Viral Pollution of the Environment

Editor

Gerald Berg, Ph.D.

Department of Civil and Environmental Engineering
College of Engineering
University of Cincinnati
Cincinnati, Ohio

CRC Press
Taylor & Francis Group
Boca Raton London New York

CRC Press is an imprint of the
Taylor & Francis Group, an **informa** business

First published 1983 by CRC Press
Taylor & Francis Group
6000 Broken Sound Parkway NW, Suite 300
Boca Raton, FL 33487-2742

Reissued 2018 by CRC Press

Library of Congress Cataloging in Publication Data
Main entry under title:
Viral pollution of the environment.

Includes bibliographies and index.
1. Viral pollution of water. 2. Water—Purification—Virus removal. I. Berg, Gerald, 1928- .[DNLM: 1. Viruses. 2. Water microbiology. 3. Water pollution. QW 80 V813]
TD427.V55V57 1983 628.1'62 82-17865
ISBN 0-8493-6245-8

A Library of Congress record exists under LC control number: 82017865

Publisher's Note
The publisher has gone to great lengths to ensure the quality of this reprint but points out that some imperfections in the original copies may be apparent.

Disclaimer
The publisher has made every effort to trace copyright holders and welcomes correspondence from those they have been unable to contact.

ISBN 13: 978-1-315-89845-2 (hbk)
ISBN 13: 978-1-351-07755-2 (ebk)

Visit the Taylor & Francis Web site at http://www.taylorandfrancis.com and the
CRC Press Web site at http://www.crcpress.com

PREFACE

Only a few decades ago, viruses could be detected only occasionally in flowing sewage and there seemed little reason for great preoccupation with the water vehicle in the transmission of viral diseases. Of course, the technology that could only intermittently detect viruses in sewage did not often detect viruses in flowing streams or in drinking waters. Thus, the pollution with viruses of the source waters for much of the modern world's drinking supplies seemed a matter of little gravity.

In the years since the end of the third decade of this century, however, techniques have been developed that are sufficiently sensitive to detect viruses frequently even at water intakes. In recent years, moreover, there have appeared growing numbers of reports of recoveries of viruses from apparently well-treated surface water supplies in technologically advanced nations. The viruses detected were probably sufficient in numbers to infect at least a proportion of those who ingested them. The size of the problem appears to grow with the advances in technology that permit the recoveries of greater numbers of viruses. The viruses have always been there. Only our ability to detect them has changed.

In the decades ahead, as virus detection technology continues to improve, we may expect greater attention to the problem of water transmission of these agents. Those who have written in this volume have been major contributors to the developing knowledge and technology for the detection and measurement of viruses in waters and wastewaters and in dealing with the problems presented by that presence. They can also be expected to be the vanguard for the technological advances that tomorrow brings.

CONTRIBUTORS

Elmer W. Akin
Chief
Microbiology Branch
Toxicology and Microbiology Division
Health Effects Research Laboratory
U.S. Environmental Protection Agency
Cincinnati, Ohio

Gerald Berg, Ph.D.
Department of Civil and Environmental Engineering
College of Engineering
University of Cincinnati
Cincinnati, Ohio

Jean-Claude Block, Ph.D.
Centre des Sciences de l'Environment
Metz, France
Laboratoire d'Hygiene
Faculte de Medecine
Vandoeuvre-Les-Nancy, France

Samuel R. Farrah
Associate Professor
Department of Microbiology and Cell Science
University of Florida
Gainesville, Florida

Brian J. Ford
Visiting Research Biologist
Department of Zoology
University College
Cardiff, Wales

Charles P. Gerba, Ph.D.
Associate Professor of Microbiology
Departments of Microbiology and Nutrition and Food Science
University of Arizona
Tucson, Arizona

Sagar M. Goyal
Veterinary Diagnostic Investigation
College of Veterinary Medicine
University of Minnesota
St. Paul, Minnesota

John C. Hoff, Ph.D.
Research Microbiologist
Microbiological Treatment Branch
Drinking Water Research Division
Municipal Environmental Research Laboratory
U.S. Environmental Protection Agency
Cincinnati, Ohio

Edward F. Landry, Ph.D.
Scientist
Virus Research Group
Department of Energy and Environment
Brookhaven National Laboratory
Upton, New York

Stephen A. Schaub
Environmental Biology Branch
Environmental Quality Division
U.S. Army Bioengineering Research and Development Laboratory
Ft. Detrick
Frederick, Maryland

John S. Slade
Virologist
Virology Unit
Thames Water Authority
London, England

Charles A. Sorber
Associate Dean
College of Engineering
University of Texas at Austin
Austin, Texas

James M. Vaughn, Ph.D.
Scientist
Head, Virus Research Group
Department of Energy and Environment
Brookhaven National Laboratory
Upton, New York

Richard L. Ward, Ph.D.
Associate Member
Christ Hospital Institute of Medical Research
Cincinnati, Ohio

TABLE OF CONTENTS

Introduction

Chapter 1

DISCHARGE TO THE ENVIRONMENT OF VIRUSES IN WASTEWATER, SLUDGES, AND AEROSOLS

John S. Slade and Brian J. Ford

TABLE OF CONTENTS

I. INTRODUCTION

The study of viruses in water and wastewater is one of the newest branches of the natural sciences. Viruses in water is also a topic of urgency, which the increasing demands of a rapidly developing global human community will continue to emphasize. In a climate of understanding that tends to take virus investigations for granted, it is easy to forget the essential novelty of the discipline. Yet a mere 20 years ago, there were no data showing that there were any viruses present in London's river water,[1] whereas it is now known that every surface water in the Thames Water Authority region which has been examined for viruses has provided a positive result in routine tests. To move from a state of total ignorance of contamination to the acceptance of a virtually 100% incidence in less than 2 decades can have few precedents, and we are faced with a growing list of problems in consequence. When we realize that parallel situations exist in the rest of the developed nations and in many of those that are still essentially primitive, it is apparent that the international magnitude of the task facing the virologist and those who are responsible for the future of water management is considerable.

Although viruses had not been demonstrated in surface waters, their presence there had clearly been anticipated. The existence of fecal contamination in river water, for instance, implies that any intestinal pathogen is likely to be present. The upsurge of scientific interest in this field mirrors the development of laboratory techniques to detect viruses, rather than a sudden rise in their occurence. Indeed, it is important to distinguish between an apparent increase revealed by novel technology and a "new" problem. When we consider the transmission of viruses via fecal and airborne routes we are misled if we conclude that these are societal phenomena, resulting from human artifice alone. Though it is undeniable that civilization has altered the way in which viruses interact with the environment, and hence with the human host, in many ways we may more clearly perceive the rationale of virus transmission when we emphasize the evolutionary significance of the process. It may be said (with the exception of such minority routes as venereal contact or direct contagion) that viruses are usually ingested, injected, or inhaled.[2] In this context we can see how important in the natural history of a virus disease may be the fecal and aerosol modes of dissemination. This will also teach us something of the ability of a virus to survive in the route. Viruses have a lineage far longer than that of mankind, without doubt, and we may assume that they are well-equipped to endure their seemingly arduous travels. Knowing the sensitivity of viruses to changes in temperature, pH, etc. we may better devise means of tackling the virus problem.

The vast communities of microorganisms that occur in natural water courses are in many ways responsible for their wholesomeness, and it is helpful to envisage the purification process *not* as the imposition of high-technology on wastewater, but rather the optimization of natural, ecological mechanisms.[3] Through this "evolutionary" approach we can appreciate how an unsophisticated society is likely to suffer a burden of virus diseases when these diseases are not controlled by the natural process of water purification.

It is possible to elaborate an understanding of the incidence of virus diseases and the control of the hazard, through basically ecological principles. We have to avoid providing encouragement to those who would prefer to ignore the problem altogether, however, since there is a school of thought which argues as follows:

1. We have elaborated a complex water supply and recycling system over several centuries without concerning ourselves with viruses, so it is unnecessary to consider them now.
2. A few viruses in the water supply are beneficial, as they lead to a valuable level of acquired immunity in a human settlement.
3. Little can be done to detect virus in a short-term testing procedure.

In consequence we have to address ourselves to the question, "do viruses matter?" — which seems academic and unrealistic to a microbiologist in the laboratory environment, but which many of us are asked when this discipline crosses the path of another, whether it is engineering, management, or future policy.[4]

II. THE EXTENT OF THE PROBLEM

A leading cause of mortality in the children of the under-developed nations is infant diarrhea, which it is claimed causes 6×10^6 deaths per year. We now recognize that viruses (notably rotaviruses) are a major cause of this syndrome. This is an international problem, a statistical datum which it is tempting to dismiss as "unqualifiable" or "imprecise". A well-documented local outbreak of viral hepatitis in New Delhi[5] is more precisely understood. There were 30,000 cases in a population with a likelihood of well-developed immunity, caused by viruses in drinking water from a modern works in which there had been no major breakdown.

The eating of shellfish has led to outbreaks of hepatitis A in Western countries, the filter-feeding of these molluscs causing them to accumulate concentrations of virus-containing material greatly in excess of those found in the water where these viruses were obtained. In an investigation of Galveston Bay, Texas, 26 seawater samples out of a total of 44 revealed enteroviruses in concentrations of up to $0{\cdot}4/\ell$; while in samples of oysters harvested from 40 pools, each containing 10 to 12 oysters, 14 of the pools gave a positive result with virus concentrations up to 224/100 g. On five occasions positive results were obtained from oysters taken from water in which viruses could not be detected.[6]

This does not make it easy to point to individual patients and to definitively assert that they have contracted a virus disease from a specified source. There are many children who act as short-term carriers of rotaviruses but who do not develop the diarrheal syndrome (the asymptomatic excreters) and without detailed serological analysis of samples from patient and source, it would not be easy to ascribe an individual case of hepatitis A to an identifiable shellfish source. But the prima facie evidence which exists is sufficient cause for concern, and it would be unwise to discount these hazards because of the relatively primitive techniques of state-of-the-art epidemiology. The existence of virus diseases in these categories is recent knowledge itself; our understanding of their extent and the interpretation of their effects is even more undeveloped.

But setting an arbitrary standard (such as $<1/1000$ gal[7]) would raise political problems and possible legal complications. It might be construed that such a proviso amounts to the authorization of the admission of viral pathogens to water. If that proved to be the case, then the development of clinical hepatitis in the consumer of drinking water from an identifiable agency would be a possible source of litigation. In that pragmatic sense, the highest standards have to be set, i.e., no detectable viruses in the water supply.

It is true that the virus population of a body of wastewater is derived from the individuals (whether human or not) who live in the catchment area and in a recycling system a cynic might argue "they are only getting back what they put in". But that is an unrealistic objection to the highest standards of purity. Though excrement is natural, sewage is not. The concentration and channelling of waste matter through man-made ducts prevents many of the ecological purification processes, to which we have alluded earlier, from functioning. In this way, a localized source of contamination can become a general health hazard.

Therefore, although we concede that the magnitude of the virus problem cannot be adequately expressed with the strictures imposed by our still-young experience in this field, there is no justification for complacency, and to nonmicrobiologists who fail to understand the potential hazard we face, it is important to emphasize the ignorance with which we are still surrounded[8] as a mirror for the real and urgent data-base we have already begun to establish.

III. VIRUSES IN DRINKING WATER

There are instinctive preferences in people, as in many animals, to drink the purest water available. For a small proportion of the global human population this takes the form of purified, piped water supplies. Here we have a product — drinking water — as innocuous and as pure as conventional treatment can provide. It is enlightening to examine the virus levels known to occur in this highly processed product, for this gives an insight into the virus intake which the world community accepts at present. This is a direct indicator of the level of contamination of the environment by viruses.

We consider three aspects of the development in our understanding of these issues:

1. The levels of viruses in waters of the less developed countries
2. The incidence of contamination of drinking water in developed nations
3. The constraints that prevent the dissemination of new findings exerted by orthodox channels of responsibility in administrative hierarchies

A. Levels in the Third World

The data-base for under-developed countries is still incomplete in many areas of the world. The authors have separately visited establishments of disparate kinds in Asia, the Far East, Africa, and the Pacific Islands. It is possible to find very low levels of contamination in areas where natural ecological forces are at work. For example, virus levels in much of India are low during the dry seasons. To a considerable extent this is attributable to the lack of sewage: piped waste disposal is often absent, so that desiccation and UV sterilization can operate from the intense sunlight, and higher ambient temperatures potentiate the destruction of viruses. It is important that we emphasize here *if modern sewerage systems were installed in such an area, it is possible that the incidence of viral contamination of these water courses would increase substantially.* Once again, we are recognizing the need to fit such developments into the cultural and environmental milieu of the region concerned. For a people who are accustomed to a life-style founded on millenia of acquired patterns of behavior, an injudicious or ill-founded modernization scheme would bring with it the risk of lowered, rather than raised, standards of community hygiene. This is the reverse of what would be experienced in a Western country, where it would be justified to assume that the introduction of a similar scheme would bring about a clear improvement in purity standards.

The novel nature of virus examination techniques makes them unavailable for most of the Third World. For this reason, little is known of the levels of virus contamination in those countries. The practical significance of this area of ignorance becomes apparent when we bear in mind that this means most of the world's population experiences an intake of viruses which remains unquantified.

However, evidence that the levels of viruses in the water of the Third World may be very high is supplied by recently published data on bacterial contamination[9] since we may justifiably assume that a raised level of contamination by fecal organisms corresponds to a parallel change in virus incidence. Data from Uganda show that fecal coliform counts up to 8000/100 mℓ may be found in river water, and even in borehole supplies the count reaches 60. Counts of many thousands are not exceptional; in the open hand-dug wells of the Gambia together with the river samples from Kenya, levels up to 100,000/100 mℓ have been found. Even more extreme examples have been obtained from Nigeria, where *total* coliforms in ponds can reach 4×10^6/100 mℓ, and canals in central Jakarta, Indonesia, where fecal coliform organisms alone have given counts as high as 3,100,000/100 mℓ. We do not know to what extent we can utilize these figures as direct indicators of viral contamination, since it is widely accepted that there is not a close and invariable relationship between bacterial counts and virus levels. Neither can we say much about virus levels (even when known) and population risk or infectivity of the pathogens, since much remains to be discovered about the epidemiology of virus transmission in communities where the virus is endemic. Though it is a truism to say that the presence of a pathogen is necessary for an infection to commence, it does not follow that the occurence of a virus will produce disease in each potentially susceptible individual. The immunological, constitutional and behavioral factors that underlie this curious selectivity are complex and in many respects unresolved.

We wish to emphasize that the use to which such findings are put should be carefully considered in the light of the cultural constraints to which we have alluded. There seems a real danger that premature broadcasting of contaminant levels may pose more difficulties than it solves. The introduction of standards in one instance[9] caused the official closure of a village water supply where 50 fecal coliforms per 100 mℓ had been found, only to force the local population to utilize instead water from irrigation canals where the count was 10,000/100 mℓ. We would wish to counsel those who obtain virus counts to relate the new data to the societal needs and values of an area and to present them in context and in a constructively pragmatic manner. The overall aim, after all, must be the removal of obstacles to real progress and not merely the substitution of alternative problems.

B. Virus Contamination of Drinking Water in Developed Nations

It is in the industrialized countries that sewers are sufficiently widespread for problems of virus distribution to occur. Additionally, the water supplies are typically treated with a range of well-proven methods to eliminate pathogens, ranging from traditional filtration to chemical sterilization with chlorine, ozone, or some other disinfectant. Even in these intensively regulated supplies, significant virus levels are found. Human enteric viruses have been reported in samples from France, Israel, Italy, Romania, South Africa, the U.S., and the Soviet Union.[10] However, these data reveal more about the distribution of virus laboratories than they do about the problem with the viruses themselves. Furthermore, the results in qualitative terms are only as broad as the testing procedures utilized, which are themselves limited. Indeed, some of the most highly significant virus contaminants (including rotaviruses and hepatitis A viruses) are not amenable to detection by routine laboratory tests.

When faced with the interpretation of data on the contamination of water by viruses in any industrialized nation, there are important theoretical limits to the understanding we can usefully derive, and some practical strictures on the acquisition of further data. It may be that the analysis of these factors will prove to be an important aspect of the future progress of water virology.

C. Establishment Constraints on the Dissemination of Knowledge

As in many other fields of specialized knowledge, new research work is largely predicated by what has been published in the journals. But not everything that is achieved reaches the stage of publication, and in a new and nascent discipline, it is inevitable that orthodoxy acts *against* the introduction of new departures. In the field of water supply and the study of contamination, we are faced with the traditional hierarchy of responsibilities in which, in order to produce an effective relationship at a controlling level between engineering and water-examination staff, a senior microbiologist and/or administrator is inevitably assigned to the task. This implies that a highly qualified officer will take the responsibility for the process of the management of water purity policy, while a junior member of staff will be appointed to investigate such new or experimental areas as, in this case, the contamination of water or the environment in general with viruses from wastes. Though it may be argued that this brings a "fresh mind to the problem", it also makes it difficult to establish channels of communication for the findings that emerge. It has been found in practice that some appointments in the field of virology are made so that a department may lay a claim to be in the forefront of knowledge. However, if the senior official has no great experience of virology then the support given to a junior member of staff may be limited, not by deliberate policy, but by the orientation of conventional attitudes towards better-known or more available disciplines. In one or two cases of which we have knowledge, the presence of viruses in some samples was greeted with disfavor since it seemed to cast a slur, or an implied criticism, on what were regarded as tried and tested methodologies closely associated with the status and reputation of the organization itself.

We do not here suggest that these emphases are a matter of conscious prejudice, but that they are an inevitable consequence of responsibilities in pyramid configurations. Further controls may be exerted by the state-of-the-art limitations of our understanding. Thus, it may be found that a raised level of viruses will not be reported because of the time which has elapsed between the taking of the sample and the reading of the results. Here, we are faced with the concept that it is, by the time the data are at hand, too late to do anything practicable about preventing the occurence. In this way, research findings can be designated as "of academic interest only" and disregarded. For these and related reasons, it will be important in the future for us to seek to give this branch of virology the status it deserves, so that the pathway to deeper levels of understanding may be smoothed, and the opportunity for new and relevant discovery facilitated. At the present time, it seems that small programs of research and individual findings which might in concert give a fuller picture of the present position are being lost in administrative files, rather than being afforded an audience. Personal discussions have suggested that in Britain (where no formal reports of viruses in tapwater have ever appeared) there have been several individually identified examples of positive results for coxsackievirus B and polioviruses, for instance. The objections to the publication of these spasmodic reports is that they may be atypical, or that they may represent false positives. However, only a full examination of the results that *are* being obtained will enable the most sensible interpretations to emerge.

Once again it is important that a discussion of these data be couched in balanced and realistic terms. A report in the media that suggested we were faced with undue health hazards from viruses which were being detected in minute concentrations would do a disservice to a specialty which is beginning to develop in order to assess, and in due course regulate, a newly recognized problem. It would be unfortunate if scientific staff were to become the victims of criticism because of their interest in tackling a novel and unquantified hazard, particularly when the risk to the population as a whole appears to be very slight, and certainly factorially smaller than many risk factors we daily take for granted.

IV. VIRUSES IN THE ENVIRONMENT

So long as waste liquids are ducted through closed sewers, their virus content poses few problems. Once at the site of treatment or disposal, however, a potential risk of virus escape exists. It is practically impossible to ensure that every part of a water-main or sewerage system is perfectly sealed, so that some contamination by direct leakage and ingress is a possibility. In London, for example, the total length of sewage and water pipeways exceeds the distance between Chicago and London, if the distance is measured the *long way* round, via the Pacific! In a situation like London's, considerable increases in traffic density and the mass of individual trucks have caused vibration and stress fracture of sewers. Water mains have suffered less, partly because of their positive pressure loading which aids structural integrity. But the leakage of sewage pipes means that health hazards which were once largely confined to sewer workers may now affect the general population.

Once discharged into the environment (for example, by release from a river-bank outfall) sewage is no longer under the control of the relevant sewerage authority. Viruses and other pathogens have been released into the environment, and from here on, official controls cease to apply. In the majority of sewage treatment processes, the solid and liquid fractions are, as far as possible, separated. This is not the case in lagoon or impoundment processing, but these are processes which have limited applications in global terms. No matter what treatment is applied, however, it is impossible to retain these products indefinitely and so there comes a time when they have to be returned to the environment. Degrees of sewage treatment vary greatly, from none (where sewage consists of collection and discharge into a sea or lake, an antiquated and dangerous approach still regretfully found in many long-standing settlements in the Western World) to highly sophisticated recycling systems designed to reuse both liquid and solid components. Even when such modern and thorough processes are in use and they are still relatively rare, it is still true to say that in the great majority of cases, effluents from sewage treatment agencies contain substantial numbers of viruses.

Finally, it is inevitable that procedures that involve the agitation and aeration of virus-containing wastes will liberate viruses in aerosol form. Contaminated riverwater or seawater, is liable to produce aerosols in, for example, spray irrigation or cooling towers, apart from natural agencies from sea-shore breakers to waterfalls and rapids. Once in aerosol form, virus contamination is almost impossible to detect or control, and prolonged distances may be covered by released viruses which are (in so many cases) well-adapted to colonization of a new host via the respiratory route. One could even speculate whether the release of wash-basin wastes from aircraft might lead to the transmission of a virus disease in some susceptible individual many miles distant from the point of contamination.

V. CATEGORIES OF VIRUSES

It is tempting to think only in terms of human enteric viruses when sewage is considered and to disregard pathogens of other forms of (nonhuman) life.

But there are some significant categories which involve very different species, from bacteriophage to pathogens of domestic and agricultural animals, and even some virus diseases of fish and plant species which have to be considered.

A. Viruses of Human Origin

In practical terms, it is the human pathogens which have to be given most urgent consideration. An understanding of the nature and extent of these agents in mankind's environment would be instrumental in leading to a considerable reduction of suffering and mortality world-wide.

1. Fecally-Excreted Enteric Viruses

More than 100 different fecally excreted enteric viruses of human origin are known. The principle types (polioviruses, coxsackieviruses, echoviruses, adenoviruses, reoviruses, and parvoviruses) have been taxonomically studied and are identified by clearly defined criteria. Many of these have been repeatedly cultured *in vitro*. Others, such as rotaviruses, have been identified by electron microscopy in fecal matter but the etiologic role of numerous other categories is obscured by the lack of reliable laboratory test procedures.

2. Other Viral Pathogens

Feces are not the only source of human pathogens, however. Water from wash-basins may contain viruses from superficial lesions, while blood in a sewage sample (perhaps from traumas or menstrual losses) could contribute any agent causing a viremia. Nose and throat secretions are a potent source of human respiratory viruses. The recently reported upsurge in genital herpes may lead to bathwater contamination, we speculate, and questions might also be raised over transmissable autoimmune disease. Therefore, although we must emphasize that the pathogenic or epidemiologic significance of any of these viruses remains imprecisely understood, it is clearly arguable that any virus pathogenic to humans could be found in sewage.

B. Viruses of Animal Origin

We are concerned here with viral pathogens that may be of economic importance (e.g., foot and mouth disease) or involve domestic pets (including distemper in dogs). In recent years, attention has focused on swine vesicular disease (SVD) which has been known to occur in imported pork products and could be transmitted through contaminated wastes. Regulations controlling the preparation of pig-swill (waste food boiled down to feed pig stocks) have been introduced in consequence. The virus, which can in certain circumstances cause an infection in humans, is similar to coxsackievirus B5. Since this latter virus is known to occur in sewage and in sewage-derived contamination, it is reasonable to assume that the virus of SVD might be similarly distributed.

1. Virus Diseases of Fish

In recent years, increased attention has centered on infectious pancreatic necrosis (IPN) which has caused losses up to 90% in the fish fry reared in trout farms and other hatcheries of salmonid species. A proportion of the survivors may act as carriers of the IPN virus. The disease was introduced into Britain during the 1970s, probably from contaminated imports from Western Europe. It has been suggested that this arose from the introduction of whole trout, but the financial incentives of this form of farming tend to encourage illicit dealing and it may be that the virus was imported with unregistered fry or eggs. In either event, a transmission route through sewage and the subsequent contamination of the environment by the virus is likely.

C. Viruses of Plant Origin

Here too we are dealing with poorly understood mechanisms. In typical plant diseases, vectors such as nematodes or aphids transmit virus particles from host to host. Although one might reasonably expect to isolate plant viruses from wastewaters, it is likely that without the vector, transmission would not occur. However (particularly in the case of aquatic nematodes), it is possible that the virus-containing vector itself is a constituent of wastewater and this could pose problems for agriculture. The use of the end-products of sewage treatment, whether sludges or partially-purified water, might act as a means of distributing plant viruses on a large scale.

D. Viruses of Bacterial Origin (Bacteriophages)

We now know that many different types of bacteriophage are found in wastewater and on occasions they are present in considerable quantities. The difficulties they pose would seem to be few, since problems such as the contamination of biotechnology establishments with viruses that could decimate the microbial population are much reduced by the systematic treatment of influents into these plants by their operators. It is possible that phages in water may serve as indicators of the presence of host bacteria. In this way, phages may become valuable indicators, which could be exploited to give a rapid test (within 6 hr) for the presence of fecal matter.

Not only are phages potentially valuable as indicators of the occurence of contaminants, but they have applications as an index of the extent to which a contaminated water-course has been cleaned up. For example, it has been claimed that coliphages are less susceptible to chlorination than the host coliforms, and therefore survive in greater numbers.[11] For this reason, coliphages might be useful for evaluating the performance of sewage treatment plants in the elimination of human pathogens.

VI. ENVIRONMENTAL CONTAMINATION

All forms of life are afflicted by viruses and some species are susceptible to more than one type of virus. It may be argued, therefore, that there are more viruses than there are species of living organisms. The environment has supported viruses, either in their hosts or in a route of transmission, for virtually as long as there have been living organisms to replicate them.

So the mere *presence* of viruses in the environment does not definitively amount to the same thing as *contamination*. We define this term in two senses: (1) when undue concentrations of viruses are produced by artificial means and (2) when any virus of economic or hygienic importance is present in the environment of the host.

This implies that, first, we must avoid the use of processing methodologies which (in a manner analogous to the filter-feeding of mussels referred to in Section II) will concentrate viruses normally present in the environment. Some once-unsuspected examples will be described below. Second, even if amounts of a given pathogen in the environment are not "excessive" in the historical or ecological sense, methods must be evolved to eliminate them. In this way, as in the case of variola, a disease once prevalent can be eliminated from human populations. In the *sensu stricto*, these are two distinct usages of the term "contamination". But the need to control pathogens for humanitarian or economic reasons means that, in practical terms, we must seek to introduce measures that will totally eliminate all viruses in wastewater and prevent the recycling of viruses by their redistribution through environmental channels. We can thus seek to apply more rigorous controls than those exerted by nature and in due course, this could have a pronounced action in promoting the health of the entire human (and animal) community.

It must be accepted that with present-day economic and technological limitations, we are faced with the more practical task of ensuring that wastes are disinfected, rather than entirely sterilized. Even here we are in contentious territory. There are many meanings of the term "sterile" which introduce complications to any analysis of this problem[12] and the concept of "disinfection", with its human-centered interpretation, is equally vague. Therefore, in proposing an ideal we have to relate to the practical realities of what may be costly or inefficient technologies, and also to the inherited orthodoxies of treatment on which present waste disposal and sewage treatment processes are based.[2]

A. Viruses in Sewage

Not only is there a considerable range of viruses in human excreta, more than 100 different types being known to date, but the amounts excreted may be very large. More than 1 million infectious viruses may be excreted per gram of feces by infected individuals, irrespective of whether they show any signs of illness.[13] In sewage, the concentrations may be as high as 100,000 infectious particles per liter, and they can survive for several months under suitable circumstances.

There are many factors which influence the concentrations of viruses found in sewage:

1. Health and age structure of the community
2. Epidemic/etiologic status of local viral pathogens
3. Physical condition of the sewage (for example, degree of dilution by nonvirus-containing effluents)
4. Relationship of ambient conditions to those best fitted to virus survival
5. Diurnal and seasonal cycles of temperature, rainfall, etc.

An early study from the U.S.[14] considered enteroviruses found in sewage samples from the state of Michigan between 1955 to 1957. East Lansing sewage specimens gave 14% positive, while samples from Lansing which had been subjected to dilution with industrial effluents showed half this figure. The highest recovery rates were in the summer season, July through November. Only 10% of the effluent samples from the East Lansing activated-sludge plant gave positive results, compared with 33% of the raw influent.

A similar investigation in Haifa, Israel, between 1972 and 1974 showed a monthly average of between 6×10^3 and 4.9×10^5 viruses per liter in sewage samples.[15] The highest value recorded was 1 million. The degree of virus contamination of sewage was studied in Ottawa in 1977[16] where 79% of sewage samples were shown to contain human pathogenic viruses. Of 72 isolates submitted to examination by serology and electron microscopy, 56 (78%) were reoviruses and the remaining 16 were enteroviruses. Of the latter category, 15 were polioviruses (five of these wild strains); the remainder were coxsackieviruses. At the time, it was wisely pointed out that the presence of wild polioviruses, occurring at a time when popular support for vaccination was waning, gave cause for concern.

Nearly 80% of the 11,500 virus particles per liter found in an Indian study[17] were polioviruses and approximately 70% of these were wild strains. In this case too, the public health aspects of virus contamination were emphasized and it was pointed out that 60% of the cases of paralytic poliomyelitis reported in India occurred during the rainy season.

A continuing study in London during 1981 involving collection of a series of samples at biweekly intervals over a period of more than of 7 months showed up to 9140 viruses per liter of raw sewage influent to a modern activated sludge plant. Every sample of influent gave a positive result. The comparable figures for treated effluents show that the average reduction in virus levels was 98.8%. Even with this reduction, 80% of effluent samples gave a positive result with concentrations per liter as high as 60; one sample (containing stormwater) reached 120/ℓ. Though the amount of data remains somewhat limited, it seems reasonable to make some general statements. Thus, all sewage contains viruses. Human sewage is, in our view, the most potent source of human pathogenic viruses capable of contaminating the environment of today's civilized communities.

B. Viruses in Sludges

Partly due to their diminutive size, viruses readily adsorb onto solid particles and tend in nature to become associated with solid matter. There is an inherent tendency for viruses to aggregate in sludges, and for this reason the removal of solid matter from a contaminated source of sewage leads to a reduction of virus levels in the effluent. A water engineer who reduces the solids of a sewage under treatment is therefore also reducing its virus content, even though it would be difficult to relate the two phenomena mathematically.

Although this is a useful property of viruses in some ways, it poses problems since solid-associated viruses tend to survive for longer than viruses in a dispersed phase. The disposal of sludge residues can act as a significant source of virus contamination of the environment. The same phenomenon which reduces virus concentrations in settling sewage, therefore, potentiates the hazard at a later stage in the cycle.

The ability of sludges to concentrate viruses implies that sewage sludges (though inevitably the most potent source) are not the only sludges in which viruses are found. When drinking water sources are treated by settling, the levels of viruses in sludge are proportionately increased. Thus, coagulation and sedimentation, while reducing viruses in the product, can produce undesirable levels in the by-product. Not only do the viruses survive for prolonged periods, but they retain their infectivity there. So (although sources of water that are to be purified for drinking are generally cleaner than raw sewage), these sludges remain a potential hazard to the environment.

1. Primary and Secondary Sludges

An additional factor to be taken into consideration with the points enumerated in Section VI. A is that virus levels vary with the point in the treatment process the sludge sample is taken. Levels in treated sludges are lower than those in raw sludges, though they remain significant.

Primary sludges may contain up to 1 million viruses per liter. Raw sludge in the Thames region has yielded virus concentrations up to 65,000/ℓ and though levels in treated sludges (particularly lime-treated, where high pH militates against viruses) may be zero, they can be as high as 1000/ℓ, and in some cases higher. Many attempts to show how long viruses survive have revealed that 12 weeks is not exceptional for types that include coxsackievirus B5, echovirus 9, and poliovirus.[18] Some idea of the rate at which inactivation proceeds may be gained from studies under varying climatic conditions[19] which show that a 1 log reduction per month during a Danish winter season may be compared with a 2 log reduction per week during the summer in Texas.

2. Ultimate Fate of Sludges

At the end of the treatment cycle, sludges may be tipped or used as land-fill, used in agriculture as fertilizer and/or soil conditioner, discharged into the sea, or utilized as an industrial raw material or as an energy source.[20] There are many possibilities here for contamination and even when sludges are combusted, it could be argued that virus particles might possibly escape in convected dust.

There can be no doubt that on a world-wide basis the greatest of these hazards lies in the use of sludge as an agricultural fertilizer. Though it seems only prudent, in a world of limited resources, that this valuable and nutrient-rich waste material should be recycled through agricultural use, it is clear that such use provides a direct route for reinfection of the population that consumes crops contaminated with viruses.

Further spread is possible through run-off into water courses, contamination of adjacent ground-water, distribution by birds and mammals, and even direct infection or contamination of agricultural workers or rural communities that have access to the land.

A review[21] has shown that viruses in sludge reused in this way can survive for 175 days and other results have given similar limits. Among the factors that encourage the inactivation of viruses are low humidity, high temperature, high solar irradiation, and a lack of organic material. Extremes of pH and an aerobic environment similarly act against viruses. Though these factors have been experimentally demonstrated, some of them (e.g., low levels of organic matter) cannot be said to apply to practical conditions of sludge utilization, but are experimental situations.

One extreme example[22] involved the injection of sewage sludge into soil. Very high recovery rates of viruses were obtained more than 6 months after primary sludge and anaerobically digested (secondary) sludge samples had been mechanically injected below ground level. The conclusion of this work, and other studies, shows beyond doubt that the sludge/soil matrix is able to support the survival of viruses for prolonged periods of time. Clearly, only the thorough and systematic disinfection of sludge at the treatment plant can eliminate the hazard that results.

C. Viruses in Aerosols

It has been said that there are many means by which virus-contaminated liquids can be sources of aerosols, whether through artificial or natural causes. Even the introduction of air into a treatment tank or the act of flushing a domestic lavatory generates aerosols. Viruses adsorb onto the air/liquid interface at the periphery of a bubble and are projected into the atmosphere when the bubble bursts.[23] This is an important means by which viruses may be liberated from liquid medium and of becoming liberated in concentrated form. The results obtained with coliphages T2 and T4 showed that a 50-fold concentration could be obtained. In a trial of the natural behavior of surf droplets thrown off by waves exhibited concentrations up to 250 times higher than those experimentally produced in the seawater and traveled at least 30 m inland.[24]

The amount of viruses released through sprinkler-irrigation systems is potentially high. One retrospective survey carried out in Israel[25] showed that the levels of infectious hepatitis were 2 to 4 times higher in communities around spray irrigation establishments than in 130 control settlements. Though no evidence of increase in influenza was demonstrated by laboratory tests, twice as many cases were clinically diagnosed in the potentially exposed communities. It was concluded that this might be due to upper respiratory diseases caused by enteric viruses being misreported as influenza.

Finally, what evidence exists of contamination hazard from the domestic lavatory? After seeding 10^8 polioviruses into a toilet bowl, flushing liberated approximately 3000 infectious units to the level of the seat.[26] Later trials with coliphage MS2 showed that, although in an unventilated bathroom 94% of aerosolized viruses settled out of the air in the first 2 hr, small numbers of viruses apparently remained suspended in the air for very much longer.

VII. SUMMARY AND CONCLUSIONS

In spite of some comprehensive reviews of this subject, which have done much to encourage interest in the contamination of the environment by aerosols,[27] this route of transmission is often considered almost as a "poor relative" of the alternatives we have considered earlier. There are some specific problems associated with aerosols that it is helpful to emphasize, however.

It is not practical to take personal-hygiene precautions against contaminated aerosols, for instance. Care over toilet use or the washing of hands before touching food are typical of the kind of precautions advocated in public health which cannot apply to the air we have to breathe. Authorities using clean-looking water in power-plant cooling towers, in agricultural applications, or even in decorative fountains in parks and gardens, may be producing an unwitting dispersal of viruses into the environment. They would not be expected to show the kind of caution as would apply to a sewage authority handling effluents that could be presumed contaminated.

Once released, virus aggregates in an aerosol might travel over considerable distances and perhaps produce widespread, if sparse, outbreaks. Furthermore, if (as may have been the case in the Israel study referred to above[25]) diagnostic confusion surrounds the epidemiologic analysis of this problem, it might prove difficult or impossible to delineate the extent of the hazard or even to recognize a minor epidemic should one occur.

In world-wide terms, it is clear that human excreta represent the most potent source of naturally produced human viruses. Mankind seems to have an inborn aversion to contact with feces, as a rule, which may be a natural trait that serves to minimize the chances for a vulnerable species to contract feces-transmitted infections.

When we collect, concentrate, distribute, and treat sewage wastes we are therefore handling a potentially hazardous material, and at any point where it or its products comes into contact with the environment, there is a clear potential risk to the health of the community. The contamination may be direct, as when a sewer leaks or a contaminated outfall overflows, or it may be indirect in aerosol form or as a constituent of an end-product (like irrigation or cooling water) that is assumed to be innocuous.

Since our knowledge of the full extent of virus diseases and their impact remains relatively undeveloped in many areas, it is impossible to quantify the hazard or to make viable recommendations about its reduction. But we should not expect to provide a product in which there may be unquantified ingredients. There are many reasons for this, some of them political and legal, many of them concerned with the basic concept of "professional pride", and all of them related to the demands of all people for a clean, pure, and wholesome environment. Esthetic considerations play their part, also, alongside the mundane questions of litigation, political maneuvering, and professional status.

The proven existence of viruses in virtually every stage of the waste treatment industry throughout the world, combined with the growing knowledge of diseases caused by viruses in their many forms, makes it important for us to study the implications of this area as a matter of urgency, to seek to regularize and reform current practices wherever that might help to better the lot of our fellows, and to make the environment safer for coming generations and more wholesome than it has been in mankind's past.

REFERENCES

1. **Poynter, S. F. B.,** The problem of viruses in water, *Water Treat. Exam.*, 17(3), 187, 1968.
2. **Slade, J. S.,** Viruses in sewage, *Viruses and Wastewater Treatment,* Pergamon Press, Oxford, 1981.
3. **Ford, B. J.,** The Microecology of Water Purification, paper presented at the Australian Water and Wastewater Association Symposium, University of Hobart, Tasmania, 1978.
4. **Ford, B. J.,** The European experience of water management (Keynote address), *Water,* 5(1), 10, 1978.
5. **Berg, G., Ed.,** Transmission of Viruses by the Water Route, *Wiley-Interscience,* New York, 1966, 14.
6. **Goyal, S. M., Gerba, C. P., and Melnick, J. L.,** Human enteroviruses in oysters and their overlying waters, *Appl. Environ. Microbiol.*, 37, 572, 1979.
7. **Melnick, J. L.,** Social impact of viruses, in *Viruses in Water,* Berg, G., Ed., American Public Health Association, Washington, D.C., 1976.
8. **Ford, J.,** Areas of Ignorance Keynote address, Inter Micro 77, King's College, Cambridge, U.K., 1977.
9. **Feachem, R.,** Bacterial standards for drinking water quality in developing countries, *Lancet,* 255, 1980.
10. World Bank (report), *Low-cost Methods of Sanitation and Excreta Disposal,* in press.
11. **Durham, D. and Wolf, H. W.,** Waste-water chlorination: panacea or placebo?, *Water Sewage Works,* 120, 67, 1973.
12. **Ford, B. J.,** in Terms and expressions, *Microbiology and Food,* Northwood, London, 1970, 167.
13. WHO Scientific Group, *Human Viruses in Water, Wastewater and Soil,* WHO Tech. Rep. Ser. 639, Geneva, 1979.
14. **Bloom, H. H., Mack, W. N., Krueger, B. J., and Mallmann, W. L.,** Identification of enteroviruses in sewage, *J. Infect. Dis.*, 105, 61, 1959.
15. **Buras, N.,** Concentration of enteric viruses in wastewater and effluent: a two year study, *Water Res.*, 10, 295, 1976.
16. **Sattar, S. A. and Westwood, J. C. N.,** Isolation of apparently wild strains of polio virus Type 1, from sewage in the Ottawa area, *Can. Med. Assoc. J.*, 116, 25, 1977.
17. **Rao, V. C., Lakhe, S. B., and Waghmare, S. V.,** Developments in environmental virology in India, *Indian Assoc. Water Pollut. Control Tech. Annu.*, 5, 1, 1978.
18. **Subrahmanyan, T. P.,** Persistence of enteroviruses in sewage sludge, *Bull. W. H. O.*, 55, 431, 1977.
19. **Damgaard-Larsen, S., Jensen, K. O., Lund, E., and Nissen, B.,** Survival and movement of enterovirus in connection with land disposal of sludges, *Water Res.*, 11, 503, 1977.
20. **Ford, B. J.,** Energy Conservation and World Resources, paper presented to the United Nations Association, New York, November 1974.
21. **Gerba, C. P., Wallis, C., and Melnick, J. L.,** Fate of wastewater bacteria and viruses in soil, *J. Irrig. Drainage Div. Am. Soc. Civil Eng.*, 101, 157, 1975.
22. **Sagik, B. P.,** Virus survival in soils injected with municipal wastewater treatment residuals, U.S. National Technical Information Service Report, #PB 287, 215, 1978.
23. **Baylor, E. R., Peters, V., and Baylor, M. B.,** Water-to-air transfer of virus, *Science,* 197, 763, 1977.
24. **Baylor, E. R., Baylor, M. B., Blanchard, D. C., Syzdek, L. D., and Appel, C.,** Virus transfer from surf to wind, *Science,* 198, 575, 1977.
25. **Katzenelson, E., Buium, I., and Shuval, H. I.,** Risk of communicable disease infection associated with wastewater irrigation in agricultural settlements, *Science,* 194, 944, 1976.
26. **Gerba, C. P., Wallis, C., and Melnick, J. L.,** Microbiological hazards of household toilets: droplet production and fate of residual organisms, *Appl. Microbiol.*, 30, 229, 1975.
27. **Berg, G.,** Water pollution, microbiology-detection, occurence, and removal of viruses, *J. Water Pollut. Control Fed.*, 50, 1395, 1978.

Methodology

Chapter 2

METHODS FOR RECOVERING VIRUSES FROM THE WATER ENVIRONMENT

Charles P. Gerba

TABLE OF CONTENTS

I. INTRODUCTION

The major preoccupation of those who study viruses in water has been with the development of methods capable of detecting viruses in large volumes of water. Because of the high infectivity of viruses, such studies have been seen as a necessity to adequately determine the public health significance of these viruses when they are waterborne. Research toward this goal has produced a large variety of methods, but the most promising for processing large volumes of water are those dependent upon viral adsorption. The major problem with other methods has been that they can only be used with limited volumes of water or they can only be used with highly treated, turbidity-free waters. However, many of these methods are suitable for processing limited volumes of heavily contaminated waters or in reconcentrating viruses initially concentrated by other methods. The purpose of this chapter is not to give an exhaustive review of all the methods that have been tested for concentrating viruses from water because this has been done previously[1,2] and because most of these methods have not seen practical application. The aim of this chapter is to focus on the most promising and extensively used methods and the principles upon which these methods are based. Concentrating viruses from water is not achieved simply by an application of methods; concentrating viruses from water is a strategy where the complexity of the environment from which the viruses must be isolated, the nature of the viruses, and their interaction with their environment are taken into consideration.

Many biochemical procedures used for concentrating and purifying proteins have been explored in the development of methods for concentrating viruses from water. Viruses are largely protein, and methods useful for concentrating macromolecular proteins such as phase partition, immunochemical, precipitation, and adsorption techniques have been applied to the concentration of viruses from water. In practice, many of these methods are highly effective in the laboratory for concentrating viruses, but their application is often limited to use with highly contaminated waters.

II. METHODS WHICH DEPEND ON VIRUS ADSORPTION

A. Mechanisms of Virus Adsorption

The most useful methods for concentrating virus from water have been those which depend upon virus adsorption to a surface and subsequent desorption by a small volume of eluent. In this regard, microporous filter adsorption-elution techniques, because of their ability to process very large volumes of water (100 to 1000 gal of widely varying quality, have seen greater application than any other method. Other adsorbents such as polyelectrolytes, clays, metal oxides, and glass beads (Table 1) have also been successfully used.

Fundamental to the basis of this methodology is an understanding of the processes involved in virus adsorption to surfaces. Virus adsorption to microporous filter surfaces is now believed to be controlled by both electrostatic and hydrophobic interactions.[3,4] Viruses are biocolloids and their adsorptive behavior by electrostatic phenomena can be described by concepts of colloidal chemistry. If the charges of a surface are of opposite signs, electrostatic attraction will allow the particles to deposit on the media. If they are of the same sign, repulsion will occur and deposition will be hindered. In neutral solutions and most natural waters, viruses are negatively charged and thus there occur strong electrostatic repulsion forces between viruses and similarly charged surfaces. These electrostatic forces can, to a large measure, by controlled by altering various environmental factors such as pH and electrolyte concentration. This phenomenon is explained by the electrical double-layer theory of colloidal chemistry.[5] A particle such as a virus immersed in an aqueous solution develops a surface charge by adsorbing ions to its surface. A fixed layer of oppositely charged ions develops around the surface of the adsorbent. To maintain the electrically neutral system, there is a

Table 1
METHODS USED FOR CONCENTRATING VIRUSES FROM WATER

Method	Initial volume of water	Applications	Remarks
Filter adsorption-elution			
Negatively charged filters	Large	All but the most turbid waters	Only system shown useful for concentrating viruses from large volumes of tapwater, sewage, seawater, and other natural waters. Cationic salt concentration and pH must be adjusted before processing
Positively charged filters	Large	Tapwater	To date these filters have only been tested with tapwaters, but they may be useful for other types of water as well. No preconditioning of waters may be necessary
Adsorption to metal salt precipitate, aluminum hydroxide, ferric hydroxide	Small	Tapwater, sewage	Have been useful as reconcentration methods
Polyelectrolytes — PE60	Large	Tapwater, lakewater, sewage	Due to its unstable nature and lot-to-lot variations in efficiency for concentrating viruses the method has not been used in recent years
Bentonite	Small	Tapwater, sewage	
Iron oxide	Small	Tapwater, sewage	
Talcum powder	Large	Tapwater, sewage	Can be used as a "sandwich" between filter paper supports to process up to 100-ℓ volumes
Gauze pad	Large	—	First method developed for detecting viruses in water, but not quantitative or very reproducible
Glass powder	Large	Tapwater	Columns containing glass powder have been made which are capable of processing 400-ℓ volumes
Organic flocculation	Small	Reconcentration	Widely used method for reconcentrating from primary filter eluents
Protamine sulfate	Small	Sewage	Very efficient method for concentrating reoviruses and adenoviruses from small volumes of sewage
Polymer two-phase	Small	Sewage	Processing is slow; method has been used to reconcentrate viruses from primary eluents
Hydroextraction	Small	Sewage	Often used as a method for reconcentrating viruses from primary eluents
Ultrafiltration			
Soluble filters	Small	Clean waters	Clogs rapidly with even low turbidity
Flat membranes	Small	Clean waters	Clogs rapidly with even low turbidity
Hollow fiber or capillary	Large	Tapwater, lake water	Up to 100 ℓ have been processed, but water is often prefiltered
Reverse osmosis	Small	Clean waters	Also concentrates cytotoxic compounds which adversely affect assay methods

diffused layer containing a sufficient number of counterions extended for some distance into the solution. If the bulk solution of counterions increases by addition of cationic salts or increasing pH, the thickness of this layer decreases because less volume is required to contain enough counterions to neutralize surface charge. The reduction of the thickness of this layer facilitates the approach of the surfaces, allowing van der Waal's forces to have an effect.[6] Reducing the pH or adding cationic salts thus reduces the electronegativity of negatively charged surfaces allowing increased adsorption to occur.

Control of pH is of considerable importance to virus adsorption. In general, proteins such as those that compose the virus surface are negatively charged above their isoelectric pH and positively charged below their isoelectric pH. They are strongly negatively charged at very high pH values and strongly positively charged at very low pH levels. We take advantage of this phenomenon when we concentrate viruses on microporous filters and other adsorbents. With negatively charged filters such as fiberglass, the water sampled is preconditioned by adjustment to pHs below 5.0 which act to reduce the net negative charge on the virus allowing for more optimal adsorption of the virus to the filter surface. Elution of the virus from the filter is achieved by passing a solution with a high pH (usually above pH 8.0) through the filter which causes the virus to become more negatively charged favoring desorption. The isoelectric points of enteric viruses vary from below pH 3.0 to above pH 7.0. This variation in isoelectric points can result in different optimal conditions for adsorption of different viruses to identical surfaces under a given set of conditions and may account to some extent for differences in efficiencies for adsorption of different viruses.

Amino acids which compose proteins may have nonpolar groups lacking an affinity for water and can serve as a source of attraction for other nonpolar groups in the environment.[6] Recent studies indicate that hydrophobic interactions are a major factor in virus adsorption to microporous filters at high pH, whereas electrostatic forces are dominant at low pH.[4] Thus, substances which weaken hydrophobic interactions such as detergents and chaotropic salts can be used to deadsorb viruses from surfaces where these types of interactions are significant.

B. Microporous Filter Adsorption Methods

1. Negatively Charged Filter Media

Metcalf[7] was the first to point out the potential application of microporous filters for virus concentration from aqueous suspensions. Cliver[8] later reported the application of Millipore® membrane filters to the concentration of enteroviruses from a variety of solutions. Wallis and Melnick[9] are credited with the development of the first practical microporous filter system for concentration of viruses from water and wastewater. Nitrocellulose membrane filters that were subject to clogging were used in early methods for virus concentration, and sample sizes were generally less than 5ℓ[8-10] Wallis et al.[11] found that virus adsorption to nitrocellulose filters could be greatly enhanced by trivalent ($AlCl_3$) and divalent ($MgCl_2$) salts and adjustment of the pH of water to 3.5. Enteroviruses adsorbed to the filters were eluted with 0.05 *M* glycine adjusted to pH 11.5 with NaOH. The authors also described a reconcentration procedure whereby viruses recovered from filters were readsorbed to smaller diameter filters and eluted with smaller volumes of eluent, thus greatly reducing the amount of eluent to be assayed.

These developments provided the basis for the first portable virus concentrator for concentrating viruses in the field.[12,13]

The original system was a continuous flow-through apparatus in which incoming water was passed through a series of five nonvirus-adsorbing clarifying textile filters followed by treatment with an anion-exchange resin to remove materials which clogged the adsorbent filters before adsorption of the virus onto 293 mm diameter cellulose nitrate disc filters. Magnesium chloride was added via injection to enhance adsorption of viruses. Viruses were eluted from the filter with 1ℓ volumes of pH 11.5 glycine buffer. The primary eluent was then reconcentrated on smaller diameter cellulose nitrate membranes from which adsorbed viruses could be eluted with small volumes (approximately 20 mℓ) of pH 11.5 glycine. The virus concentrator was capable of recovering up to 80% of exogenously added viruses during laboratory processing of 300 gal (1134 ℓ) of tapwater. Further refinement of the portable virus concentrator resulted in a unit in which the water was first clarified through a series of orlon or polyester spun fiber cartridge depth filters, and the viruses adsorbed to a fiberglass or cellulose acetate spun fiber cartridge depth filter.[14] In addition to these changes, $AlCl_3$ (0.005 *M*)was substituted for $MgCl_2$ to enhance virus adsorption, since a 100-fold less concentration was needed. This system was used successfully in field studies for the isolation of naturally occurring viruses in wastewater and seawater.[14,15] Based on the prototype virus concentrator developed by Wallis and Melnick, the Carborundum Company (Niagara Falls, N.Y.) made commercially available a self-contained virus concentrator referred to as the "Aquella® Virus Concentrator."

These early systems concentrated viruses from large volumes of tapwater and smaller volumes of sewage and seawater, but several problems soon became evident. Suspended matter was so great in wastewater and other turbid waters that it tended to clog the filters and greatly reduce the flow rate. Clarifying filters placed in front of the adsorbing filters reduced the magnitude of this problem, but often decreased the efficiency of virus recovery because they removed solids-associated virus.[14,15]

Further study resulted in a modified version of the portable virus concentrator.[16] In this system, tapwater was acidified to pH 3.5 with 1 *N* HCl and then passed through a virus adsorber consisting of a fiberglass cartridge depth filter (K-27) and a 142 mm diameter, 0.65 μm pore size Cox® membrane filter in series. With this method, small quantities of poliovirus 1 in 100 gal (378 ℓ) volumes of tapwater were concentrated nearly 40,000-fold with average recoveries of 77%. This methodology was also useful for concentrating viruses from 50 to 100 gal amounts of seawater[17] and of wellwater beneath a wastewater land treatment site.[18]

Unfortunately, this system could not be effectively scaled up to process larger volumes at reasonable flow rates. Because of the limited surface area of the flat-disc adsorbent filters, maximum flow rates of only 3 gal (approximately 11 ℓ)/min could be achieved with finished tapwater. Moreover, humic acids and other organic compounds were also concentrated along with the viruses.[19] These substances seriously interfered with reconcentration of viruses in the initial eluate on membrane filters[19] when more than 100 gal of tapwater or smaller volumes of seawater and sewage were processed. To overcome these limitations, Farrah et al.[20] tested a variety of membrane filters and found that Filterite® fiberglass membrane filters (Duo-Fine® series) were far less easily clogged than Cox® (series AA), acrylonitrile polyvinyl chloride copolymer filters (Acropor® series), and nitrocellulose (Millipore®) of approximately the same rated pore size. All filters adsorbed greater than 90% of poliovirus added to tapwater at pH 3.5. Filterite® Duo Fine® filters are manufactured as 10-in. (about 25.4 cm) long, pleated cartridges whose surface areas are 280 times that of a 47 mm diameter disc filter. Flow rates of up to 37.8ℓ/min (10 gal/min) were obtained with the pleated membrane filter cartridges. Large volumes of tapwater can be processed without a prefilter, but a prefilter (a 3 μm pore size Filterite® or a K-27 spun fiberglass depth filter) spares the 0.25 or 0.45 μm final adsorbing filter when sewage and marine waters are processed. The 0.45 μm nominal pore size filter was used with sewage and seawater to reduce problems with clogging. With these filters, seeded polioviruses could be recovered from 472 to 1900 ℓ (125 to 500 gal) of tapwater, 378 ℓ (100 gal) of seawater, and 19 to 190 ℓ (5 to 50 gal) of secondarily treated sewage with an average efficiency of 52, 53, and 50%, respectively. Because humic acids and other organics eluted from the filters in the initial concentration step clogged filters, reconcentration was accomplished by a combination of aluminum flocculation followed by hydroextraction.[21]

Another advantage of this system is that the filters can be reused several times (autoclaving or soaking overnight in a concentrated solution of NaOH between uses) without loss of concentration efficiency, thus reducing operational costs.[27]

Tube filters (Balston®) have also been commonly used for concentrating viruses from tapwater. The use of these epoxy-fiberglass filters was first described by Jakubowski et al.[23] The method has been described in detail also in the 15th edition of *Standard Methods for the Examination of Water and Wastewater.*[37] Virus adsorption to the filters was optimized by pH adjustment to 3.5 and addition of $AlCl_3$.

Elution was accomplished with pH 11.5 glycine buffer. The recovery of seeded polioviruses from 100 gallon (378 ℓ) volumes of tapwater on 8 μm pore size tube filters ranged from 42 to 57%.

Unfortunately, tube filters suffer from many of the drawbacks inherent in flat-disc membrane filters. The Balston® tube filters clog more readily than pleated membrane filters and cannot be used even with moderately turbid water. They cannot be used above an in-line pressure of 25 $lb/in.^2$ and thus high flow rates cannot be achieved. This limits their practical use to only finished tapwater of good quality. In contrast, pleated filters can operate with an in-line pressure of up to 100 $lb/in.^2$ Also, systems that utilize flat and tube filters can process tapwater at only about 5 to 10 ℓ (2.5 gal) /min;[16,17,23] pleated cartridge filters can process almost 40 ℓ (10 gal) in the same period of time.[20]

2. Positively Charged Filter Media

Positively charged microporous filters have been used recently to recover small amounts of seeded poliovirus and bacteriophages from drinking water. These filters are composed of cellulose-diatomaceous earth-"charge modified" resin mixtures and are referred to as series S, Zeta-plus® filters (AMF/CUNO, Meriden, Conn.). These filters remain positively charged up to about pH 6 and can absorb >99% of seeded polioviruses from water at pH 7.5 without the addition of salts. About 80% of the adsorbed viruses were eluted from the filters with a small volume of 0.05 *M* glycine (pH 9.5 to 10). The volume of the eluate was further reduced by adsorption of viruses to smaller filters at pH 6 and then elution with a still smaller volume of glycine at pH 10. Small numbers of polioviruses in 380 mℓ of water were concentrated in a final volume of 25 mℓ of this two-stage procedure; average efficiency of recovery was 22.5%.[3] It is important that the investigators used monodispersed viruses for testing, which probably accounts for the overall apparent lower recovery of viruses, i.e., many elution procedures break-up aggregates giving high apparent recovery of viruses. These studies clearly demonstrated that positively charged filters would have major advantages over the previously used negatively charged filter systems. Viruses could be adsorbed from most waters without prior conditioning of the water and elution could proceed at lower pHs thus reducing the chance of viral inactivation due to pH effects.

With these major advantages in mind, AMF/CUNO (Meriden, Conn.) developed a positively charged filter especially designed for the concentration of viruses from water. This media, 1MDS Virozorb®, was tested by Sobsey and Glass[24] for its efficiency in concentrating polioviruses from tapwater. Samples of tapwater (100 gal) were passed through double-layered, pleated-sheet cartridges of a similar design to the Filterite® negatively charged fiberglass media.[20] In a comparative study, virus adsorption from tapwater between pH 3.5 and 7.5 was more efficient with electropositive filters than with Filterite® filters. Beef extract-glycine at pH 9.5 was an effective eluent for recovering polioviruses from 1MDS® filters. In paired comparative studies, 1MDS® filters, with adsorption at pH 7.5 and no added polyvalent cation salts, gave less variable results in virus concentration efficiencies than did Filterite® filters with adsorption at pH 3.5 plus added $MgCl_2$. Recovery of polioviruses from 1000 ℓ (264 gal) tapwater volumes was approximately 30% efficient with both 1MDS® Virozorb and Filterite® pleated cartridge filters, but the former were much simpler to use. Electropositive filter media appear to offer great promise in simplifying concentration procedures from large volumes of tapwater and perhaps from other types of water. To date, these filters have only been optimized for tapwater. Further research is needed to evaluate their potential with waters containing high concentrations of organics and saline waters, both of which could alter the effectiveness of this type of media.

3. Elution of Viruses from Filters

Many substances have been used to elute adsorbed viruses from filter surfaces. High pH buffers proved useful for some groups of viruses, but could not be used with other groups that were inactivated at pHs above 10.0. Moreover, high pH buffers eluates had to be rapidly neutralized to limit virus inactivation. As a replacement for high pH, basic amino acids,[25] casein,[26] tryptose phosphate broth,[27] nutrient broth,[28] bovine serum albumin,[29] chaotropic agents,[4] beef extract,[10] urea,[31] etc. have been tested.

The most commonly used eluent in recent years has been solutions of 3% beef extract adjusted to pH 9.0 to 10.5.[10,30,32] Not only can the beef extract be used at a lower pH, but reconcentration of the primary eluent is more easily achieved by a bioflocculation procedure originally described by Katzenelson and co-workers.[30]

4. Reconcentration of Viruses from Filter Eluents

For detection of viruses in small amounts of fluids, a one-step adsorption-elution procedure is often used. With larger volumes, however, a second-step reconcentration method must be incorporated. The addition of second-step reconcentration to the virus concentration system may reduce virus recovery, but it reduces the eluate to a manageable volume for subsequent assays.

Techniques that have been used for reconcentration include: two-phase separation,[33] hydroextraction,[21] precipitation with inorganic salts,[11] continuous-flow ultracentrifugation,[34] readsorption to and elution from successively smaller diameter membrane filters,[16] and bioflocculation.[30]

The use of glycine as an eluent enables one to reconcentrate the eluate on another set of smaller diameter filters. Thus, the sample is adjusted to pH 3.5, cations are added, and the sample is passed through the filters. The adsorbed viruses are subsequently eluted with a high pH buffer such as 0.05 *M* glycine at pH 11.0 to 11.5. The pH of this eluate is readjusted to 3.5, cations are added, and viruses are adsorbed to smaller membrane filters which are subsequently eluted with a still smaller volume of eluent.

These procedures do not work, however, when large volumes of tapwater[20] or turbid estuarine water[21,48] are sampled, because organic compounds and metal ions in water adsorb to membrane filters along with the viruses. Elution of viruses from these filters results in elution of the organic compounds as well.[70] Subsequent acidification of the eluate for sequential adsorption-elution results in the production of organic flocs that clog the smaller diameter filters used for the second concentration (reconcentration) step.

These components also form an insoluble precipitate when the eluate is neutralized that interferes with reconcentration on smaller filters when greater than 100 gal of tapwater or smaller volumes of seawater and sewage are processed.[35] By removing organic compounds with activated carbon and ion-exchange resins, reconcentration can be performed with membrane filters.[36] However, large amounts of carbon and resin are required to treat eluates that contain large amounts of organics, and a high percentage of virus adsorption to the columns occurs. Farrah and Bitton[31] reported an alternate reconcentration procedure in which they used a 3% lysine solution at pH 9 to elute viruses adsorbed to Filterite® filters. When this eluate (approximately pH 8.8) was passed through a 25 mm diameter Zeta-plus® C-10 filter, 85 to 90% of the viruses adsorbed to the Zeta-plus® filter. The filter was then eluted with 4 mℓ of 3% beef extract solution at pH 9. With this procedure, they were able to concentrate a seeded poliovirus from 4 ℓ of tapwater to a final volume of 4 mℓ with an average recovery of 67%. Although this method seems to work well with small volumes, experiments need to be done with large volumes of tapwater to make the method a feasible field procedure.

When a protein solution such as beef extract is used to elute viruses adsorbed to filters, the sequential membrane adsorption-elution scheme does not work for reconcentration of the eluates because proteins interfere with virus adsorption to filters.[14] To alleviate this problem, Katzenelson et al.[30] described an organic flocculation method that does not involve the use of membrane filters. In this method, viruses are eluted from filters by 3% beef extract solution at pH 9,[13,41,78] and the pH is then lowered to pH 3.5. This results in flocculation of proteins. The flocs are recovered by centrifugation. This sediment obtained from flocculation of the beef extract is then solublized in 0.15 *M* Na_2HPO_4 (pH 9.0).

5. Recommended Microporous Methods for Concentrating Viruses from Water

Currently, it is difficult to suggest any standard filter methodology for the concentration of viruses from water because of the rapid development of new techniques and filter media. A number of tentative schemes for a standard method for virus concentration from finished tapwaters are given in the 15th edition of *Standard Methods for the Examination of Water and Wastewater.*[37] A flow diagram of recommended strategies is shown in Figure 1. Differences in the strategies occur depending upon the type of adsorbent filter, type of eluent, and choice of reconcentration method.

Studies comparing the various suggested filter media for tapwater[24,38-40] have indicated that the Filterite® and 1 MDS Virozorb® give the best overall efficiencies of virus recovery when beef extract is used as the primary eluent. Table 2 lists the various types of media which have been used for concentrating viruses from water.

Because of the complexity of the methodology for concentrating viruses from large volumes of water, an evaluation of the efficiency of the method with a vaccine strain of poliovirus should first be attempted by anyone not familiar with the technique before attempts are made to isolate naturally occurring viruses. Moreover, because virus concentration methodology is so dependent on the nature of the water being tested, it is beneficial to first evaluate recovery efficiency with test waters artificially seeded with viruses to obtain an idea of the efficiency of virus recovery from the water being examined.

Because the nature and type of suspended matter varies from one type of water to another, one filter system may be more useful than another. The size of the volume processed also influences the type of filter to be used. For example, small diameter cellulose nitrate filters may be adequate for processing 1 or 2 ℓ of tapwater, but not the same volume of sewage or riverwater. Of the negatively charged filter media, the Filterite® fiberglass filters and the Nucleopore® D39 (very fine fiberglass prefilter) are the most resistant to clogging.[21,41]

The 1MDS Virosorb® is also resistant to clogging, but to date has only seen use wth polished tapwaters. The choice of filter combination is based on experience, cost, and availability. Certain modifications are often made in methodology depending upon the virus being concentrated. A list of the different animal viruses for which membrane filter concentration has been reported is given in Table 3.

C. Inorganic Precipitates

A number of procedures have been described for concentrating viruses from waters in which the viruses were either precipitated (coagulated) by or adsorbed to preformed precipitates of polyvalent metal salts. Efficient adsorption usually requires controlled pH and ionic conditions, suggesting that electrostatic forces are involved.[1] Wallis and Melnick[42] first studied the use of preformed precipitates of aluminum phosphate, aluminum hydroxide, and calcium phosphate for concentrating viruses from water. They found that aluminum hydroxide and calcium phosphate efficiently adsorbed enteroviruses and adenoviruses, but not reoviruses, while aluminum phosphate did not adsorb any of these viruses. Wallis and Melnick were successful in concentrating viruses from 1 gal (3.8 ℓ) volumes of sewage by adding 1 g of preformed aluminum hydroxide and collecting the floc by filtration through a Millipore® AP-20 prefilter. The floc was then resuspended in a 10% solution of fetal bovine sera. A similar procedure with ferric hydroxide flocs has also been described.[43]

Aluminum hydroxide flocs formed *in situ* have been used to concentrate viruses from

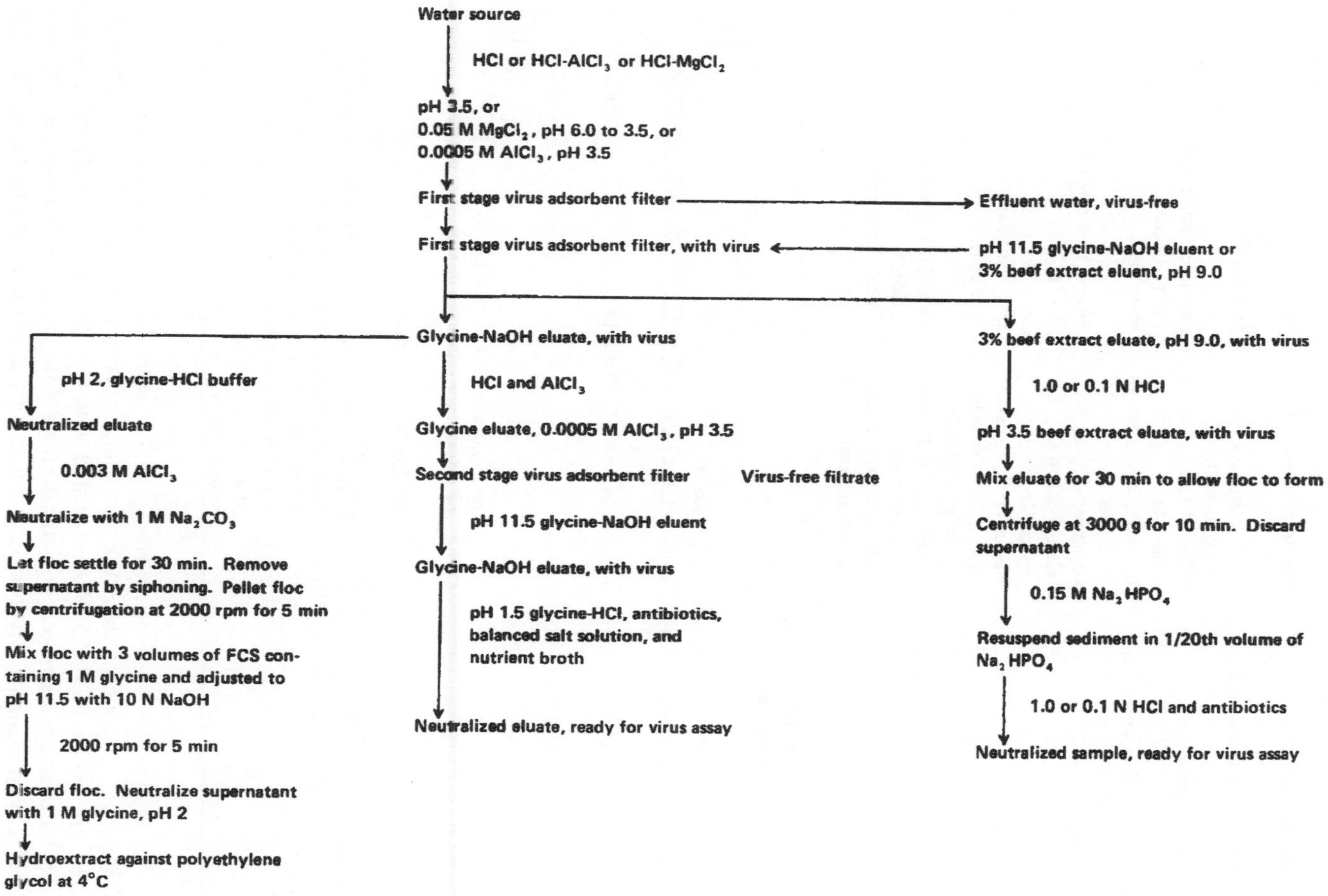

FIGURE 1. Methods for concentrating enteroviruses from large volumes of water. (Modified from *Standard Methods for the Examination of Water and Wastewater*, 1st ed., American Public Health Association, Washington, D.C., 1978.)

Table 2
FILTERS COMMONLY USED FOR CONCENTRATION OF VIRUSES[a]

Composition	Manufacturer	Code
Nitrocellulose	Millipore® Corp., Bedford, Mass.	HA
Epoxy-fiberglass-asbestos	Cox® Instrument Corp., Detroit, Mich.	Series AA[b]
Epoxy-fiberglass	Filterite® Corp., Timonium, Md.	Duo-Fine® series (cartridge filters)
Borosilicate glass microfiber-epoxy resin	Balston®, Inc., Lexington, Mass.	Filter tubes
Fiberglass	Commercial Filter Div., Carborundum Co., Lebanon, Ind.	Wound fiber depth filter (K-27)
Fiberglass	Nucleopore®, Pleasanton, Calif. (formerly Johns-Manville)	D39 (very fine prefilter)
Charge modified cellulose inorganic filter aid	AMF®, CUNO Division, Meriden, Conn.	Zeta-plus® (S and C grades)
Charge modified cellulose	AMF®, CUNO Division, Meriden, Conn.	1-MDS Virozorb®

[a] Modified from Reference 77.
[b] No longer manufactured with asbestos.

Table 3
METHODS FOR CONCENTRATING VIRUSES WITH MEMBRANE FILTERS

Viruses	Remarks	Ref.
Enteroviruses (polioviruses, coxsackieviruses, echoviruses)		24,30,35
Adenoviruses		39,78
Reoviruses	The membrane filter method has not been tested for efficiency of recovery, but isolation of reoviruses from sewage has been reported	16,79
Rotaviruses		8
Hepatitis A virus	This virus can be concentrated with the methods described for enteroviruses	4,24,81
Parvoviruses	Minute virus of mice (MVM) was adsorbed to Zeta-plus® filters at pH 7.5 and eluted at pH 10 with beef extract with an average recovery of 80%	79
Bacteriophages		82

eluates of filters used to concentrate viruses from large volumes of water.[21] In this procedure, a neutralized glycine eluate (0.05 *M*) is adjusted with 0.003*M* aluminum chloride, which results in a pH drop to approximately 4. Neutralization of the sample with a sodium carbonate solution produces an aluminum hydroxide floc that is an efficient virus adsorbent. The floc is collected by centrifugation and mixed with 1*M* glycine dissolved in fetal calf serum, pH 11.5. After mixing, the sample is centrifuged and the supernatant neutralized.

None of the methods involving inorganic precipitates have been used to process more than a few liters of sample since they cannot be scaled up to process large volumes of water.

D. Polyelectrolytes

Viruses have been concentrated successfully from sewage and water by adsorption onto and elution from insoluble polyelectrolytes. The polymer which has been used most extensively is referred to as PE60, produced by Monsanto®. The compound PE60 is a cross-linked copolymer of isobutylene/maleic anhydride containing diloweralkyl-aminoloweralkylimide groups.[44] According to Cookson,[44] the PE60 adsorption sites for viruses are carboxyl groups and ammonium radicals which are involved in hydrogen bonding and quaternary ammonium groups which electrostatically interact with the negatively charged virus surface. Elution of viruses from PE60 occurs at pH 8 to 9 in isotonic solutions. Elution is also enhanced in the presence of protein solutions such as fetal calf serum, since they compete with viruses for adsorption sites. PE60 can be used both in batch processes and sandwiched between filter pads to concentrate viruses from water.[45,46] PE60 was effective for concentrating viruses from water and sewage,[45-47] but has not been used in recent years because of variability in its efficiency in virus concentration.[1] PE60 is an unstable polymer and with increasing storage time, it becomes chemically altered and less efficient as a virus adsorbent. Moreover, different production lots of PE60 have varied in their chemical characteristics, including the optimum pH for virus adsorption. The efficiency of PE60 is not identical for different enteric viruses.[46]

E. Minerals and Clays

Talcum powder, a hydrous magnesium silicate, has been used for concentrating enteric viruses from water by a "sandwich" technique similar to that developed for PE60.[48,49] This technique has been used extensively by Sattar and co-workers[49-51] for enteric virus isolation from tapwater, riverwater, and sewage. In practice, water samples are adjusted to pH 6.0, Earle's balanced salt solutions added to a final concentration of 1:100, and the sample is passed through a talc-celite layer. The layer is composed of three parts talc and one part celite sandwiched between two coarse filter papers. Viruses are eluted from the talc with 10% fetal calf serum or 3% beef extract adjusted to pH 9.0. The talc-celite concentration method is able to recover efficiently small numbers of viruses from up to 1000 ℓ of tapwater. The clay mineral, bentonite, has been used for virus concentration from small volumes (0.5 to 2 ℓ) of wastewater.[52] The method consists of adsorbing viruses to bentonite clay in the presence of 0.01 M $CaCl_2$ and allowing a sufficient length of time for virus adsorption to the clay. The virus-clay complex is then removed by centrifugation and eluted with tryptose phosphate broth.

Magnetic and nonmagnetic mineral iron oxides have also been used as adsorbents in the concentration of viruses from water. This mineral has been used in batch studies similar to those in which bentonite has been used[53] as a filter sandwich,[54] or in packed columns.[54] In the batch procedure developed by Bitton and Mitchell,[55] $CaCl_2$ was added to obtain optimal adsorption of coliphase T7 to magnetic iron oxide (Fe_3O_4). The magnetic iron oxide-virus complex was collected by these investigators by passage of the mixture through stainless steel wool placed in a background magnetic field. The magnetite, along with the viruses, were retained by the coarse filter. Viruses were eluted from the matrix with either beef extract[59] or isoelectric casein[53] adjusted to pH 8 to 9.

F. Glass Beads

Small glass beads and glass powder have been used as adsorbents by several French investigators for concentration of viruses from tapwater.[56,57] As with negatively charged microporous filters, adsorption of viruses to the glass beads was enhanced by adjustment of the pH of water to 3.5 and addition of $AlCl_3$ to a final concentration of 0.0005M. Viruses were eluted from the beads with pH 11.5, 0.005M glycine buffer. Large volumes of water can be processed by passage of the water sampled through columns of glass beads.[58] With such a system, Schwartzbrod and Lucerna-Gutierrez[58] recovered poliovirus[1] (LSc 2ab) from 50 ℓ volumes of tapwater, with a mean efficiency of 55%. The processing of 50 ℓ of tapwater necessitated the use of 150 g of glass beads. The flow rate through the columns was 80ℓ/hr. Systems using glass columns capable of processing 400 ℓ volumes of tapwater have been constructed.

G. Protamine Sulfate

Protamines are basic proteins rich in arginine with an isoelectric point near 12 and are very useful precipitants near neutral pH for negatively charged macromolecules of high molecular weight such as viruses. England[59] developed and tested procedures utilizing the sulfate salt of protamine for concentrating enteric viruses from clarified sewage and treated effluent. Bovine albumin was added to the sample as a precipitate enhancer along with the protamine sulfate at pH 7.5 to 7.8, and the precipitate collected by filtration. The precipitate was then dissolved in a small volume of NaCl and diluted to isotonicity before assay. Reoviruses and adenoviruses in experimentally contaminated wastewater were recovered with efficiencies of 80 to 100%, but enterovirus recoveries were variable and depended upon the specific virus tested.

III. OTHER METHODS FOR CONCENTRATING VIRUSES FROM WATER

A. Ultrafiltration and Reverse Osmosis

Ultrafiltration refers to the passage of solutions through membranes, usually of cellulosic material, with pore sizes that permit the passage of water and low molecular weight materials but prevent the passage of viruses and macromolecules. The viruses and macromolecules are then concentrated on the membrane or in the liquid portion that does not pass through the membrane. Because of the small pore size, these membrane filters are subject to rapid clogging and have usually been used only to concentrate viruses from distilled water.[60,61] This clogging problem can be greatly reduced by flowing water tangentially across the membrane as is the case with hollow fiber membranes.

Ultrafilters consisting of aluminum alginate gel containing lanthanum ions have been used to concentrate viruses from 1 to 10 ℓ volumes of preclarified sewage and tapwater.[1,62] These ultrafilters are soluble in sodium citrate. The filters clog easily and most waters need to be prefiltered resulting in potential loss of virus. Aluminum alginate gel ultra filters have been used recently to concentrate viruses recovered by a filter adsorption-elution procedure from treated wastewater.[63] A number of workers have reported the application of anisotropic membranes to the concentration of viruses from water.[64-66] Nupen and Stander[64] used polymeric membranes with a molecular weight "cut-off" of 30,000 (Amicon® Type PM30) to concentrate viruses from 10 ℓ effluent samples from the Windhoek wastewater reclamation plant, South-West Africa. Viruses were eluted from the membrane with a salt solution containing 10% fetal calf serum and 0.5% lactalbumin hydrolysate. The average recovery of viruses was 70%. Foliguet et al.[66] reported that poliovirus 1 could be concentrated in a battery of flat-sheet, cellulosic ultrafilter elements in an tangential flow system to a volume of 350 mℓ from 20-ℓ volumes of tapwater and from 50-ℓ volumes of filtered or raw riverwater with recovery efficiencies of 50 to 100%.

Belfort et al.[65] used polysulfone asymmetric hollow fiber membranes to concentrate poliovirus 1 from 50-ℓ volumes of tapwater to a final volume of 250. The average efficiency was 42%. They found that backwashing of the membranes was necessary. They also found that polysulfone membranes were superior to cellulose acetate membranes for virus recovery and in general chemical and biological stability. A time period of 170 to 260 min was required to process 50-ℓ volumes of tapwater.

Recently, capillary (1.5 mm diameter capillaries) rather than hollow fibers have been used to concentrate viruses from 100-ℓ volumes of tapwater.[65] The water was continuously recirculated through the concentrating unit until the volume was reduced to approximately 1 ℓ. This concentration required about 86 min. During processing of the sample, the virus adsorbed to the membrane. Viruses were eluted from the membrane by backwash with 1 ℓ of 1% beef extract and further reconcentrated by organic flocculation.[30] With this procedure, poliovirus 1 could be concentrated from 100 ℓ to approximately 35 mℓ with an average virus recovery efficiency of 35%.

Although tangential flow ultrafiltration appears a promising alternative to filter adsorption-elution approaches to concentrating viruses from water, the high cost of such ultrafiltration and the tendency of ultrafilters to clog has limited their present application to laboratory studies. Belfort et al.[66] found that larger ultrafiltration units with reusable membranes could concentrate viruses from 1900 ℓ of tapwater in about 3 hr.

Reverse osmosis (RO) membranes are of a finer porosity than conventional ultrafilters and retain microsolutes which are within one order of magnitude of the solvent. They generally operate at high pressures. Sweet and co-workers[60] used asymmetric cellulosic RO membranes to recover viruses, and although recoveries of viruses from distilled water were good, recoveries from tapwater were poor.

B. Hydroextraction

Hydroextraction is a simple method for concentrating viruses in which the sample is placed in a semipermeable membrane such as dialysis tubing and exposed to a macromolecular hydroscopic material such as polyethylene glycol, which results in the transfer of water but not macromolecules of viruses across the membrane to the hydroscopic material. This method has been used to concentrate viruses directly from water samples[69] and as a second-stage concentration step for recovering viruses from filter eluents.[21,35] Hydroextraction is a simple method, but it is only useful for samples of 1 ℓ or less. It may also result in the concentration of materials toxic to cell cultures.[70]

C. Two-Phase Separation

When two different organic polymers are dissolved in water, two liquid phases may be produced which are not compatible. Sometimes, viruses and macromolecules can be partitioned between the two immiscible aqueous phases that are produced.[71] Partitioning of viruses depends upon the type of polymers and their molecular weights and the ionic composition and strength of the partitioning system. By selecting the appropriate polymers and controlling ionic composition, ionic strength, and pH, viruses can be partitioned largely into one of two phases. If the volume of the virus-containing phase is small in relation to the original fluid volume, then a considerable degree of concentration is achieved. The polymer mixtures usually used for concentrating viruses from water are dextran-polyethylene glycol or dextran sulfate-polyethylene glycol.[72-74] Two-phase separation has been used to concentrate viruses from sewage effluents,[72] and tapwater,[73] and in the reconcentration of eluates from filters.[28]

Grindrod and Cliver[75] found that dextran sulfate inhibited certain enteroviruses and interfered with their detectability in cell cultures. These investigators recommended replacing dextran sulfate with dextran.[76] Shuval et al.[73] used a two-step phase separation procedure to achieve concentration factors of 250 to 500 for viruses in 2 to 7 ℓ of raw sewage or tapwater; 7ℓ is about the largest volume size that can be sampled by two-phase separation.

V. SUMMARY AND CONCLUSIONS

Methods involving virus adsorption-elution from microporous filters continue to be the most promising and useful methods for concentrating viruses from large volumes of water.

The recent development of positively charged microporous filter media offers further simplification of current methodology. Methods involving microporous filters are the only ones which have been shown useful for concentrating viruses from large volumes of tapwater, sewage, seawater, and other types of natural waters. Glass beads and talc "sandwiches" may be used for concentrating viruses from 100- to 1000-ℓ volumes of tapwater but offer no distinct advantage over microporous filter methods. Ultrafiltration may offer some future alternative to methods dependent on adsorption but the high cost of the necessary capital equipment and lack of current application of the method to only clean waters limits its current usefulness.

Currently, there are a number of methods particularly well-suited for concentration of viruses from small volumes of water (i.e., less than 4 ℓ) Some such methods are bioflocculation, adsorption to clays and inorganic precipitates, and hydroextraction. In summary, reasonably reliable methods are currently available for concentrating viruses from water. Improvements, of course, are needed to increase the efficiency and general application of these methods to various environmental waters.

REFERENCES

1. **Sobsey, M. D.,** Methods for detecting enteric viruses in water and wastewater, in *Viruses in Water,* Berg, G., Bodily, H. L., Lennette, E. H., Melnick, J. L., and Metcalf, T. G., Eds., American Public Health Association, Washington, D.C., 1976, 89.
2. **Hill, W. F., Jr., Akin, E. W., and Benton, W. H.,** Detection of viruses in water: a review of methods and application, *Water Res.*, 5, 967, 1971.
3. **Sobsey, M. D. and Jones, B. L.,** Concentration of poliovirus from tapwater using positively charged microporous filters, *Appl. Environ. Microbiol.*, 37, 588, 1979.
4. **Farrah, S. R., Shah, D. O., and Ingram, L. O.,** Effects of chaotropic and antichaotropic agents on the elution of poliovirus adsorbed to membrane filters, *Proc. Natl. Acad. Sci. U.S.A.*, 78, 1229, 1981.
5. **Hou, K., Gerba, C. P., Goyal, S. M., and Zerda, K. S.,** Capture of latex beads, bacteria, endotoxin, and viruses by charge-modified filters, *Appl. Environ. Microbiol.*, 40, 892, 1980.
6. **Mix, T. W.,** The physical chemistry of membrane-virus interaction, *Dev. Ind. Microbiol.*, 15, 136, 1974.
7. **Metcalf, T. G.,** Use of membrane filters to facilitate the recovery of virus from aqueous suspensions, *Appl. Microbiol.*, 9, 376, 1961.
8. **Cliver, D. O.,** Enterovirus detection by membrane chromatography, in *Transmission of Viruses by the Water Route,* Berg, G., Ed., Wiley-Interscience, New York, 1967, 139.
9. **Wallis, C. and Melnick, J. L .,** Concentration of viruses from sewage by adsorption on Millipore membranes, *Bull. W.H.O.*, 36, 219, 1967.
10. **Berg, G., Dahling, D. R., and Berman, D.,** Recovery of small quantities of viruses from clean waters on cellulose nitrate membrane filters, *Appl. Microbiol.*, 22, 608, 1971.
11. **Wallis, C., Henderson, M., and Melnick, J. L.,** Enterovirus concentration on cellulose membranes, *Appl. Microbiol.*, 23, 46, 1972.
12. **Wallis, C. and Melnick, J. L.,** A portable virus concentrator for use in the field, in *Advances in Water Pollution Research,* Jenkins, S. H., Pergamon Press, Elmsford, N.Y., 1973, 119.
13. **Wallis, C., Homma, A., and Melnick, J. L.,** A portable virus concentrator for testing water in the field, *Water Res.*, 6, 1249, 1972.
14. **Homma, A., Sobsey, M. D., Wallis, C., and Melnick, J. L.,** Virus concentration from sewage, *Water Res.*, 7, 945, 1973.
15. **Metcalf, T. G., Wallis, C., and Melnick, J. L.,** Environmental factors influencing isolation of enteroviruses from polluted surface waters, *Appl. Microbiol.*, 27, 920, 1974.
16. **Sobsey, M. D., Wallis, C., Henderson, M., and Melnick, J. L.,** Concentration of enteroviruses from large volumes of water, *Appl. Microbiol.*, 26, 529, 1973.
17. **Sobsey, M. D., Gerba, C. P., Wallis, C., and Melnick, J. L.,** Concentration of enteroviruses from large volumes of turbid estuary water, *Can. J. Microbiol.*, 23, 770, 1977.
18. **Gilbert, R. G., Gerba, C. P., Rice, R. C., Bouwer, H., Wallis, C., and Melnick, J. L.,** Virus and bacteria removal from wastewater by land treatment, *Appl. Environ. Microbiol.*, 32, 333, 196.

19. **Farrah, S. R., Goyal, S. M., Gerba, C. P., Wallis, C., and Shaffer, P. T. B.,** Characteristics of humic acid and organic compounds concentrated from tapwater using the Aquella virus concentrator, *Water. Res.*, 10, 897, 1976.
20. **Farrah, S. R., Gerba, C. P., Wallis, C., and Melnick, J. L.,** Concentration of viruses from large volumes of tapwater using pleated membrane filters, *Appl. Environ. Microbiol.*, 31, 221, 1976.
21. **Farrah, S. R., Goyal, S. M., Gerba, C. P., Goyal, S. M., Wallis, C., and Melnick, J. L.,** Concentration of enteroviruses from estuarine water, *Appl. Environ. Microbiol.*, 33, 1192, 1977.
22. **Farrah, S. R., Gerba, C. P., Goyal, S. M., Wallis, C., and Melnick, J. L.,** Regeneration of pleated filters used to concentrate enteroviruses from large volumes of tapwater, *Appl. Environ. Microbiol.*, 33, 308, 1977.
23. **Jakubowski, W., Hoff, J. C., Anthony, N. C., and Hill, W. F., Jr.,** Epoxy-fiberglass adsorbent for concentrating viruses from large volumes of potable water, *Appl. Microbiol.*, 28, 501, 1974.
24. **Sobsey, M. D. and Glass, J. S.,** Poliovirus concentration from tapwater with electropositive adsorbent filters, *Appl. Environ. Microbiol.*, 4, 201, 1980.
25. **Farrah, S. R. and Bitton, G.,** Elution of poliovirus adsorbed to membrane filters, *Appl. Environ. Microbiol.*, 36, 982, 1978.
26. **Bitton, G., Feldberg, B. N., and Farrah, S. R.,** Concentration of enteroviruses from seawater and tapwater by organic flocculation using non-fat dry milk and casein, *Water, Air Soil Pollut.*, 12, 187, 1979.
27. **Farrah, S. R., Goyal, S. M., Gerba, C. P., Conklin, R. H., Wallis, C., Melnick, J. L., and Dupont, H. L.,** A simple method for concentration of enteroviruses and rotaviruses from cell culture harvests using membrane filters, *Intervirology*, 9, 56, 1978.
28. **Hill, W. F., Jr., Akin, E. W., Benton, W. H., and Metcalf, T. G.,** Virus in water. II. Evaluation of membrane cartridge filters for recovering multiplicities of poliovirus in water, *Appl. Microbiol.*, 23, 880, 1972.
29. **Goyal, S. M., Hanssen, H., and Gerba, C. P.,** A simple method for the concentration of influenza virus from allantoic fluid on microporous filters, *Appl. Environ. Microbiol.*, 39, 500, 1980.
30. **Katzenelson, E., Fattal, B., and Hostovesky, T.,** Organic flocculation: an efficient second-step concentration method for the detection of viruses in tapwater, *Appl. Environ. Microbiol.*, 32, 638, 1976.
31. **Farrah, S. R. and Bitton, G.,** Low molecular weight substitutes for beef extract as eluents for poliovirus adsorbed to membrane filters, *Can. J. Microbiol.*, 25, 1945, 1979.
32. **Landry, E. F., Vaughn, J. M., Thomas, M. Z., and Vicale, T. V.,** Efficiency of beef extract for the recovery of poliovirus from wastewater effluents, *Appl. Environ. Microbiol.*, 36, 544, 1978.
33. **Hill, W. F., Jr., Akin, E. W., Benton, W. H., Mayhew, C. J., and Jakubowski, W.,** Apparatus for conditioning unlimited quantities of finished waters for enteric virus detection, *Appl. Microbiol.*, 27, 1177, 1974.
34. **Scutt, J. E.,** Virus retention by membrane filters, *Water Res.*, 5, 183, 1971.
35. **Gerba, C. P., Farrah, S. R., Goyal, S. M. Wallis, C., and Melnick, J. L.,** Concentration of enteroviruses from large volumes of tapwater, treated sewage and seawater, *Appl. Environ. Microbiol.*, 35, 540, 1978.
36. **Farrah, S. R., Wallis, C., Shaffer, P. T. B., and Melnick, J. L.,** Reconcentration of poliovirus from sewage, *Appl. Environ. Microbiol.*, 32, 653, 1976.
37. American Public Health Association, *Standard Methods for the Examination of Water and Wastewater*, 15th ed., American Public Health Association, Washington, D.C., 1980.
38. **Jakubowski, W., Hill, W. F., and Clarke, N. A.,** Comparative study of four microporous filters for concentrating viruses from drinking water, *Appl. Environ. Microbiol.*, 31, 58, 1975.
39. **Sobsey, M. D., Glass, J. S., Carrick, R. J., Jacobs, R. R., and Rutala, W. A.,** Evaluation of the tentative standard method for enteric virus concentration from large volumes of tapwater, *J. Am. Water Works Assoc.*, 72, 292, 1980.
40. **Sobsey, M. D., Glass, J. S., Jacobs, R. R., and Rutala, W. A.,** Modifications of the tentative standard method for improved virus recovery efficiency, *J. Am. Water Works Assoc.*, 72, 350, 1980.
41. **Payment, P. and Trudel, M.,** Efficiency of several microfiber glass filters for recovery of poliovirus from tapwater, *Appl. Environ. Microbiol.*, 38, 365, 1979.
42. **Wallis, C. and Melnick, J. L.,** Concentration of viruses on aluminum and calcium salts, *Am. J. Epidemiol.*, 85, 459, 1967.
43. **Schaefer, E.,** Experimentelle Untersuchungen zur quantitativen Erfassung von Polio-Impfvirus Typ II, in *Oberflachenwasser*, GWF-Wasser/Abwasser, 112, 109, 1971.
44. **Cookson, J. T.,** The chemistry of virus concentration by chemical methods, *Dev. Ind. Microbiol.*, 15, 160, 1974.
45. **Grinstein, S., Melnick, J. L., and Wallis, C.,** Virus isolations from sewage and from a stream receiving effluents of sewage treatment plants, *Bull. W.H.O.*, 42, 291, 1970.
46. **Wallis, C., Grinstein, S., Melnick, J. L., and Fields, J. E.,** Concentration of viruses from sewage and excreta on insoluble polyetectrolytes, *Appl. Microbiol.*, 18, 1007, 1969.

47. **Hawley, H. B., Morin, D. P., Geraghty, M. E., Tomkow, J., and Phillips, C. A.,** Coxsackievirus B epidemic at a boys' summer camp, *JAMA*, 226, 33, 1973.
48. **Kalter, S. S. and Millstein, C. H.,** Efficacy of methods for the detection of viruses in treated and untreated sewage, in *Virus Survival in Water and Wastewater Systems*, Malina, J. F., Jr. and Sagik, B. P., Eds., Center for Research in Water Resources, The University of Texas at Austin, 1974, 33.
49. **Sattar, S. A. and Westwood, J. C. N.,** Comparison of talc-celite and polyelectrolyte 60 in virus recovery from sewage: development of techniques and experiments with poliovirus (type 1, Sabin)-contaminated samples, *Can. J. Microbiol.*, 22, 1620, 1976.
50. **Sattar, S. A. and Westwood, J. C. N.,** Viral pollution of surface waters due to chlorinated primary effluents, *Appl. Environ. Microbiol.*, 36, 427, 1978.
51. **Sattar, S. A. and Ramia, S.,** Use of talc-celite layers in the concentration of enteroviruses from large volumes of potable waters, *Water Res.*, 13, 637, 1979.
52. **Moore, B. E., Sagik, B. P., and Sorber, C. A.,** Procedure for the recovery of airborne human enteric viruses during spray irrigation of treated wastewater, *Appl. Environ. Microbiol.*, 38, 688, 1979.
53. **Farrah, S. R. and Bitton, G.,** Methods (other than microporous filters) for concentration of viruses from water, in *Methods in Environmental Virology*, Gerba, C. P. and Goyal, S. M., Eds., Marcel Dekker, New York, 1982, 117.
54. **Rao, V. C., Sullivan, R., Read, R. B., and Clarke, N. A.,** A simple method for concentrating and detecting viruses, *J. Am. Water Works Assoc.*, 60, 1288, 1968.
55. **Bitton, G. and Mitchell, R.,** The removal of *Escherichia coli* bacteriophage T7 by magnetic filtration, *Water Res.*, 8, 549, 1974.
56. **Sarrette, B. A., Danglot, C. D., and Vilagines, R.,** A new and simple method for recuperation of enteroviruses from water, *Water Res.*, 11, 355, 1978.
57. **Hugues, B., Plisser, M., Andre, M., Pagliardini, D., and Laurent, A.,** Evaluation de la charge viral dans les eaux d'une station d'epuration biologique au moyen de deux methodes de concentration de virus par adsorption-elution sur poudre de verre ou sur microfibre de verre, *Water Res.*, 13, 1117, 1979.
58. **Schwartzbrod, L. and Lucena-Gutierrez, F.,** Concentration des enteroviruses dans les eaux par adsorption sur poudre de verre: proposition d'un appareillage simplifié, *Microbia*, 4, 55, 1978.
59. **England, B.,** Concentration of reovirus and adenovirus from sewage and effluents by protamine sulfate (salmino) treatment, *Appl. Microbiol.*, 24, 510, 1972.
60. **Sweet, B. A., Ellender, R. D., and Leong, J. K. L.,** Recovery and removal of viruses from water utilizing membrane techniques, *Dev. Ind. Microbiol.*, 15, 143, 1974.
61. **Berman, D., Rohr, M. E., and Safferman, R. S.,** Concentration of poliovirus in water by molecular filtration, *Appl. Environ. Microbiol.*, 40, 426, 1980.
62. **Gartner, H. and Witt, G.,** Virologische Stuhl-und Abwasser-unter-suchungen, ein Beitrag zur Praktischen Erfassung der Poliomyelitis-Sitaution, *Arch. Hyg. Bacteriol.*, 150, 189, 1966.
63. **Selna, M. W. and Miele, R. P.,** Virus sampling in wastewater-field experiences, *J. Environ. Eng. Div., Am. Soc. Civ. Eng.*, 103, 693, 1977.
64. **Nupen, E. M. and Stander, G. V.,** The virus problem in the Windhoek wastewater reclamation plant, in *Advances in Water Pollution Research*, Jenkins, S. H., Ed., Pergamon Press, Elmsford, N.Y., 1972, 133.
65. **Belfort, G., Rotem, Y., and Katzenelson, E.,** Virus concentration using hollow fiber membranes — II, *Water Res.*, 10, 279, 1976.
66. **Foliguet, J. M., Lavillaurex, J., and Schwartzbrod, L.,** Viruses in water. II. General review of the methods available to detect viruses in water, *Rev. Epidemiol. Med. Soc. Sante Publique*, 21, 185, 1973.
67. **Rotem-Borensztajn, Y., Katzenelson, E., and Belfort, G.,** Virus concentration by capillary ultrafiltration, *J. Environ. Eng. Div. Am. Soc. Civ. Eng.*, 105, 401, 1979.
68. **Belfort, G., Rotem-Borensztajn, Y., Katzenelson, E.,** Virus concentration from waters by hollow fiber ultrafiltration: the importance of hydrodynamics, in *Proceedings Water Reuse Symposium*, Vol. 3, American Water Works Association Research Foundation, Denver, Colo., 1979, 1925.
69. **Wellings, F. M., Lewis, A. L., and Mountain, C. W.,** Demonstration of solids-associated virus in wastewater and sludge, *Appl. Environ. Microbiol.*, 31, 354, 1976.
70. **Shuval, H. I., Cymbalista, S., Fattal, B., and Goldblum, N.,** Concentration of enteric viruses in water by hydro-extraction and two-phase separation, in *Transmission of Viruses by the Water Route*, Berg, G., Ed., John Wiley & Sons, New York, 1967, 45.
71. **Albertsson, P.,** Two-phase separation of viruses, in *Methods in Virology*, Maramorosch, K. and Koprowski, H., Eds., Academic Press, New York, 1967, 303.
72. **Lund, E. and Hedstrom, C. E.,** The use of an aqueous polymer phase system for enterovirus isolations from sewage, *Am. J. Epidemiol.*, 84, 287, 1966.
73. **Shuval, H. I., Fattal, B., Cymbalista, S., and Goldblum, N.,** The phase-separation method for the concentration and detection of viruses in water, *Water Res.*, 3, 225, 1969.
74. **Nupen, E. M.,** Virus studies on the Windhoek wastewater reclamation plant (South-West Africa), *Water Res.*, 4, 661, 1970.

75. **Grindrod, J. and Cliver, D. O.,** Limitations of the polymer two-phase system for detection of viruses, *Arch. Ges. Virusforsch.*, 28, 337, 1969.
76. **Grindrod, J. and Cliver, D. O.,** A polymer two-phase system adapted to virus detection, *Arch. Ges. Virusforsch.*, 31, 365, 1970.
77. **Goyal, S. M. and Gerba, C. P.,** Membrane filters in virology, in *Membrane Filtration: Applications, Techniques and Problems*, Dutka, B. J., Ed., Marcel Dekker, New York, 1981, 219.
78. **Fields, H. A. and Metcalf, T. G.,** Concentration of adenovirus from seawater, *Water Res.*, 9, 357, 1975.
79. **Gilbert, R. G., Gerba, C. P., Rice, R. C., Bouwer, H., Wallis, C., and Melnick, J. L.,** Virus and bacteria removal from wastewater by land treatment, *Appl. Environ. Microbiol.*, 32, 333, 1976.
80. **Smith, E. M. and Gerba, C. P.,** Survival and detection of rotaviruses in the environment, in *Proc. 3rd Int. Symp. Neonatal Diarrhea*, Veterinary Infectious Disease Organization, Saskatoon, Saskatchewan, Canada, 1981, 67.
81. **Hejkal, T. W., Keswick, B., LaBelle, R. L., Gerba, C. P., Sanchez, Y., Dressman, G., Hafkin, B., and Melnick, J. L.,** Viruses in a community water supply associated with an outbreak of gastroenteritis and infectious hepatitis, *J. Am. Water Works Assoc.*, 74, 318, 1982.
82. **Goyal, S. M., Zerda, K. S., and Gerba, C. P.,** Concentration of coliphages from large volumes of water and wastewater, *Appl. Environ. Microbiol.*, 39, 85, 1980.

Removal of Viruses from Waters and Sludges by Treatment Processes

Chapter 3

REMOVAL OF VIRUSES FROM WASTEWATER AND EFFLUENT BY TREATMENT PROCESSES

Charles A. Sorber

TABLE OF CONTENTS

I. INTRODUCTION

Historically, sewers were constructed primarily to carry stormwater from urban areas. With the advent of water carriage waste disposal systems, the storm drains acted as combined sewers. The combined sewers were directed to the nearest watercourse over the shortest route. When it became mandatory to treat sanitary wastewater, collector sewers were constructed to intercept the dry-weather flow of the combined sewers. This interceptor carried the sanitary wastewater to a central point for treatment.

Although many older urban areas are still served by some combined sewers, newer communities have separate sanitary and storm sewers. Obviously, it would be very costly to replace all combined sewers with separate sewer systems. Thus, there has been a gradual improvement in the quality of wastewater collection systems. Nevertheless, many municipalities have some combined sewers and, during periods of precipitation, discharge a portion of their sanitary wastewater directly to receiving watercourses without benefit of treatment. This fact should not be lost in a discussion of the removal of viruses from wastewater and the establishment of viral standards for wastewater treatment plant effluents.

Wastewater conveyed to a treatment facility comes from households and industries. Furthermore, wastewater can include significant quantities of infiltration of runoff or groundwater into the sewer system. Water quality criteria, established for receiving waters, have been used to help define the amount of treatment required prior to discharge to a watercourse. Effluent standards dictate the degree of treatment required for a wastewater of some specific quality and quantity. The minimal treatment required under law is secondary treatment. Some municipalities on major waterways, however, still provide only primary treatment; others have installed advanced wastewater treatment systems. For examples of some of the wastewater treatment schemes commonly used see Figure 1.

Conventional wastewater treatment facilities are designed to reduce the amount of organic material (as measured by the biochemical oxygen demand test) and suspended material discharged to natural waters. Only the disinfection process is intentionally included for the reduction of concentrations of microorganisms in the treatment train effluent. Nevertheless, some wastewater treatment unit processes do impact the population of microorganisms in raw sewage.

Unfortunately, the overwhelming majority of the information available on virus removal by treatment processes has been developed with laboratory strains of viruses and model or laboratory-scale treatment units. These laboratory systems are not subjected to normal variations in hydraulic and organic loading and other variables found under field conditions. Furthermore, exogenously added test viruses may not respond to removal/inactivation as do indigenous viruses. Equally important is the fact that results obtained from the evaluation of a few strains of human viruses or bacteriophages cannot be unequivocally extrapolated to all human viruses of concern. Consequently, much of the data represent an overestimate of viral removal efficiencies. Thus, much of the published data must be interpreted with caution as to its reproducibility or applicability to field conditions.

II. VIRUS REMOVAL BY TREATMENT PROCESSES

The efficiency of various wastewater treatment processes for the removal of viruses has been discussed by Berg[1,2] and Sproul.[3] Treatment process effectiveness can have a marked effect on the performance of virus concentration/detection methodologies. Furthermore, the actual wastewater treatment process train employed dictates to a large degree the disposition of the viruses.

A. Primary Treatment

Most conventional wastewater treatment facilities consist of some preliminary processes

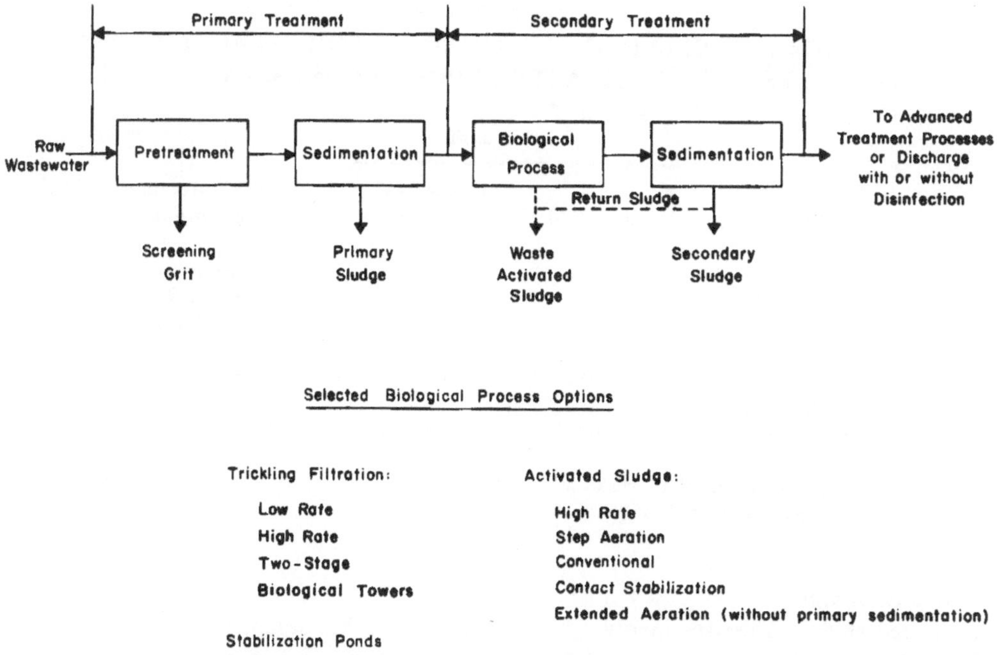

FIGURE 1. Selected wastewater treatment options.

such as pumping, screening, and grit removal. For all practical purposes these processes should not affect the levels of viruses in the wastewater.

The first major process in a conventional wastewater treatment facility is that of primary treatment. Primary treatment is sedimentation and it is designed to remove heavy solids and floatable materals. Primary sedimentation with overflow rates of 600 gal/day/ft^2 can remove 40 to 60% of the suspended solids and perhaps up to 35% of the 5-day biochemical oxygen demand (BOD_5) found in raw municipal wastewater.

Primary sedimentation provides minimal virus removal. Malherbe and Strickland-Cholmley[4] reported removals up to 69% for poliovirus 1. In studies involving polioviruses 1, 2, and 3 under field conditions, England and co-workers[5] reported removals of between 0 to 12%.

Sproul,[3] in a review of the effectiveness of conventional wastewater treatment processes on virus removal, indicated that primary sedimentation provided incomplete removal for all organisms and occasionally increases in the numbers of viruses were detected. The latter phenomenon has been attributed to fecal solids breaking up in the sedimentation tank thereby releasing viruses to the wastewater stream.

B. Secondary Treatment

Secondary wastewater treatment is a biological process involving aeration to metabolize and flocculate colloidal and dissolved organics. In addition to the biological processes (generally some form of activated sludge or trickling filtration), final clarification (or sedimentation) of the wastewater is accomplished. This part of the process separates the solids generated through biological metabolism from the liquid stream.

Satisfactorily operating secondary wastewater treatment processes remove 90% or more of the BOD_5 and suspended solids originally present in the raw wastewater. Generally, this results in BOD_5 and suspended solids levels of less than 30 mg/ℓ in the process effluent.

Table 1
GENERAL OPERATIONAL VARIABLES FOR SELECTED TYPES OF ACTIVATED SLUDGE PROCESSES

	Aeration basin		
Type process	Loading (F/M[a])	Detention time (hours)	MLSS[b] in Aeration basin (mg/ℓ)
High rate	0.4—1.5	0.5—2	4,000—10,000
Step aeration	0.2—0.4	5—7	2,000—3,500
Conventional	0.2—0.4	4—8	1,500—3,000
Contact stabilization			
Contact basin	0.2—0.6	0.5—1	1,000—3,000
Stabilization basin	—	3—6	4,000—10,000
Extended aeration	0.05—0.15	18—36	3,000—6,000

[a] Pounds of BOD_5/day/lb of MLSS in the aeration basin.
[b] Mixed liquor suspended solids.

1. Activated Sludge

Activated sludge processes involve the suspension of microorganisms in a wastewater. Air is added to provide an aerobic environment and to ensure adequate mixing. The aeration basin (reaction chamber) is the heart of the activated sludge system. Its design and operation is governed by a number of variables that include organic loading, detention time, and the concentration of microorganisms in suspension. The range of general operating variables for selected types of activated sludge processes are shown in Table 1.

The variables shown in Table 1 clearly indicate that there is a wide range in operating conditions available for the different types of activated sludge processes. Furthermore, these are typical design ranges which are based on average daily BOD_5 in the influent to the process and average daily flow rates, conditions which exist for very short periods of time during any given day.

In evaluating the performance of activated sludge processes (or any treatment process, for that matter), it is important to describe the type of process involved, variations in flow and organic loading, and actual operating conditions when they deviate from design assumptions. Flow-composite sampling is essential to accommidate daily variations in flow and microbiological loading. Thus, laboratory (bench) studies rarely simulate conditions that exist at field sites.

The activated sludge process is effective for the removal of large numbers of viruses from the liquid portion of wastewater. In one study, Moore et al.[6] evaluated a 10 million gal/day (MGD) contact stabilization plant located in Austin, Texas. The plant operated with a contact time of 20 to 30 min followed by a 4-hr sludge stabilization period. With grab sampling techniques, indigenous enteric virus levels detected in raw wastewater entering the plant were shown to remain fairly constant at around 1000 PFU/ℓ (plaque-forming units per liter) and to decrease to 250 PFU/ℓ during a low flow period. Heavy rains resulted in what appeared to be a flush of the wastewater collection system and the enteric virus levels reaching 1500 PFU/ℓ as the plant flow increased to 15 MGD.

Virus levels in the effluent of the contact stabilization process ranged from a low of approximately 10 PFU/ℓ to a high of 300 PFU/ℓ. Sludge-associated virus levels consistently ranged five- to tenfold greater than suspended virus levels reach 900 PFU recovered from the solids of a 1-ℓ volume of mixed liquor. Under normal operating conditions, greater than 90% of the detectable incoming viruses were removed from the liquid fraction of the wastewater.

Table 2
ENTEROVIRUS DISTRIBUTION IN ACTIVATED SLUDGE AERATION BASINS

Treatment plant location	MLSS (mg/ℓ)	Viruses detected (PFU/ℓ)		Solids associated (%)
		Liquid	Solids[a]	
Austin, Tex.	1800	90	950	93
	2100	40	600	95
	1700	70	450	91
Chicago, Ill.		40	202	83
		2	190	99
		5	279	98
Butte, Mont.	1560	57	720	93
Portland, Ore.	2780	34	500	94

[a] Total PFU eluted from the collected solids of 1-ℓ mixed liquor sample.

Adapted from Moore, B. E., Sagik, B. P., and Sorber, C. A., *Risk Assessment and Health Effects of Land Application of Municipal Wastewater and Sludges*, Sagik, B. P. and Sorber, C. A., Eds., Center for Applied Research and Technology, The University of Texas at San Antonio, San Antonio, 1978, 154.

More recently, a study was conducted at the John E. Egan Water Reclamation Plant in Schaumburg, Ill.[7] The first biological treatment stage of this 15 MGD plant employed conventional activated sludge. Most of the enteric viruses detected were mixed liquor suspended solids (MLSS) — associated on every sampling day but one. Even considering that single occasion when only 49% of the enteric viruses were found to be MLSS-associated the mean values for the 6 sampling days indicated that 93% of the enteric viruses isolated were solids-associated.

Field studies encompassing a wide range of plants employing some form of activated sludge treatment have indicated that generally greater than 90% of the enteroviruses recovered in the aeration basin were solids-associated. These data are reported in Table 2.

Data on indigenous virus removal from an oxidation ditch system (essentially, extended aeration) are provided in Table 3. With flow composite sampling techniques, virus removals ranged from 61 to 99.9% over the 10 sampling days. Despite a wide range of influent virus levels (5.6 to 910 PFU/ℓ), a rather narrow range of effluent concentrations were observed (0.3 to 5.7 PFU/ℓ). For more details on this system, see Trickling Filtration, below.

2. Trickling Filtration

Trickling filtration is a process whereby biological growth is attached to a fixed medium. The medium may be natural (stone) or synthetic (plastic). Sludge recycling is generally not required and the amount of sludge produced is far less than that produced by the activated sludge process. The quantity of biological growth developed is controlled by the available food (organic material), hydraulic loading of the medium (flow rate), the type of media, temperature, filter depth, and a number of other factors.

As with activated sludge there exist a number of trickling filter systems with different design characteristics. Trickling filters are classified as to hydraulic loading, organic loading, depth, medium, and recirculation. A partial listing of these classifications can be found in Figure 1.

Table 3
EFFECTIVENESS OF TWO BIOLOGICAL PROCESSES IN REDUCING ENTEROVIRUS NUMBERS (SELECTED SAMPLING DAYS) KERRVILLE, TEX.

	Enteroviruses recovered (PFU/ℓ, 5 day) and treatment effectiveness (%) Sample Number									
Sampling point	**1**	**2**	**3**	**4**	**5**	**6**	**7**	**8**	**9**	**10**
Raw wastewater	9.1	6.8	31	8.0	7.0	5.6	70	780	910	34
Oxidation ditch effluent	0.7	0.3	2.0	1.3	0.5	2.2	1.8	1.0	5.7	1.2
% Reduction	92	96	94	84	93	61	97	99.9	99.4	96
Trickling filter effluent	2.5	0.7	1.6	1.1	2.1	1.9	21	320	140	7.0
% Reduction	73	90	95	86	71	66	70	59	85	79
Flow (% of Mean Flow)	105	102	99	94	109	84	103	105	105	97

From Moore, B. E., Sagik, B. P., and Sorber, C. A., *J. Water Pollut. Control Fed.*, 53, 1492, 1981. With permission.

Generally, trickling filtration is less effective than activated sludge in removing viruses. Data compiled in a recent field study illustrate this point (Table 3). In this case, the activated sludge process was an oxidation ditch. Both the oxidation ditch and the trickling filter were hydraulically under-loaded and well-operated. The results are comparable because the same raw wastewater served as the influent to both unit processes. Mean virus removal efficiency was 77.4% for the trickling filter compared to 91.2% for the oxidation ditch.

Sproul[3] noted that several studies have shown virus removals in trickling filters of up to 85%. The increase in virus concentrations occasionally observed in trickling filter effluents may result from the breakup of fecal solids in the filters, the low biomass available for viral adsorption, and the relatively short contact time between the wastewater and the biological growth on the filter medium.

3. *Stabilization Ponds*

The removal of viruses from wastewater by stabilization ponds has been studied by relatively few workers. Most of these report reductions, but not elimination, of enteroviruses from wastewater treated in ponds.[4,10-17] Malherbe and Strickland-Cholmley[4] studied a number of ponds in South Africa for their abilities to lower virus concentrations in wastewater. In the effluent of the first pond of a four-cell system with a total holding time of 38 days, seeded polioviruses were recoverable 56 days after addition. The second system was a three-cell maturation pond with a total detention time of 7 days. This system reduced, but did not eliminate, reovirus and enterovirus levels. The third system consisted of a trickling filter followed by four maturation ponds in series. The detention time was 14 days. Viruses were detected in 11 of 13 samples of trickling filter effluent but in only 2 of 16 samples of pond effluent.

In a recent study of poliovirus survival in model wastewater holding ponds, Funderburg and co-workers[14] found that up to 10% of the viruses added to the ponds were deposited into the pond sludge. The inactivation rate for these virions was slower than that for the viruses in the overlying pond water. Virus inactivation was found to be primarily a function of detention time and temperature of the pond. Primary wastewater effluent in ponds was more conducive than secondary effluent to prolonged virus survival. Elevated pH levels of the pond waters, due to the metabolic activity of algae, correlated with a more rapid inactivation of the polioviruses. Other studies have attributed virus reduction in stabilization ponds to detention time, sunlight,[15] biological factors,[16] and adsorption to solids.[17]

C. Advanced Wastewater Treatment

Advanced wastewater treatment (AWT) involves the use of processes that remove more contaminants than are usually removed by conventional wastewater treatment processes. More often than not, the goal of AWT is to remove refractory compounds. Many of these compounds adversely affect aquatic life in streams or lakes and preclude the reuse of the liquid stream for higher purposes.

There are a number of unit processes which are popular for advanced wastewater treatment. Included are chemical coagulation with sedimentation, granular filtration, and microstraining for suspended solids removal; adsorption on activated carbon or extended biological oxidation for organics removal; chemical coagulation and biochemical precipitation and clarification for phosphorous removal; and nitrification-denitrification, ammonia-air stripping, breakpoint chlorination, and ion exchange for nitrogen removal. In addition, land application of wastewater on cropland has been effective in removing many refractory compounds. Membrane processes (reverse osmosis and ultrafiltration) have been highly effective in removing most refractory compounds, also. This discussion will focus on those processes most commonly used in AWT. Land application of wastewaters is beyond the scope of this chapter and can be found elsewhere in this book.

1. Coagulation

Most studies have not separated the virus removal effectiveness of the coagulation and the filtration processes. This is particularly true when the studies have been conducted on indigenous viruses at naturally occurring levels. More often than not, sufficiently high viral levels were not observed to permit this degree of discrimination.

On the other hand, laboratory studies have shown the coagulation process (combined with sedimentation) to be quite effective for the removal of test viruses. Guy and co-workers[18] demonstrated an average removal of 99.9% for polioviruses and 99.7% for coliphage T4 by the coagulation-clarification process. Results of similar studies have been summarized by Sproul.[19]

It is generally accepted that a properly operated coagulation-sedimentation process can account for up to 99% removal of indigenous viruses.[20] In a pilot-scale study of tertiary treatment of wastewater, removals exceeding 99% have been demonstrated.[21] After alum coagulation and clarification, no polioviruses (<0.5 PFU/mℓ were observed in clarified effluent. Initial virus concentrations ranged from 86 to 156 PFU/mℓ, representing a removal in excess of 99.7%.

Use of coagulation in association with direct filtration should result in similar virus reductions. In any case, care must be taken to separate the sludge (with viruses adsorbed) for handling other than by direct return to the primary treatment train.

2. Filtration

Where efforts have been made to discriminate between the effectiveness of the coagulation and filtration processes, virus removal by filtration alone has not been consistent. In an early study, Gilcreas and Kelly[22] examined the movement of coxsackievirus A5, Theiler virus (Strain 4, 727D, or GD VII), and coliphage T4 in spring water by sand filtration following flocculation with aluminum sulfate. Slow filtration was very effective in removing the viruses while rapid filtration reduced enterovirus titers by about 10%.

Robeck and co-workers[23] studied virus removal during various filtration regimens. Filtration of tapwater containing from 1.0×10^6 PFU/mℓ to 6.0×10^6 PFU/mℓ of poliovirus 1 (Mahoney) through 15 in. of an unsaturated California dune sand produced from about 95% to over 99% virus removal during 98 days of periodic loading. Virus removal was greatly decreased at higher flow rates. Almost complete retention of polioviruses applied at similar concentrations was seen in two saturated sands of differing particle size, also. During 6 weeks of use, no saturation point for virus retention was reached. Most virions were retained within the top 1- to 2-ft depth of the infiltration beds. When the sand filters were treated with alum, poliovirus removal exceeded 98% even at rapid flow rates.

Removal of poliovirus 1 (LSc) from lime-flocculated secondary effluent by rapid sand filtration (2.25 gal/min/ft^2) was examined by Berg and co-workers.[24] The flocculated effluents ranged in pH from 9.3 to 11.3. The virus removals observed ranged from 82% to greater than 99.8%.

Nestor and Costin[25] emphasizing the possible role of accumulating organic and inorganic material in enterovirus removal during drinking water treatment, used sand taken from operating plants in addition to clean unused sand in their experimental studies. Passage of coxsackievirus A4 diluted in sterile tapwater was monitored. These authors noted that the virus was more effectively removed by clean sand that was wet at the time of virus application than by clean sand that was dry. Sand from operating water treatment plant filters retained the virus more effectively than either wet or dry clean sand. The sand in operating filters was slightly more acidic than the clean sand, thus suggesting that the lower pH might be responsible for the greater retention of virions by the sand from operating filters.

3. Microstraining

Microstraining has not been evaluated for its effectiveness in virus removal. However, the nature of the process as well as the type of processes which are most effective for virus removal suggest that microstraining will be less effective than rapid sand or mixed media filtration. Its major advantage may lie in its ability to reduce the suspended solids concentration of wastewater thereby improving the virucidal effectiveness of the disinfection process.

Fabrics for microstraining are usually made of woven stainless steel containing 2 to 3% molybdenum. Several aperture sizes are available in the area of 23, 35, and 60 μm. In removal of suspended solids from an activated sludge effluent the 23 μm and the 35 μm screens have been reported to be 89 and 73% efficient, respectively.[26]

4. Membrane Processes

Rejection of viruses by membranes used in the reverse osmosis process has not been studied extensively. This is due, in part, to the size of viruses and the generally accepted membrane transport theory which holds that no viruses should appear in the product water.

One extensive study was conducted to test this theory. Sorber et al.[27] concluded that a small number of viruses can penetrate asymmetrical cellulose acetate membranes but that the penetration was a random phenomenon. The penetration which was observed was associated with individual viral particles and was thought to have occurred either through a mechanical defect in the membrane (i.e., a very small hole which would be a defect in the membrane manufacturing process) or through large pores resulting from random imperfections in the cross-linkage of the cellulose acetate.

More recently, Cooper and Straube[28] used a reverse osmosis system to evaluate virus removals at an operating wastewater reclamation facility in San Diego, Calif. Cellulose acetate membranes were employed in the test unit. Seeded poliovirus 1 (LSc) and naturally occurring bacteriophage were removed from the wastewater in numbers up to 7.3 orders of magnitude.

Thus, it would appear that of those unit processes discussed here, reverse osmosis would represent the most effective process for virus reductions from wastewater. Furthermore, the product stream of a reverse osmosis process may be readily disinfected and easily monitored for its physical integrity. Unfortunately, process capital and operating costs are relatively high.

D. Disinfection

Although disinfection is discussed more extensively elsewhere in this book, that subject will be discussed here also as it pertains to wastewater. It is generally agreed that despite the effectiveness of a single or even most multiple unit processes for the removal of viruses

Table 4
VIRUS INACTIVATION WITH CHLORINE

Test virus	Conditions	Inactivation (%)
Poliovirus 1	0.5 mg/ℓ FAC[a] for 16 min; 2°C; pH 7.8	99.99
Poliovirus 1	1 mg/ℓ FAC for 2 min; 5°C; pH 6	99
Coxsackievirus A9	1 mg/ℓ FAC for 2 min; 5°C; pH 6	99

[a] Free available chlorine.

Adapted from Sproul, O. J., *Risk Assessment and Health Effects of Land Application of Municipal Wastewater and Sludges*, Sagik, B. P. and Sorber, C. A., Eds., The Center for Applied Research and Technology, The University of Texas at San Antonio, San Antonio, 1978, 282.

from wastewater, the level of reduction achieved is not adequate to provide the margin of safety required for many uses or discharges of wastewater. The major process employed to minimize the numbers of viruses in treated wastewater is disinfection. In the U.S., disinfection is practically synonymous with chlorination for most applications.

1. Chlorination

A number of studies have demonstrated the variability in resistance to chlorine of different viruses. Generally, hypochlorous acid (HOCl) is the major chlorine species that occurs at lower pH in clean waters. HOCl is a more effective virucide than the major chlorine species found at high pH in such waters (hypochlorite ion [OCl^-]). Further, combined chlorine species, which occur in waters that contain NH_3 and contain organics, are less virucidal than either of the dominant free chlorine species (HOCl and OCl^-).

Most experimental work on the inactivation of viruses with chlorine has been conducted in clean systems (essentially no suspended material) with free viruses added to the test systems. Under these conditions, the inactivation of viruses is reasonably fast when free chlorine residuals are present. (The data in Table 4 provide an example.) The more extensive data provided in Table 5 permit comparison of survival rates of different viruses in river water containing 0.5 mg/ℓ free chlorine residual.

Recent field studies have shown significant differences in the effectiveness of chlorine inactivation of indigenous fecal coliforms and bacteriophages.[30] Specifically, at a chlorine dose of 5 mg/ℓ at pH 7.4 to 7.8, temperatures of 22 to 30°C, and a contact time of 3 min, a 2.5 $\log_{10}$ loss was observed for fecal coliforms and a 0.8 $\log_{10}$ loss for coliphages. Similarly, at a chlorine contact time of 30 min, a 4.0 $\log_{10}$ loss was observed for fecal coliforms while a 1.2 $\log_{10}$ loss was observed for coliphages.

Unfortunately, none of the chlorine inactivation studies have included the waterborne virus of most concern, hepatitis A virus, due to laboratory limitations in handling this virus. Further, there has developed an increasing concern that chlorine disinfection (or any disinfection, for that matter) may not be as effective for naturally occurring viruses as it is with laboratory-cultivated viruses. Data supporting this contention result from studies which consider that a large part of the naturally occuring viruses are solids-associated and that solids are protective of viruses during disinfection. Sproul[3] suggests that under actual field conditions viruses can be protected by particulate matter and may require longer times or higher chlorine residuals for inactivation. Under such conditions, it may be possible for viruses to pass the disinfection process intact, regardless of the chlorine concentration or contact time. In any case, chlorination will be most effective when wastewater is relatively

Table 5
TIME REQUIRED FOR 99.99% INACTIVATION OF VARIOUS HUMAN VIRUSES[a] WITH 0.5 MG/ℓ FREE CHLORINE

Virus	Time (min)
Reovirus 1	2.7
Reovirus 3	<4.0
Reovirus 2	4.2
Adenovirus 3	<4.8
Coxsackievirus B2	6.5
Coxsackievirus A9	6.8
Coxsackievirus B4	7.0
Echovirus 7	7.1
Echovirus 5	8.0
Coxsackievirus B1	8.5
Echovirus 9	12.0
Adenovirus 7a	12.5
Echovirus 8	13.0
Echovirus 11	14.0
Poliovirus 1	16.2
Echovirus 29	20.0
Adenovirus 12	23.5
Echovirus 1	27.0
Poliovirus 3	30.0
Coxsackievirus B3	35.0
Coxsackievirus A5	35.3
Coxsackievirus B5	39.5
Poliovirus 2	40.0
Echovirus 12	>60.0
Coxsackievirus A6	>120.0

[a] In samples of Potomac River water at 2°C and pH 7.8

From Mahdy, M. S., *J. Am. Water Works Assoc.*, 71, 445, 1979. With permission.

free of turbidity, the chlorine is mixed effectively so that the active species comes in contact with the viruses, and a free chlorine residual is insured for at least 30 min of contact time.

2. *Ozonation*

Ozone is widely used in Europe as a disinfectant for water and wastewater. It may find increased use in the U.S. as concern grows over potentially carcinogenic halogenated hydrocarbons resulting from reactions between chlorine and selected organics in water and wastewater. Ozone is a potent virucide and under proper conditions may be more effective than chlorine in the inactivation of viruses.

Several studies have shown ozone to be an effective disinfectant. Interference in the disinfection process by suspended solids and organics appears to be more severe with ozone than with chlorine, however. Data from two recent studies illustrate this point.

Venosa and co-workers[31] showed that filtration of secondary effluent prior to ozonation significantly improved the disinfection process as measured by reductions in coliform levels. These researchers suggested that the removal of organics as measured by the chemical oxidation demand test may be at least as important as the removal of suspended solids.

Howser and associates[32] demonstrated the protection of fecal matter for coliform bacteria, poliovirus 1 (Sabin), and porcine picornavirus (Type 3) at a turbidity of 5 NTU (Nephelometric Turbidity Unit) for initial ozone concentrations of between 0.073 and 0.096 mg/ℓ. Complete inactivation of non-solids-associated poliovirus 1, coxsackieviruses, porcine picornaviruses, *Escherichia coli*, and *Streptococcus fecalis* was achieved at similar initial ozone concentrations within 10 sec. However, protection of Hep-2 cell-associated poliovirus and coxsackieviruses was demonstrated at initial ozone concentrations of 2.8 and 4.7 mg/ℓ, respectively.

Both of these studies emphasize that treatment of wastewater by coagulation and filtration enhances the bactericidal and virucidal effectiveness of ozone. Contact time has been reported to be a less important consideration in ozone disinfection than in chlorine disinfection.[33]

Unfortunately, definitive studies with ozone in naturally occurring waters such as those shown in Table 5 for chlorine have not been conducted. This represents a major gap in the store of knowledge on the effectiveness of ozone for dependable use as a virucide. Furthermore, the lack of a stable residual and the requirement for on-site ozone generation by relatively sophisticated equipment has tended to favor chlorination over ozonation for most wastewater disinfection applications.

3. Chlorine Dioxide

Chlorine dioxide (ClO_2) has been used extensively in Europe as a disinfectant for water. In the U.S., it has been used primarily for the oxidation of manganese and for taste and odor control.[34] Increased interest in the use of ClO_2 for water and wastewater disinfection has been stimulated in the U.S. for two reasons: (1) it does not react in aqueous solutions to produce detectable concentrations of trihalomethanes and (2) it does not appear to react to any significant degree with ammonia.

ClO_2 is an effective virucide in wastewater. Tifft and co-workers[35] demonstrated approximately four and five $\log_{10}$ reductions of exogenous poliovirus 1 after 2 min of contact in wastewater at chlorine dioxide doses of 12 and 16 mg/ℓ, respectively. Comparable results with chlorine were achieved by approximately twice the disinfectant dose.

Similarly, Walters[36] reported about a four $\log_{10}$ reduction of coliphage f2 seeded into effluent from a secondary wastewater treatment facility after 30 sec of contact. Less than one $\log_{10}$ inactivation of the test virus was observed with chlorine at the same doses (5.0 and 7.5 mg/ℓ) and contact time.

Field studies with indigenous coliphages showed similar results.[30,37] Unfortunately, there has not been much work reported with indigenous human viruses.

Although less ClO_2 than chlorine is required to provide comparable disinfection, ClO_2 is considerably more expensive than chlorine.[38] This, combined with the practical requirement for generating ClO_2 onsite, makes ClO_2 less attractive than chlorine at present. Perhaps, with time, the obvious advantages of ClO_2 as a wastewater disinfectant will result in its more extensive use and thereby permit its more thorough evaluation.

III. SUMMARY AND CONCLUSIONS

It should be obvious from the preceding discussion that the wastewater treatment train employed or the unit processes selected significantly affect the degree of virus removal from wastewater. An example of this variability is shown in Figure 2.

Perhaps more importantly, most modern conventional wastewater treatment processes employ some form of activated sludge for secondary treatment. While these processes are quite effective for the removal of viruses from the wastewater stream, they result in the concentration of viruses in the primary and secondary sludges. Similarly, viruses are concentrated in the sludges or concentrates of the waste streams of most advanced wastewater

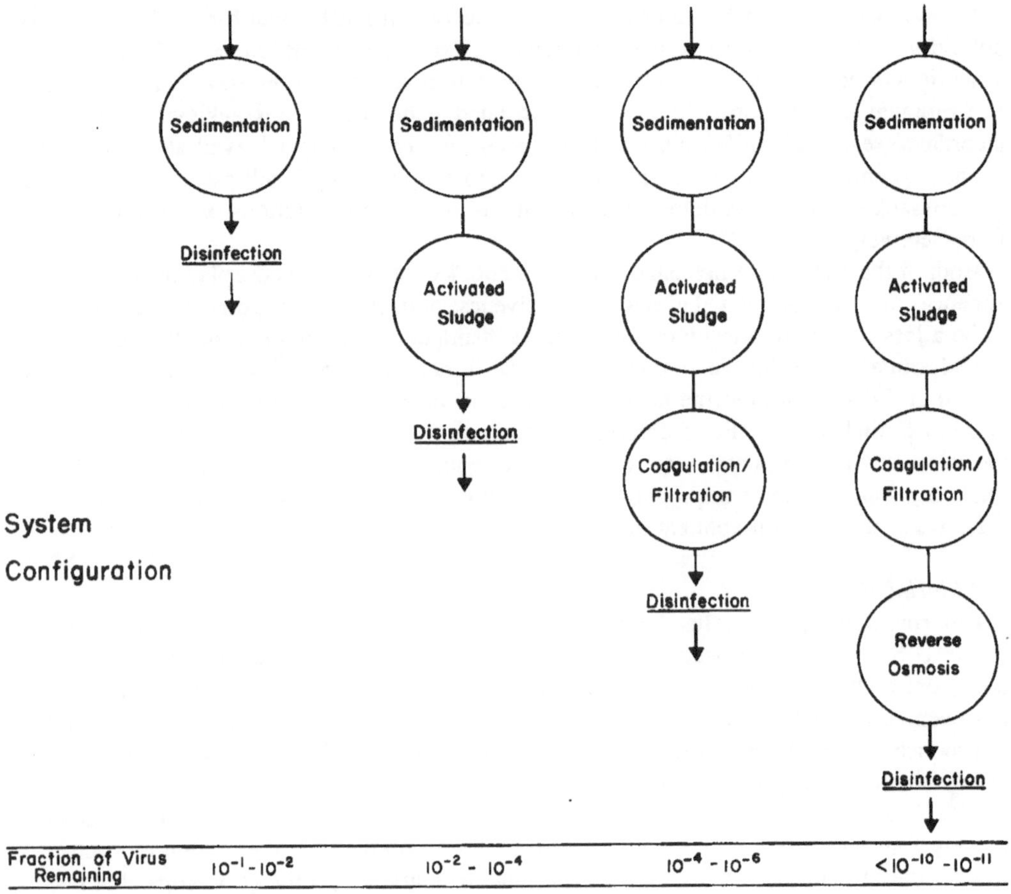

FIGURE 2. Effectiveness of various treatment schemes for virus removal from wastewater.

treatment processes such as coagulation, filtration, and reverse osmosis or ultrafiltration. Therefore, the ultimate control or destruction of the majority of viruses in wastewater is most often accomplished in the sludge treatment/processing phase of wastewater treatment. The effect of these processes on viruses is beyond the scope of this chapter and can be found elsewhere in this book.

An often overlooked aspect of wastewater treatment is daily operation of the in-place facilities. To derive the optimal benefits of any unit process requires proper operation and maintenance. With the advent of federal construction grants programs, many municipalities installed relatively sophisticated wastewater treatment facilities. Operational costs are high and wastewater treatment often suffers in municipalities which have economic problems.

A recent policy statement of the Water Pollution Control Federation makes this point quite emphatically.[39] The statement indicates that only 50% of the public wastewater treatment facilities in the U.S. are performing properly. Among the reasons cited for this poor performance are inadequate manpower, inept or untrained operators, as well as inadequate supplies, budgets, and parts inventory. It seems that many communities have neglected to plan for the resources required to adequately operate those facilities they were so eager to construct, particularly with federal grant programs providing 75% of the construction costs. Unfortunately, there is no federal grant program for the operation and maintenance of wastewater treatment facilities. Similarly, optimal operation of many unit processes is not possible in older communities with combined sewers or sewer systems with high infiltration rates. During periods of precipitation hydraulic induced upset of the treatment processes can occur, thereby greatly reducing treatment efficiency.

Several advanced wastewater treatment processes are particularly effective for removing viruses which pass conventional secondary treatment. Unfortunately, there is little economic justification for the installation of these processes at most wastewater treatment facilities. Thus, proper operation of conventional secondary wastewater treatment facilities, including effective disinfection, provides the most practicable virus removal scheme for the majority of municipalities.

REFERENCES

1. **Berg, G.,** Removal of viruses from sewage, effluents, and waters, I. A review, *Bull. W. H. O.,* 49, 451, 1973.
2. **Berg, G.,** Removal of viruses from sewage, effluents, and waters, 2. Present and future trends, *Bull. W. H. O.,* 49, 461, 1973.
3. **Sproul, O. J.,** The efficiency of wastewater unit processes in risk reduction, in *Risk Assessment and Health Effects of Land Application of Municipal Wastewater and Sludges,* Sagik, B. P. and Sorber, C. A., Eds., The Center for Applied Research and Technology, The University of Texas at San Antonio, San Antonio, 1978, 282.
4. **Malherbe, H. H. and Strickland-Cholmley, M.,** Quantitative studies on viral survival in sewage purification processes, in *Transmission of Viruses by the Water Route,* Berg, G., Ed., Wiley-Interscience, New York, 1967, 379.
5. **England, B., Leach, R. E., Adame, B., and Shiosaki, R.,** Virological assessment of sewage treatment at Santee, California, in *Transmission of Viruses by the Water Route* Berg, G., Ed., Wiley-Interscience, New York, 1967, 401.
6. **Moore, B. E., Funderburg, L., Sagik, B. P., and Malina, J. F., Jr.,** Application of viral concentration techniques to field sampling, in *Proc. Center Res. Water Resources Symp. No. 7,* Malina, J. F. and Sagik, B. P., Eds., The University of Texas, Austin, 1974, 3.
7. **Moore, B. E., Sagik, B. P., and Sorber, C. A.,** An assessment of potential health risks associated with land disposal of residual sludges, in *Proc. 3rd Nat. Conf. Sludge Manage., Disposal, Utilization,* Information Transfer, Rockville, Md., 1977, 108.
8. **Moore, B. E., Sagik, B. P., and Sorber, C. A.,** Land application of sludges: minimizing the impact of viruses on water resources, in *Risk Assessment and Health Effects of Land Application of Municipal Wastewater and Sludges,* Sagik, B. P. and Sorber, C. A., Eds., Center for Applied Research and Technology, The University of Texas at San Antonio, San Antonio, 1978, 154.
9. **Moore, B. E., Sagik, B. P., and Sorber, C. A.,** Viral transport to groundwater at a wastewater land application site, *J. Water Pollut. Control Fed.,* 53, 1492, 1981.
10. **Bopardikar, M. V.,** Microbiology of a waste stabilization pond, in *Advances in Water Pollution Research, Proc. 4th Int. Conf.,* Prague, Jenkins, S. H., Ed., Pergamon Press, Oxford, 1969, 595.
11. **Kott, Y.,** Hazards associated with the use of chlorinated oxidation pond effluents for irrigation, *Water Res.,* 7, 853, 1973.
12. **Shuval, H. I.,** Detection and control of enteroviruses in the water environment, in *Developments in Water Quality Research, Proc. Jerusalem Int. Conf. Water Quality Pollut. Res.,* Ann Arbor-Humphrey Science, London, 1969.
13. **Slanetz, L. W., Bartley, C. H., Metcalf, T. G., and Nesman, R.,** Survival of Enteric Bacteria and Viruses in Oxidation Pond Systems, Research Rep. No. 6, Water Resources Research Center, University of New Hampshire, Durham, 1972.
14. **Funderburg, S., Moore, B. E., Sorber, C. A., and Sagik, B. P.,** Survival of poliovirus in model wastewater holding ponds, in *Progress in Water Technology* Vol. 10, Pergamon Press, Oxford, 1978, 619.
15. **Ullrich, A. H.,** Use of wastewater stabilization ponds in two different systems, *J. Water Pollut. Control Fed.,* 39, 965, 1967.
16. **Horning, W. B., Porges, R., Clarke, H. F., and Cooke, W. B.,** Waste stabilization pond study, Lebanon, Ohio, USPHS Publ. 999-WP-16, 1964.
17. **Christie, A. E.,** Virus reduction in the oxidation pond, *Water Pollut. Control,* 105, 45, 1967.

18. **Guy, M. D., McIver, J. C., and Lewis, M. J.,** The removal of virus by a pilot treatment plant, *Water Res.*, 11, 421, 1977.
19. **Sproul, O. J.,** Standards for viruses in effluents, sludges, and ground and surface waters, in *Virus Aspects of Applying Municipal Waste to Land,* Baldwin, L. B., Davidson, J. M., and Gerber, J. F., Eds., Eds., University of Florida, Gainesville, 1976, 131.
20. American Water Works Association Committee Report, Viruses in drinking water, *J. Am. Water Works Assoc.*, 71, 441, 1979.
21. **Wolf, H. W., Safferman, R. S., Mixson, R. A., and Stringer, C. E.,** Virus inactivation during tertiary treatment, *J. Am. Water Works Assoc.*, 66, 526, 1974.
22. **Gilcreas, F. W. and Kelly, S. M.,** Relation of coliform-organism test to enteric-virus pollution, *J. Am. Water Works Assoc.*, 47, 683, 1955.
23. **Robeck, G. G., Clarke, W. A., and Dostal, K. A.,** Effectiveness of water treatment processes in virus removal, *J. Am. Water Works Assoc.*, 54, 1275, 1962.
24. **Berg, G., Dean, R. B., and Dahling, D. R.,** Removal of poliovirus 1 from secondary effluent by lime flocculation and rapid sand filtration, *J. Am. Water Works Assoc.*, 60, 193, 1968.
25. **Nestor, I. and Costin, L.,** Presence of certain enteroviruses (Coxsackie) in sewage effluents and in river waters of Roumania, *J. Hyg. Epidemiol. Immunol.*, 20, 137, 1976.
26. **Bodien, D. G. and Stenburg, R. L.,** Microscreening effectively polishes activated sludge plant effluent, *Water Waste. Eng.*, 3, 74, 1966.
27. **Sorber, C. A., Malina, J. F., and Sagik, B. P.,** Virus rejection by the reverse osmosis-ultrafiltration processes, *Water Res.*, 6, 1377, 1972.
28. **Cooper, R. C. and Straube, D.,** Reverse osmosis reduces viral count in the influent stream, *Water Sewage Works*, R162, 1979.
29. **Mahdy, M. S.,** Viruses in the water environment: an underestimated problem, *J. Am. Water Works Assoc.*, 71, 445, 1979.
30. **Longley, K. E., Moore, B. E., and Sorber, C. A.,** Comparison of chlorine and chlorine dioxide as disinfectants, *J. Water Pollut. Control Fed.*, 52, 2098, 1980.
31. **Venosa, A. D., Meckes, M. C., Opatken, E. J., and Evans, J. W.,** Disinfection of filtered and unfiltered secondary effluent in two ozone contactors, presented at the 52nd Annu. Water Pollut. Control Fed. Conf., Houston, 1979.
32. **Howser, D. M., Emerson, M. A., Buke, C. E., Walsh, D. S., and Sproul, O. J.,** Ozone inactivation of cell and fecal associated viruses and bacteria, presented at the 51st Annu. Water Pollut. Control Fed. Conf., Anaheim, Calif., 1978.
33. **Stover, E. L. and Jarnis, R. N.,** Obtaining high level wastewater disinfection with ozone: water quality considerations, presented at the 52nd Annu. Water Pollut. Control Fed. Conf., Houston, 1979.
34. **Symons, J. M.,** Ozone, chlorine dioxide, and chloramines as alternatives to chlorine for disinfection of drinking water, Water Supply Research Div., Municipal Environmental Research Lab, Cincinnati, Ohio, 1977.
35. **Tifft, E. C., Moffa, P. E., Richardson, S. L., and Field, R. I.,** Enhancement of high rate disinfection by sequential addition of chlorine and chlorine dioxide, *J. Water Pollut. Control Fed.*, 49, 1652, 1977.
36. **Walters, G. E.,** Chlorine Dioxide and Chlorine: Comparative Disinfection, M. S. thesis, The Johns Hopkins University, Baltimore, 1976.
37. **Longley, K. E.,Moore, B. E., and Sorber, C. A.,** Disinfection efficiencies of chlorine and chlorine dioxide in a gravity flow contactor, *J. Water Pollut. Control Fed.*, 54, 140, 1982.
38. **Roberts, P. V.,** Chlorine dioxide: a state-of-the-art, in *Proc. of Wastewater Disinfection Alternatives State-of-the-Art Workshop,* Water Pollution Control Federation, Washington, D.C., 1979, 138.
39. *Highlights*, Water Pollution Control Federation, Washington, D.C., 17, 1, 1980.

Chapter 4

REMOVAL OF VIRUSES FROM RAW WATERS BY TREATMENT PROCESSES

John C. Hoff and Elmer W. Akin

TABLE OF CONTENTS

I. INTRODUCTION

Man has long associated the purity of water with its clarity. The earliest recorded knowledge of water treatment dating back to 2000 B.C. includes descriptions of methods for clarifying water, including filtration and the use of mineral and vegetable substances to enhance the removal of solids from water. Although great advances have been made in understanding and applying this technology for removing solids during drinking water treatment, the basic principles used have remained essentially the same.

The water treatment unit processes to be considered in this chapter include coagulation, flocculation, sedimentation, and filtration and combinations of these processes (e.g., direct filtration which involves addition of a coagulant followed by immediate filtration with no intermediate settling process). The major objective of all of these unit processes is to remove the wide variety of particulate materials present in water. Other substances including certain dissolved organic compounds may also be removed to some degree by some of the processes. The solid material suspended in water is extremely diverse, both in size and in chemical and physical characteristics. Included are particles ranging in diameter from about 50 to 100 μm to those three to four orders of magnitude smaller. They include silt, clays, inorganic and organic particles from domestic and industrial wastes, decaying organic matter, and living forms such as algae bacteria, and viruses. Although the viruses present in raw water contribute virtually nothing to the turbidity, they, along with other colloidal materials, are removed with varying degrees of efficiency by these unit processes. Removal, of course, does not imply destruction or loss of viability of the virions although this also may occur as a consequence of the removal process used, e.g., softening under high pH conditions.

II. SEDIMENTATION PROCESSES

The term sedimentation, as employed in water treatment, is the process by which suspended matter separates from water by subsidence and deposition due to the force of gravity. Major factors that influence sedimentation rates of large particles include particle size, shape, and specific gravity. Also, as particle size decreases, the ratio of surface area to mass increases. For colloidal material, the ratio is very high and surface properties such as electric charges and inorganic groups, which increase the resistance of the particles to settling, become dominant. The nature of these effects is shown in Table 1. It is evident that particles in the size range of the enteroviruses and even particles 10- to 100-fold larger cannot be efficiently removed by gravity sedimentation.

A. Coagulation and Flocculation

The purpose of the closely tied coagulation and flocculation processes is to destabilize the forces that maintain colloidal particles in a dispersed state allowing them to aggregate and associate with the coagulating agent into particles sufficiently large for gravity sedimentation to be effective. Coagulation is the process that effects a reduction in the forces that maintain the particles in a dispersed state. Coagulation reactions occur almost instantaneously upon mixing a coagulant with water. Flocculation is the process in which destabilized colloidal particles aggregate into floc networks through formation of chemical bridges. This is a much slower process usually requiring 10 to 60 min.

The reactions of hydrophilic colloids with coagulants differ from those of hydrophobic colloids. The stability of hydrophobic colloids is mainly dependent on electrostatic repulsion while hydrophilic colloids are stabilized by both electrostatic forces and elements of hydration or nonpolar groups that have an affinity for water. The successful removal of hydrophilic colloids may depend more on interparticle bridging than on surface charge neutralization.[2] Some differences in the characteristics of coagulation of hydrophobic and hydrophilic colloids are shown in Table 2.

Table 1
RELATIONSHIP OF PARTICLE SIZE TO SURFACE AREA AND SETTLING TIME

Diameter of particle, μm	Description	Total surface area[a]	Time required to settle[b]
100	Fine sand	48.7 $in.^2$	38 sec
10	Silt	3.38 ft^2	33 min
1	Bacteria	33.8 ft^2	55 hr
0.1		3.8 $yard^2$	230 days
0.01[c]	Colloidal particles	0.7 acres	6.3 year
0.001		7.0 acres	63 year minimum

[a] Area for particles of indicated size produced from a particle 10 mm in diameter with a specific gravity of 2.65.
[b] Calculations based on sphere with a specific gravity of 2.65 to settle 1 ft.
[c] Enterovirus diameters range from about 0.03 to 0.08 μm.

Adapted from Powell, S. T., *Water Conditioning for Industry*, McGraw-Hill, New York, 1954, 22. With permission.

Table 2
CHARACTERISTICS OF COAGULATION OF DILUTE CLAY SUSPENSIONS AND HUMIC SUBSTANCES WITH ALUMINUM SULFATE

Dilute clay[a] suspensions	Humic substances[b]
Optimum pH 6.5—7.5	Optimum pH 5—6
Minimum residual turbidity independent of pH	Minimum residual color dependent on pH
Increased concentration slightly reduces coagulant dose	Increased concentration increases coagulant dose
Dose and optimum pH changed in presence of humic substance	Dose and optimum pH independent of presence of clay

[a] Hydrophobic colloid.
[b] Hydrophilic colloid.

From Hall, E. S. and Packham, R. F., *J. Am. Water Works Assoc.*, 57(9), 1965. With permission. Copyright 1965, The American Water Works Association.

Much of the information on theory and mechanisms of sedimentation, coagulation and flocculation and their application to water treatment has been developed in the last 20 years and a number of excellent reviews are available.[4-7] Because of the many different types of colloidal particles present in water and the influences of other chemical and physical characteristics of natural waters, it still is not possible to predict the conditions required for optimum particle removal from different raw waters. The most practical method for determining optimum conditions for coagulation remains the jar test, a trial-and-error laboratory procedure that has been used for many years.[5]

Many studies on virus removal by coagulation and flocculation processes have been conducted. An excellent critical review of the available information was prepared recently by Sproul[8] and this section will draw heavily on this review.

B. Metallic Coagulants

The overall results of a number of investigations of virus removal from water by metallic coagulants are shown in Tables 3 and 4. The studies included a number of different enteroviruses (Table 3) and bacteriophages, both tailed and spherical (Table 4). As indicated by Sproul,[8] several early reports[22-24] were excluded because the experimental conditions used made the results inapplicable to water treatment practices.

Overall, enterovirus removal ranged from 48 to 99.999%, but the great majority of the results showed removals ranging upwards from about 90%. This occurred with a wide variety of viruses, virus concentrations, water conditions, and coagulants. Of the studies in which markedly lower levels were reported, one was conducted in distilled water with no added turbidity, and poor coagulation occurred.[10] In the other study poor removal occurred when only a small amount of turbidity caused by clay was present.[12] Poor floc formation and settling often occurs in low-turbidity waters.[25]

Data extracted from Tables 3 and 4 are shown in Table 5. Except for the data of Chang et al.[14] there is no clear apparent relationship between coagulant dosage and virus removal efficiency. The data further indicate that aluminum sulfate and ferric salts provide equivalent virus removal. The dosage for optimum removal appears to be related to the characteristics of the suspending water.

1. Relationship of Virus Removal to Turbidity Removal

Sproul[8] observed that in water treatment practice, coagulant dosages are normally adjusted to optimize turbidity reduction. He also noted that a number of investigations had provided evidence that increased turbidity removals were accompanied by increased virus removals.[11,17-21] Closer examination of certain of the data[17] indicated that at low pH (5.24) maximum turbidity removal occurred at slightly lower $Al_2(SO_4)_3$ dosages than those at which maximum virus removal was reached. This was not noted, however, when coagulation was conducted at pH 6.17, 7.0, and 8.3. Shelton and Drewry[20] showed that maximum turbidity reduction and virus reduction occurred at equivalent $Al_2(SO_4)_3$ dosages but that slightly higher dosages of $FeCl_3$ were required for optimum removal of viruses as compared to optimal dosages for turbidity removal. Data presented by York[18] and York and Drewry[19] also show a correlation between virus and turbidity removals although there were indications from some of the data that optimum coagulant dosages for maximum virus removal were slightly higher than those required for maximum turbidity removal.[18] Both differences in dosage required and differences in virus and turbidity removals at equivalent coagulant dosages were small. In addition, when expressed as percent reductions, virus removals were usually greater than turbidity removals.

2. Effects of Organic Matter and Inorganic Salts on Removal of Viruses

The following material from Sproul's review[8] summarizes the available data on the effects of various kinds of organic matter on virus removal by coagulation. Chang et al.[15] reported that 20 ppm of gum arabic reduced the removal of coxsackievirus A2 from 97 to 17% when 15 ppm of $Al_2(SO_4)_3$ was used. Concurrently, the removal of *Micrococcus pyogenes* phage was reduced from 92 to 0%. The gum seriously interfered with coagulation since no floc was produced. Manwaring et al.[21] reported a reduction in the removal of coliphage MS2 from 99 to 67% when 200 mg/ℓ of wastewater effluent was added to a clear suspending medium and 60 mg/ℓ of $FeCl_3$ was added subsequently. This was accompanied by a parallel decrease in turbidity removal. No decrease was observed when up to 50 mg/ℓ of bovine serum albumin was added. Chaudhuri and Engelbrecht[17] reported reductions from 99.8 to 94% in the removal of coliphage MS2 by 50 mg/ℓ of $Al_2(SO_4)_3$ from water contaminated with 200 mℓ/ℓ of wastewater effluent. Parallel reductions in turbidity removal were also noted. Reductions of only 2 to 3% were noted in the coagulation of coliphage T4 with 50

Table 3
REMOVAL OF ENTEROVIRUS BY METAL COAGULANTS

Virus		Coagulant		Condition of coagulation					Removal		
Type	Initial concentration	Type	Dose mg/ℓ or ppm	Type of water	Turbidity	Temp. (°C)	pH Start/ end	Type study	Virus (%)	Turbidity % or turbidity units	Ref.
Poliovirus 1 (Sabin)	3—7 × 10^{4} [a]	$Al_2(SO_4)_3$	10	Spiked DW[b]	50 mg/ℓ clay	Room	NS/6.8	L[c]	90	97	9
Poliovirus 1 (Sabin)	1—3 × 10^{4} [a]	$Al_2(SO_4)_3$	20	DW	None	22.5	7.0/7.0	L	70	—	10
Poliovirus 1 (Mahoney)	$10^{6.3}$ to[d] $10^{8.37}$ [a]	$FeCl_3$	66 66	Spiked demineralized Spiked demineralized	45—178 TU 45—178	13—21 13—21	7.0/ 6.6 7.7—6.9	P[e]	99.7—99.999	0.6—3.8TU	11
Poliovirus 1 (Mahoney)	1.4 × 10^{6} [f]	$FeCl_3$	60	Spiked demineralized clay	25—500 mg/ℓ	15—17	5—8	P	48—99.7	NS[g]	12
Poliovirus 1,2,3 (oral vaccine)	9.2 × 10^{7} [h]	$FeCl_3$	40	Polluted river	NS	NS	NS	P	99.8	NS	13
Coxsackievirus A2	2.25 × 10^{5} [i]	$Al_2(SO_4)_3$	40	Spiked DW	0.4 mℓ SiO_2	25	6.2/6.2	L	86	NS	14
Coxsackievirus A2	2.25 × 10^{5} [i]	$Al_2(SO_4)_3$	60	Spiked DW	0.4 mℓ SiO_2	25	6.2/6.2	L	96	NS	14
Coxsackievirus A2	2.25 × 10^{5} [i]	$Al_2(SO_4)_3$	80	Spiked DW	0.4 mℓ SiO_2	25	6.2/6.2	L	97	NS	14
Coxsackievirus A2	2.25 × 10^{5} [i]	$Al_2(SO_4)_3$	100	Spiked DW	0.4 mℓ SiO_2	25	6.2/6.2	L	99	NS	14
Coxsackievirus A2	4.5 × 10^{3} [i]	$FeCl_3$	20	Spiked DW	0.4 mℓ SiO_2	25	6.2/6.2	L	97	NS	14
Coxsackievirus A2	4.5 × 10^{3} [i]	$FeCl_3$	40	Spiked DW	0.4 mℓ SiO_2	25	6.2/6.2	L	98	NS	14
Coxsackievirus A2	NS	$Al_2(SO_4)_3$	25	Ohio River	16—240 TU	25	NS/7.3	L	99	1—5	15
Coxsackievirus A2	NS	$Al_2(SO_4)_3$	25	Ohio River	140—255 TU	25	NS/7.4	L	95	1—5	15
Coxsackievirus A2	NS	$Al_2(SO_4)_3$	25	Ohio River	40—135 TU	15	NS/7.4	L	96	1—5	15
Coxsackievirus A2	NS	$FeCl_3$	15	Ohio River	5—10 TU	5	NS/8.4	L	95	0.1	15
Naturally occurring (mainly Coxsackievirus B3 and B5)	0.004[a]	$Fe_2(SO_4)_3$	40	Polluted river	NS	NS	NS/NS	P	>88.3		13

Table 3 (continued)
REMOVAL OF ENTEROVIRUS BY METAL COAGULANTS

Virus		Coagulant		Condition of coagulation					Removal		
Type	Initial concentration	Type	Dose mg/ℓ or ppm	Type of water	Turbidity	Temp. (°C)	pH Start/ end	Type study	Virus (%)	Turbidity % or turbidity units	Ref.
Poliovirus I	17—173[a]	$Al_2(SO_4)_3$	40	Natural river	10—69 TU	NS	7.2—7.6/ 7.1—7.5	P	>90	>60%	16
Coxsackievirus B5	17—100[a]	$Al_2(SO_4)_3$	40	Natural river	10—69 TU	NS	7.2—7.6/ 7.1—7.5	P	>90	> 60%	16
ECHO-9	17—56[a]	$Al_2(SO_4)_3$	40	Natural river	10—69TU	NS	7.2—7.6/ 7.1—7.5	P	>90	>60%	16

[a] PFU/mℓ.
[b] Distilled water.
[c] Laboratory.
[d] CPE_{50}/mℓ.
[e] Pilot plant.
[f] MPNCU/mℓ.
[g] Not stated.
[h] Total dose ($TCID_{50}$)
[i] LD_{50}/mℓ.

Table 4
REMOVAL OF BACTERIOPHAGES BY METAL COAGULANTS

Virus		Coagulant		Conditions of coagulation					Removal		
Type	Initial concentration PFU/mℓ	Type	Dose mg/ℓ or ppm	Type of water	Turbidity	Temp. (°C)	pH Start/end	Type study	Virus %	Turbidity % or TU[a]	Ref.
Coliphage MS2	2.25 × 10^5	$Al_2(SO_4)_3$	20—26	NS[b]	120 mg clay per liter	24—25	6.0/6.0	L[c]	99.9	97	17
Micrococcus pyogenes phage	5 × 10^4	$Al_2(SO_4)_3$	80—100	Spiked D.W.[d]	NS	25	6.2/6.2	L	99—99.8	NS	14
M. pyogenes phage	3 × 10^6	$Al_2(SO_4)_3$	25	Ohio River	16—240 TU	25	6.7/7.3	L	97	1—5	15
M. pyogenes phage	NS	$Al_2(SO_4)_3$	25	Ohio River	140—255 TU	15	NS/7.3	L	89	1—5	15
M. pyogenes phage	NS	$Al_2(SO_4)_3$	25	Ohio River	40—135 TU	5	NS/7.4	L	84	1—5	15
Coliphage f2	1—8 × 10^5	$Al_2(SO_4)_3$	25/21	Natural lake	2—4 TU	23—25	8.1/7.4	L	99.9	96	18,19
Coliphage f2	NS	$Al_2(SO_4)_3$	15/15	Natural surface	1.1—1.2 TU	NS	NS/6.8	L	99.4	96	20
Coliphage T4	2.5—5 × 10^5	$Al_2(SO_4)_3$	20—26	NS	120 mg clay per liter	24—25	5.2/5.2	L	98	98	17
Coliphage MS2	3.9 × 10^5	$FeCl_3$	50—60	NS	12.5 TU	24—25	5.0	L	99.5	98	21
M. pyogenes phage	2—6.4 × 10^5	$FeCl_3$	20—40	Spiked D.W.	NS	25	6.2/6.2	L	99.3—99.9	NS	14
M. pyogenes phage	2—6.4 NS	$Al_2(SO_4)_3$	15	Ohio River	60—100 TU	25	NS/7.4	L	86	5—10	15
Coliphage f2	0.7—1 × 10^6	$FeCl_3$	50/23	Natural lake	0.4—3 TU	20—24	8.2/6.9[f]	L	99.4	92	18,19
Coliphage f2	7—9 × 10^5	$Fe_2(SO_4)_3$	50/49[e]	Natural lake	2 TU	22—24	8.3/7.5[f]	L	92	89	18,19
Coliphage f2	1 × 10^6	$Fe_2(SO_4)_3$	36	Natural lake	1—2 TU	24—25	8.2/8.6[f]	L	94	Minor or increase	18,19
Coliphage f2	NS	$FeCl_3$	40/34[a]	Natural surface	1.1—1.2	NS	NS/6.8	L	99.1	95	20
Coliphage f2	NS	$Fe_2(SO_4)_3$	62/35[a]	Natural surface	1.1—1.2	NS	NS/7.2	L	99.91	90	20
Coliphage T4	2.5—5 × 10^5	$Al_2(SO_4)_3$	20—26	NS	120 mg clay per liter	24—25	5.2/5.2	L	98	98	17
Natural phage	54.2	$Fe_2(SO_4)_3$	40	Polluted river	NS	NS	NS/NS	P[g]	93	NS	13

Table 4 (continued)
REMOVAL OF BACTERIOPHAGES BY METAL COAGULANTS

Virus		Coagulant		Conditions of coagulation					Removal		
Type	Initial concentration PFU/mℓ	Type	Dose mg/ℓ or ppm	Type of water	Turbidity	Temp. (°C)	pH Start/ end	Type study	Virus %	Turbidity % or TU[a]	Ref.
Coliphage T4	1.5—3 × 10^{12}[h]	$Fe_2(SO_4)_3$	40	Polluted river	NS	NS	NS/NS	P[g]	99.6	NS	13

[a] Turbidity units.
[b] Not stated.
[c] Laboratory.
[d] Distilled water.
[e] Optimal coagulant dose for virus removal/optimal coagulant dose for turbidity removal.
[f] pH is optimal value for virus removal.
[g] Pilot plant.
[h] Total dose.

Table 5
VIRUS REMOVAL BY SEDIMENTATION VERSUS METAL COAGULANT DOSAGE[a]

Coagulant Type	Coagulant Dosage mg/ℓ	Type of virus	Water system	Removal (%)	Ref.
$Al_2(SO_4)_3$	10	Poliovirus 1	Spiked DW[b]	90	9
	20	Poliovirus 1	DW	70	10
	40	Poliovirus 1	Polluted river	99.8	13
	40	Poliovirus 1	Natural surface	≥90	16
	40	Coxsackievirus A2	Spiked DW	86	14
	60	Coxsackievirus A2	Spiked DW	96	14
	80	Coxsackievirus A2	Spiked DW	97	14
	100	Coxsackievirus A2	Spiked DW	99	14
	25	Coxsackievirus A2	Ohio River	95—99	15
	40	Coxsackievirus B5	Natural surface	≥83	16
	40	Echovirus 9	Natural surface	≥90	16
	20—26	Coliphage MS2	NS[c]	99.9	17
	20—26	Coliphage T4	NS	98	17
	15	Coliphage f2	Natural surface	99.4	20
	25	Coliphage f2	Natural surface	99.4	18, 19
	40—100	*Micrococcus pyogenes* phage	Spiked DW	94—99.8	14
	25	*M. pyogenes* phage	Ohio River	84—95	15
	15	*M. pyogenes* phage	Ohio River	86	15
$FeCl_3$	66	Poliovirus 1	Spiked DW	99.7—99.999	11
	60	Poliovirus 1	Spiked DW	48—99.7	12
	20	Coxsackievirus	Spiked DW	97	14
	40	Coxsackievirus	Spiked DW	98	14
	15	Coxsackievirus	Ohio River	95	15
	50—60	Coliphage MS 2	NS	99.5	21
	50	Coliphage f2	Natural lake	99.4	18, 19
	40	Coliphage f2	Natural surface	99.1	20
	20—40	*M. pyogenes* phage	Spiked DW	99.3—99.9	14
$Fe_2(SO_4)_3$	40	Poliovirus 1, 2, 3	Polluted river	99.8	13
	50	Coliphage f2	Natural lake	92	18, 19
	62	Coliphage f2	Natural surface	99.91	20
	40	Coliphage T4	Polluted river	99.6	13

[a] Data extracted from Tables 3 and 4.
[b] Distilled water.
[c] Not stated.

mg/ℓ of $Al_2(SO_4)_3$ in the presence of 50 mg/ℓ of egg and bovine serum albumin and 200 mℓ/ℓ of wastewater effluent.

The effects of inorganic salts on the removal of viruses and turbidity were studied by Thorup et al.[9] Increasing concentrations of $CaCl_2$ from 4 to 4000 mg/ℓ resulted in increased removal of both poliovirus 1 and turbidity (suspended clay particles). In studies on the effects of other salts and the influence of coagulant aids, Throup et al.[9] consistently noted that the quality of the floc produced was a good indicator of removal efficiencies of both turbidity and polioviruses. Chang et al.[15] speculated that the presence of calcium and magnesium ions in raw water interfered with virus removal by reducing the rate of coagulant-metal ion-virus protein complex formation. Chaudhuri and Engelbrecht[17] showed that Ca^{++} or Mg^{++}, in concentrations up to 50 mg/ℓ, did not interfere with coliphage T4 removal by $Al_2(SO_4)_3$. Manwaring et al.[21] reported similar results for coliphage MS2 with $FeCl_3$ as the coagulant.

C. Flocculation Aids

As the name implies, flocculation aids, also referred to as coagulant aids, are materials that improve the quality of floc formation in waters in which coagulation-flocculation is poor. The elements of floc quality improved by flocculation aids include floc toughness, increased size, and increased settling rates. The materials used as flocculant aids fall into four categories: oxidants, adsorbents-weighting agents activated silica, and polyelectrolytes.[25] The latter group consists of natural organic compounds, such as starch derivatives, polysacharide gums, and proteinaceous materials, and synthetic materials designed to have certain electrical charge characteristics. The major benefit sought from the use of these compounds in water treatment is an increase in floc size with concomitant increases in settling rate. The use of synthetic polymers for this purpose is a comparatively recent development, the advantage being that they can be tailored, particularly with regard to molecular weight, electrical charge, and ionizable groups, for specific applications.

Although polyelectrolytes until recently were not intended as primary coagulants for water treatment, considerable research has been conducted on their efficiency in virus removal. This research apparently was stimulated by observations in 1967 indicating that polyelectrolytes effectively removed animal and plant viruses from aqueous suspension[26] and by concern about the possible presence of viruses in potable water. The available information on the use of polyelectrolytes as primary coagulants, as summarized by Sproul,[8] is shown in Table 6. Thorup et al.[9] concluded that the cationic polyelectrolyte studied performed more effectively than the anionic and nonionic types, but that the cationic polyelectrolyte did not increase virus removal beyond levels obtained with unaided conventional metal coagulants. Chaudhuri and Engelbrecht[17] also observed that anionic polyelectrolytes did not adsorb coliphages MS2 and T4. The phages were adsorbed by cationic polylelctrolytes. Overall, the results in Table 6 again show a general correlation between quality of floc formed and virus removal and a positive correlation between virus removal and turbidity removal.

In several of the studies discussed above, the use of polyelectrolytes as coagulant aids was investigated. Thorup et al.[9] observed little enhancement in virus removal by polyelectrolytes when optimum levels of $Al_2(SO_4)_3$ were used. At less than optimum levels of $Al_2(SO_4)_3$, cationic polyelectrolytes substantially increased the amount of coliphage T2 removal, but anionic and nonionic polyelectrolytes did not improve removal. Foliguet et al.[27] reported that anionic and nonionic polyelectrolytes used as aids with $FeCl_3$ caused slight reductions in removal of poliovirus 1. Use of a cationic polymer did not increase removal beyond that obtained using $FeCl_3$ alone. Chaudhuri and Engelbrecht,[17] York,[18] and York and Drewry[19] also observed little effect of polyelectrolyte coagulant aids on virus removal. Overall, the results indicate that while some cationic polyelectrolytes may be useful for virus removal, the anionic and nonionic polyelectrolytes are not useful for this purpose.

III. WATER SOFTENING PRECIPITATION PROCESSES

Water hardness is caused by the presence in solution of divalent cations, usually calcium

Table 6
VIRUS REMOVAL WITH POLYELECTYROLYTES AS PRIMARY COAGULANTS[a]

Virus		Coagulant		Conditions of coagulation				Removal (%)			
Type	Initial concentration PFU/mℓ	Type	Dose mg/ℓ or ppm	Type water	Turbidity	Temp. (°C)	pH Start/end	Virus	Turbidity	Remarks	Ref.
Poliovirus 1 (Sabin)	3.7×10^4	Cationic Purifloc C32	1	Spiked DW[b]	None	NS[c]	6.8/6.8	0—36	—	Poor to no floc	9
Poliovirus 1 (Chat)	NS	Cationic Catfloc	0.5—2.0	Ohio River	NS	NS	4.7/NS	0—40	NS	Poor to no floc	22
Coxsackievirus B3	NS	Cationic Catfloc	0.5—2.0	Ohio River	NS	NS	4.7/NS	0	NS	Poor to no floc	22
Coliphage T2	10^6—10^7	Anionic Hercules CMC	1	Spiked DW	[d] [e]	NS	7.0/7.0	0—41	NS	Poor to no floc	9
Coliphage T2	10^6—10^7	Nonionic Nacolyte 605	1	Spiked DW	[d] [e]	NS	7.0/7.0	0—37	NS	Poor to no floc	9
Coliphage T2	10^6—10^7	Cationic Nacolyte 110	1	Spiked DW	[d] [e]	NS	7.0/7.0	32—96	NS	Poor to no floc	9
Coliphage f2	1×10^6	Cationic Drewfloc 21	1.5	Lake	3.1	26	8.2/8.6	77	35	Fair floc	18, 19
Coliphage f2	1×10^6	Cationic Catfloc	1—3.5	Lake	5.3	19—24	8.2/8.2	99	40—60	Fair floc	18, 19
Coliphage f2	NS	Cationic Catfloc	2.2	Surface	1.1—1.2	NS	NS/7.5	92	Poor	Poor floc	20
Coliphage MS2	2.8×10^5	Cationic Primafloc C7	7.5	NS	12.5	24—25	5.9/5.9 6.0/6.0	99.1	98		17
Coliphage MS2	2.8×10^5	Cationic Catfloc	10	NS	12.5	24—25	5.9/5.9 6.0/6.0	99.6	97		17
Coliphage T4	5×10^5	Cationic Primfloc C7	7.5	NS	12.5	24—25	5.2/5.2 5.5/5.5	99.9	99		17

Table 6 (continued)
VIRUS REMOVAL WITH POLYELECTYROLYTES AS PRIMARY COAGULANTS[a]

Virus		Coagulant		Conditions of coagulation				Removal (%)			
Type	Initial concentration PFU/mℓ	Type	Dose mg/ℓ or ppm	Type water	Turbidity	Temp. (°C)	pH Start/end	Virus	Turbidity	Remarks	Ref.
Coliphage T4	5×10^5	Cationic Catfloc	12.5	NS	12.5	24—25	5.2/5.2 5.5/5.5	99.5	99.1		17

[a] All data from lab-scale studies.
[b] Distilled water.
[c] Not stated.
[d] 5 mg/ℓ clay.
[e] 5 mg/ℓ infusorial earth.

and magnesium, and occasionally to a lesser degree by strontium, ferrous iron, and manganese. Removal of these ions can be accomplished by precipitating them as insoluble compounds, such as $CaCO_3$ or $Mg(OH)_2$, with appropriate chemicals. In this process, frequently called lime softening, water pH is increased to at least 9.5 and usually to a value above 11 when excess lime treatment is used. The bactericidal effect of water softening precipitation processes was noted in the early 1900s. Subsequent studies showed that in general, the higher the pH attained, the greater the percent kill of bacteria present.

It was not until the 1960s that effects of lime softening precipitation processes on viruses were investigated. Sproul[8] recently reviewed the literature on virus inactivation and removal by water softening precipitation processes. He concluded that reduction of the number of viruses present results from two effects: (1) irreversible inactivation due to the high pH and (2) physical removal of the viruses by incorporation into the material precipitated. He tabulated the results of eight investigations. In general, there was a direct relationship between the pH attained and the degree of virus reduction observed in these studies. At pH 10 or lower, virus reductions of 90% or less were attained, while at higher pH values, virus reductions were correspondingly higher. Sproul[8] noted that satisfactory explanations for the mechanism of virus removal by $CaCO_3$ and $Mg(OH)_2$ precipitates had not been developed. He emphasized that for virus inactivation by high pH to occur, it is necessary for the virus to be exposed to the pH condition and that for viruses embedded in solid particles this would not occur. Such particles would require removal by sedimentation or other separation processes. Sproul concluded that virus removal or inactivation of 90% or more should be expected in a typical excess-lime-softening treatment plant, and that the process could be adequately controlled by monitoring pH and turbidity.

IV. FILTRATION PROCESSES

Filtration can be defined as the passage of a liquid through a porous medium to remove suspended materials. The three major types of filtration used in drinking water treatment are: rapid sand filtration, slow sand filtration, and diatomaceous earth or diatomite filtration. Some general characteristics of each of these processes are shown in Table 7. There are major differences in the bed depths, filter media sizes, and water flow rates of these processes. Conventional filter media used in sand filtration include silica sand, garnet sand, crushed quartz, and crushed or alluvial coal. The latter are used in dual media rapid filtration. Diatomaceous earth is also siliceous in nature, consisting of the fossil remains of diatoms processed to yield graded size ranges.

Several mechanisms are operative in the removal of particles by filtration. The importance of the particular mechanism involved depends on the type of filter used and the nature of the suspended material. In general, physical mechanisms including straining, gravity, interception, and hydroelectric forces control the removal of larger particles and those having densities significantly greater than that of water. Removal of submicroscopic particles is influenced mainly by surface forces including electrical double layer and Van der Waals forces.[28] The physical, chemical, and biological mechanisms involved in filtration have been the subjects of extensive research and a number of reviews are available.[28-34]

A. Rapid Sand Filtration

1. Description

As indicated in Table 7, rapid sand filtration as applied in drinking water treatment involves the passage of water through approximately 60 to 75 cm of sand or mixed media such as sand and coal with an effective particle size range of 460 to 1600 μm. The filter media are usually supported on a bed consisting of layers of larger graded gravel of various sizes. Conventionally, water flow rates of 5 to 15 m/hr are employed and water passage through the filter occurs in 9 min or less.

Table 7
SOME CHARACTERISTICS OF WATER FILTRATION PROCESSES

Type of filter	Bed depth (cm)	Particle size (um)	Flow rate (range, m/hr)	Empty bed contact time[a] (range, in min)
Slow sand	60—120	150—350	0.1—0.4	90—720
Rapid sand	60—75	450—1600	5—15	2—9
Diatomaceous earth	0.15—0.3	10—50	1.3—5	0.02—0.15

[a] Assumes no filter media present. Contact time would be decreased and flow rate would be increased with filter media present.

Rapid sand filtration is not effective for clarifying water that has not received pior treatment. Because of this, coagulation and flocculation are almost always used prior to rapid sand filtration.[25] In most cases, sedimentation also is used following flocculation to reduce the quantity of solids applied to the filter. A modification of this process called direct filtration involves immediate filtration after coagulation with or without flocculation and with no sedimentation step. In both of these processes, the sand operates as a depth filter, the unsedimented flocculated materials being gradually carried beneath the surface of the filter to depths of 5 to 10 cm during the course of a filter run. The depth to which floc material is carried is a function of the size of the interstices among the sand grains, size and toughness of the floc, and floc rate or filtration rate. Headloss to which the filter is operated before backwash also influences penetration depth. Total flow through such filters is dependent on the characteristics of the filter, the amount of flocculated material applied, and characteristics of the water filtered. Cleaning of the filters by backwashing is initiated upon decrease in flow rate or increase in hydraulic head loss to certain levels. Thus, as indicated by Robeck et al.[35] and Foliguet and Doncoeur,[12] coagulation, flocculation, and rapid sand filtration should be considered as parts of one process rather than as separate processes because of their interdependence. This was indicated in the early investigations of Carlson et al.[23] and Kempf et al.[24] who showed that little virus removal occurred during rapid sand filtration. However, they showed that impregnation of the filter medium with alum floc resulted in improved virus removal efficiency.

2. Virus Removal Efficiency

Data on several investigations of virus removal by rapid sand filtration are shown in Table 8. The results of several of these investigations confirmed that rapid sand filtration, without a preceding coagulation-flocculation process, was inefficient for viral removal.[13,36,37] With flocculation preceding rapid sand filtration, removals of 90 to 99% generally were attained in the filtration step. The 76% removal rate reported by Foliguet and Doncoeur[12] occurred in the presence of 100 mg/ℓ of skim milk added as organic material. They also attained removals of 96% or more with lower concentrations of organic matter present, but showed that by adding large amounts of organic matter (serum + cell extract + 10X Eagle medium) massive releases of retained viruses could be brought about. Guy et al.[13] showed much lower overall removals of seeded polioviruses and coliphage T4 as well as naturally occurring phages. The authors indicated that the calculated removal of naturally occurring phages was derived from data insufficient for truly quantitative analysis. Their seeded studies were done by batch inoculation of a flash mixer prior to flocculation and sedimentation rather than by continued addition of the contaminants. Removals by flocculation were calculated by interpolation from contaminant concentrations found in samples of the appropriate effluents

Table 8
REMOVAL OF VIRUSES FROM WATER BY RAPID SAND FILTRATION

Virus		Filtration conditions				Water			Removal (%)		
Type	Initial concentration (PFU/mℓ)	Type water	Temp. (°C)	pH	Turbidity (TU)	Type study	Velocity (m/hr)	Prefiltration flocculation	Virus	Turbidity	Ref.
Coxsackievirus A5	NS[a]	Spring	22—23	NS	NS	L[b]	5	None	10	NS	36
Coxsackievirus A5	NS	Spring	22—23	NS	NS	L	5	Al[c]	90	NS	36
Coliphage T4	NS	Spring	22—23	NS	NS	L	5	None	35	NS	36
Coliphage T4	NS	Spring	22—23	NS	NS	L	5	Al	97	NS	36
Poliovirus 1 (Mahoney)	1×10^4	River	15—26	7.7—8.1	NS	P	5—14	None	1—50	NS	35
Poliovirus 1 (Mahoney)	1×10^4	River	15—26	7.7—8.1	10—50	P	5—14	Al	90—99	>90	36
Coliphage MS2	NS	Ground	NS	7.8—8.2	70—90	L	5	None	96—98	99+	37
Coliphage MS2	NS	Ground	NS	7.8—8.2	70—90	L	10	None	92—93	99	37
Poliovirus (Mahoney)	1.4×10^5	Demineralized	15—17	NS		P	4	Fe	76—99	NS	12
Phages (natural)	4.0	River	NS	NS	NS	P	NS	Fe	37.5	NS	13
Polioviruses 1—3 (vaccine or Sabin)	4.5×10^4	River	NS	NS	NS	P	NS	Al	19—38	NS	13
Coliphage T4	1×10^9	River	NS	NS	NS	P	NS	Al	0—87	NS	13

[a] Not stated.
[b] L = Lab scale; P = pilot scale
[c] Al = $Al_2(SO_4)_3$; Fe = $FeCl_3$

centered around the calculated nominal residence time. They noted that the peak titers did not occur at the nominal residence times and that the interpolation graphs did not show normal distribution of the contaminants. However, sampling time errors in this type of experimentation would tend to shift the results so as to indicate higher removals rather than lower. Removals by flocculation were consistent with other findings (Table 4), but the filtration results were closer to those expected with rapid sand filtration without coagulation-flocculation than with coagulation-flocculation.

B. Slow Sand Filtration

1. Description

According to Huisman and Wood,[34] development of slow sand filtration processes as currently used began in the early 1800s. In contrast to rapid sand filtration, coagulation is not used prior to slow sand filtration. For the latter process, the water is often clarified by storage in reservoirs before filtration. Although largely superseded by rapid filtration methods in many countries, slow sand filtration is still in use in some highly industrialized cities as well as in smaller communities and rural areas. As indicated in Table 7, the major physical differences between slow sand filtration and rapid sand filtration are in terms of thickness of the filter bed, rate of water flow, and sand size. However, the water purification mechanisms involved are very different. In contrast to filter runs of one to a few days and consequent frequent cleaning required for rapid sand filtration, slow sand filters operate for periods of weeks or months before cleaning becomes necessary. This allows the development of microbial communities at the filter surface and within the filter bed. The surface coat, called the schmutzdecke or filter skin, consists of algae, plankton, diatoms, protozoa, rotifers, and bacteria. The microorganisms constituting this filter skin are active in trapping, digesting, and breaking down organisms and other organic matter in raw water and in mechanically straining out inert suspended particles. This biological activity extends into the filter bed, operating to depths of about 40 cm where, in addition to the biological activity, physical surface forces involving electrostatic attraction and Van der Waals forces operate. Also, development of sticky gelatinous films of biological origin on the surface of sand grains improves adhesion of particles to sand.

Cleaning of slow sand filters involves scraping off the filter skin rather than backwashing as with rapid sand filters. The biological communities within the bed are undisturbed and a new filter skin forms quickly upon resumption of filtration.

2. Virus Removal Efficiency

The results of a number of studies of virus removal by slow sand filtration are shown in Table 9. The only studies in which less than 90% removal was attained were those of Robeck et al.[35] They pointed out that their results were obtained with clean sand. Thus, the biological activity that is an important part of slow sand filtration was not involved. Although Gilcreas and Kelly[36] reported 98% removal of both coxsackievirus A5 and coliphage T4 by slow sand filtration, their experimental conditions did not duplicate slow sand filtration. Filter retention time was only 2 to 4 min and there was no indication that their filters were biologically active.

Poynter and Slade[38] studied removal of poliovirus 1 and coliphage T7 from water "spiked" with these viruses in pilot scale slow sand filters over a period of several years. The sand depths employed were 30 to 60 cm or only half the depth described as typical by Huisman and Wood.[34] This depth was sufficient for development of the range of biological activity described by Huisman and Wood.[34] The data of Poynter and Slade[38] indicated a pronounced reduction in poliovirus removal efficiency at lower water temperatures. At a flow rate of 0.2 m/hr, poliovirus numbers were reduced by an average of 99.997% at 16 to 18°C, but somewhat less (an average of 99.68%) at 5 to 8°C. Coliform bacteria and the 37°C plate

Table 9
REMOVAL OF VIRUSES FROM WATER BY SLOW SAND FILTRATION

Virus		Filtration conditions								
Type	Initial concentration (PFU/mℓ)	Type water	Temp. (°C)	pH	Turbidity	Type study	Water velocity m/hr	Prefiltration treatment	Virus removal (%)	Ref.
Coxsackievirus	NS[a]	Spring	22—23	NS	NS	L[b]	0.5	None	98	36
Coliphage T4	NS	Spring	22—23	NS	NS	L	0.5	None	98	36
Poliovirus 1 (Mahoney)	1×10^4	Demineralized	15—26	6.0—7.5	>0.1	P[b]	0.08	None	22—96	35
Poliovirus 1 (Mahoney)	$4—6 \times 10^2$	River	5—18	NS	NS	P	0.1—0.5	Settling	95.88—99.999	38
Coliphage T7	$4—6 \times 10^2$	River	5—18	NS	NS	P	0.1—0.5	Settling	90.91—99.996	38
Natural enteroviruses	0.13—5.8 (PFU/ℓ)	River	6—11	8.2	NS	F[b]	0.5—0.18	Settling	97—>99.8	39

[a] Not stated.
[b] L = Lab scale; P = pilot scale; F = field scale.

count group of bacteria evidenced similar effects although overall removals were lower. Increasing filter rates from 0.2 m/hr to 0.4 or 0.5 m/hr resulted in about a tenfold decrease in virus removal efficiency. Mature filters removed polioviruses more efficiently than filters that had been operated for only a short period.[38]

Overall, slow sand filters consistently removed greater numbers of coliforms and plate count bacteria than polioviruses and coliphage T7.[38] However, the possible effects of the virus assay procedure on these results were not taken into account. Although seeded virus concentrations were sufficiently high to allow direct assay of the influent water, alginate membrane and membrane adsorption concentration methods were necessary for assaying viruses and phages in filter effluent samples. Since the latter methods are less than 100% efficient, calculated removals would tend to be erroneously higher. Poynter and Slade[38] concluded that slow sand filtration is a highly efficient means for removing enteroviruses from drinking water and that the similarity in removal efficiency for bacteria and viruses suggests that bacteriological parameters can be used to indicate the enterovirus removal capacity of slow sand filters.

In a subsequent study involving a full-scale slow-sand filter, Slade[39] investigated the removal of naturally occurring viruses by slow sand filtration. As indicated in Table 9, virus removals of 97 to >99.8% were attained. These results were based on 16 pairs of influent and effluent samples collected over a period of 3 months. Corresponding coliform reductions in the samples ranged from 73 to >96%. Because of the low virus levels present, concentration of the viruses for assay was necessary for both influent and effluent samples, thereby reducing the type of sampling errors encountered in the previous study.[38] In these studies, viruses were present in two influent samples and in one effluent sample in which *Escherichia coli* was not detected. Slow sand filtered water has often been regarded as fit for consumption with disinfection added only as a final safeguard, but the finding of viruses in slow sand filtered water indicated that chlorination was essential. Comparison of the types of viruses found before and after filtration showed no indication that selective removal occurred.[38,39]

C. Diatomaceous Earth Filtration

1. Description

Diatomaceous earth (DE) filters for filtration of drinking water were developed by the U.S. Army during World War II mainly for the removal of amoebic dysentery cysts from water.[40] DE filtration is used extensively for filtration of swimming pool water but has found limited application in drinking water treatment. Usually these units are designed for small supplies although treatment employing DE filtration was recently chosen for a large municipality that produced an average of 610 mℓ/day (160 million gal/day).[41] As indicated in Table 7, DE filtration characteristics differ significantly from rapid and slow sand filtration. Although filtration rates are only slightly lower than those used in rapid sand filtration, DE particle sizes are nearly two orders of magnitude smaller than the sand particles used in rapid sand filtration. DE bed thickness is more than two orders of magnitude less than that for rapid sand filters, and contact time with DE is reduced by a similar factor.

DE filtration involves precoating a permeable septum with DE suspended in water pumped through the filter. DE filters act primarily as surface filters rather than as depth filters, as in the case of sand filtration. During filtration, additional DE is added as "body feed" to the unfiltered water. This continuously adds new surface for filtration, thereby delays clogging of the filter surface, and contributes to longer filter runs. Flow rates are regulated to a constant rate and filter runs are ended when the headloss reaches a limiting value. Backwashing is accomplished in a variety of ways, depending on the design of the equipment used. Used DE is discarded and new DE is added for each filter run.

2. Virus Removal Efficiency

In studies on the removal or inactivation of infectious hepatitis virus by water treatment,

Neefe et al.[42] used coagulation, settling, and DE filtration for virus removal. Water contaminated with feces believed to contain infectious hepatitis virus and treated by coagulation, settling, and diatomaceous earth filtration caused disease in 43% of the volunteers who ingested it.

Chaudhuri et al.[43] and Amirhor and Engelbrecht[44] concluded that DE alone was not effective for removing coliphages MS2 and T4 from water. The DE used in their studies was at the upper limits of the particle size range and perhaps less efficient in virus removal than smaller sized DE. Efficient removal of viruses was attained with DE filter media coated with a water soluble cationic polyelectrolyte or by adding cationic polyelectrolyte to the virus suspending medium. The presence of organic matter reduced the virus removal efficiency of coated DE.

Brown et al.[45,46] studied the removal of coliphage T2 and poliovirus 1 by DE filtration. They reported removal of >90% of coliphage T2 and the poliovirus initially with uncoated DE, but that efficiency declined over a period of 2 hr. They also showed that DE coated with cationic polyelectrolyte or ferric hydrate removed >98% of the viruses added. Virus removal was significantly reduced when water pH was reduced from pH 9.5 to 6.5.

V. INTERPRETATION OF RESULTS AND THEIR IMPLICATIONS FOR DRINKING WATER TREATMENT

A. Relationships of Viruses to Suspended Solids in Water

In many of the investigations cited, the virus preparation used to study removal efficiency consisted of purified virus freely suspended in water. Although some of the viruses present in natural waters may exist in this state, most probably exist as portions of virus-solid complexes. Because the enteroviruses of concern are produced in large numbers in epithelial cells lining the intestinal tract and are excreted with feces, it is likely that many of them are attached to or embedded in fecal material and other components of wastewaters. The results of several studies indicate that this is the case.[47,48] In addition, viruses adsorb readily to a wide variety of inorganic and organic particles.[49-51] Thus, in considering the virus removal efficiency of water treatment processes, assessment of the removal of freely suspended viruses only may not provide adequate information.

B. Recovery and Infectivity of Solids-Associated Viruses

Under current state-of-the-art conditions, virus concentration and recovery methods are more efficient when applied to waters containing relatively few suspended solid materials. The presence of large amounts of extraneous suspended particles interferes with both the volume that can be sampled and with the efficiency with which the viruses are recovered. Since detection methods rely on the ability of the viruses to demonstrate their presence by entering living cells and initiating the infectious process, the ability of viruses associated with solids to initiate infection is an important consideration. Infectivity of solids-associated viruses has been reported for viruses associated with clay minerals and natural solids.[52,53] The results of studies by Ward and Ashley[54] indicate that viruses associated with sludge must contact and be bonded to specific cell surface receptors for infection to occur. Their results indicated that viruses encapsulated in sludge solids did not utilize mechanisms other than receptor-mediated entry into susceptible cells as had been demonstrated previously for poliovirus encapsulated in lipid vesicles.[55]

C. Viruses as Colloids

The greatest number of the particles suspended in most natural waters are colloidal in nature. Colloidal particles range down in size from about 1 μm in diameter to large organic molecules. Colloids suspended in water may be divided into two groups, hydrophilic and hydrophobic, based on their affinity for water. Clays and many inorganic particles behave as hydrophobic colloids while most organic particles behave as hydrophilic colloids. A range of particles exhibiting characteristics of both groups is also present. Bacteria and viruses appear to act as hydrophilic colloids.

Recent studies by Murray and Parks[56] provide direct evidence that viruses suspended in water behave as hydrophilic colloids. Their results establish a basis for predicting virus adsorption in extra-host systems and the influence of system components on virus adsorption to surfaces. The results of their studies indicate that polioviruses adsorb to surfaces in a manner consistent with the DLVO-Lifshitz* theory of colloid stability. Murray and Parks provided experimental evidence that the adsorption characteristics of polioviruses for a wide range of material was in agreement with adsorption as predicted by the theory as follows:

Metals (strong adsorption) $>$ Sulfides $\geqslant$ Transition Metals
Oxides $>$ SiO_2 $>$ Organics (weak adsorption)

These investigators also cited experimental results from a number of investigations that indicated that the influence of ionic strength, pH, and dissolved components such as organics on virus adsorption is consistent with DLVO-Lifshitz theory. In other studies, Murray and Laband[57] reported that polioviruses were inactivated by physical disruption of the viruses when adsorbed to metal (Al) and certain metal oxides (MnO_2, Al_2O_3, CuO). Polioviruses adsorbed on SiO_2 were not inactivated.

D. Turbidity Reduction as a Measure of Virus Removal

It appears that application of the DLVO-Lifshitz theory offers great potential for design and optimization of adsorption-filtration processes for virus removal and inactivation in water treatment (see Table 2). In view of the continuing reliance on disinfection as the major control step for viruses in drinking water, however, it is unlikely that optimization of treatment processes for virus removal will be considered. It is possible that, in some cases, such alterations may be made to optimize water treatment processes for the removal of other hydrophilic colloids and dissolved organic materials because of their toxicological significance (either directly or as precursors of trace organics). Such changes could result in concomitant improved removal of viruses.

In contrast to other physical, chemical, and microbiological water quality parameters, turbidity is a measurement dependent only on the scattering of light by particles in suspension. Despite this, turbidity measurements that assess the removal of particulate materials by water treatment processes have had and continue to have a major role in water treatment process control. A turbidity Maximum Contaminant Level (MCL) of one Nephelometric Turbidity Unit (NTU) is included in the current federal drinking water regulations.[58]

Overall, the data summarized in this report indicate that the efficiency with which turbidity is removed by sedimentation and filtration processes in drinking water treatment is indicative of the efficiency with which viruses are removed. Recent studies on the removal of asbestos fibers[59] and *Giardia* cysts[60] by coagulation-flocculation followed by filtration showed a similar picture. That is, although neither the asbestos fibers nor the *Giardia* cysts contributed significantly to the turbidity, turbidity reduction served as a surrogate to indicate the efficiency of asbestos and cyst removal. Relatively small increases in effluent turbidity corresponded with increases in asbestos particle counts or detection of cysts. While there are some indications that coagulant doses for optimum virus removal may differ from those needed for optimum turbidity removal, the differences in removal efficiencies are small.

* Theory of colloid stability, the elements of which were developed between 1941 and 1961 by Derjaguin, Landau, Verwey, Overbeek, and Lifshitz.

The major factor that determines overall removal efficiency of coagulation-flocculation is how well the process is controlled, which in turn, depends on: (1) providing the optimum coagulant, coagulant dose, and flocculation time, (depending on water pH); (2) composition and ionic strength of dissolved inorganic salts; (3) amount and nature of turbidity, and (4) temperature. The efficiencies of filtration processes used in conjunction with coagulation-flocculation are also influenced by coagulation-flocculation process control. In addition, filtration efficiencies vary during the course of filter runs; removal of viruses usually is lowest at the beginning, immediately after backwashing, and at the end of the run which is often signalled by increases in effluent turbidity brought about by floc breakthrough. The variability in particle removal efficiency of coagulation-flocculation and filtration processes points up the importance of maintaining efficient disinfection as a final step in water treatment.

Although turbidity appears to be a good measure of the removal of a wide variety of particles in water treatment, recent studies indicate that particle counts may be an even more sensitive indicator. Hannah et al.[61] reported that particle counting provided a more quantitative evaluation of the suspended matter in finished water than turbidity measurement. Kavanaugh et al.[62] used particle counts and turbidity measurements in pilot plant studies of direct filtration. Particle counts in the filter effluent began to increase several hours before increased effluent turbidity was observed. Also, numbers of smaller particles (2.5 to 10 μm) in the effluent began to increase before increases in the numbers of larger particles occurred. It is likely that particle counts will be used increasingly to evaluate solids/liquid separation processes in drinking water treatment.

VI. SUMMARY AND CONCLUSIONS

The results of studies on the removal of viruses by individual drinking water treatment unit processes indicate that both sedimentation processes employing coagulants and various types of filtration processes usually remove more than 90% of the viruses present in water. Results of pilot and field-scale studies were similar to the results of laboratory-scale studies. Because of the limited number of investigations conducted and the low and inconsistent levels of viruses present, the results of investigations with naturally occurring viruses are less quantitative. Coagulation-flocculation processes are optimized for the removal of hydrophobic colloids, which constitute the major part of turbidity in water. It is likely that optimization of these processes for removal of hydrophilic colloids would result in more efficient removal of viruses. Process control is the primary factor that determines the efficiency of removal processes. The variability in removal efficiency of coagulation-flocculation and filtration processes points up the need for maintaining adequate disinfection in addition to the removal process. Overall, the efficiency with which turbidity is removed by sedimentation and filtration processes is indicative of the efficiency of virus removal.

REFERENCES

1. **Powell, S. T.,** *Water Conditioning for Industry,* McGraw-Hill, New York, 1954, 22.
2. **van Duuren, F. A.,** Water purification unit operations: fact and fiction, *J. Inst. Water Pollut. Control,* 77, 31, 1978.
3. **Hall, E. S. and Packham, R. F.,** Coagulation of organic color with hydrolyzing coagulants, *J. Am. Water Works Assoc.,* 57, 1149, 1965.
4. **Rao, M. N.,** Studies on coagulation of water — a review (ii) basic mechanism of coagulation process, *Environ. Health,* 10, 1, 1968.
5. **Packham, R. F. and Sheiham, I.,** Developments in the theory of coagulation and flocculation, *J. Inst. Water Eng. Sci.,* 31, 96, 1977.

6. **A.W.W.A.** Committee Report, State of the Art of Coagulation, *J. Am. Water Works Assoc.*, 63, 99, 1970.
7. **Miller, D. G.,** Sedimentation: a review of published work, *Water Works Eng.*, 68, 52, 1964.
8. **Sproul, O. J.,** Critical review of virus removal by coagulation processes and pH modifications, EPA-600/2-80-004, U.S. Environmental Protection Agency, Cincinnati, Ohio, 1980.
9. **Thorup, R. T., Nixon, F. P., Wentworth, D. G., and Sproul, O. J.,** Virus removal by coagulation with polyelectrolytes, *J. Am. Water Works Assoc.*, 62, 97, 1970.
10. **Boardman, G. D. and Sproul, O. J.,** Absorption as a protective mechanism for poliovirus, *Proc. 97th Am. Water Works Assoc. Conf.*, Part 1, Denver, Colo., 1977.
11. **Foliguet, J. M. and Michelet, S.,** The elimination of enteroviruses during the treatment of the public water supply, (in French), *Rev. Hyg. Med. Soc.*, 17, 175, 1969.
12. **Foliguet, J. M. and Doncoeur, F.,** Elimination of enteroviruses during treatment of drinking water by coagulation-flocculation-filtration, (in French), *Water Res.*, 9, 953, 1975.
13. **Guy, M. D., McIver, J. D., and Lewis, M. J.,** The removal of virus by a pilot treatment plant, *Water Res.*, 11, 421, 1977.
14. **Chang, S. L., Stevenson, R. E., Bryant, A. R., Woodward, R. L., and Kabler P. W.,** Removal of coxsackie and bacterial viruses in water by flocculation, I. Removal of coxsackie and bacterial viruses in water of known chemical content by flocculation with aluminum sulfate of ferric chloride under various testing conditions, *Am. J. Public Health,* 48, 51, 1958.
15. **Chang, S. L., Stevenson, R. E., Bryant, A. R., Woodward, R. L., and Kabler, P. W.,** Removal of coxsackie and bacterial viruses in water by flocculation. II. Removal of coxsackie and bacterial viruses and the native bacteria in raw Ohio River water by flocculation with aluminum sulfate and ferric chloride, *Am. J. Public Health,* 48, 159, 1958.
16. **Godbole, S. H., Illavia, S. J., Kulkarni, A. L., and Kulkarni, R. R.,** Pilot plant studies on the removal of enteroviruses during water treatment, *Environ. Health,* 10, 29, 1968.
17. **Chaudhuri, M. and Engelbrecht, R. S.,** Removal of viruses from water by chemical coagulation and flocculation, *J. Am. Water Works Assoc.*, 62, 563, 1970.
18. **York, D. W.,** Virus Removal by Chemical Coagulation, M. S. Thesis, University of Tennessee, Knoxville, 1970.
19. **York, D. W. and Drewry, W. A.,** Virus removal by chemical coagulation, *J. Am. Water Works Assoc.*, 66, 711, 1974.
20. **Shelton, S. P. and Drewry, W. A.,** Tests of coagulants for the reduction of viruses, turbidity and chemical oxygen demand, *J. Am. Water Works Assoc.*, 65, 627, 1973.
21. **Manwaring, J. F., Chaudhuri, M., and Engelbrecht, R. S.,** Removal of viruses by coagulation and flocculation, *J. Am. Water Works Assoc.*, 63, 298, 1971.
22. **Chang, S. L., Isaac, P. C. G., and Baine, N.,** Studies on destruction of bacterial viruses. III. Dynamics of the removal of bacterial virus (bacteriophage against *Micrococcus pyogenes.* var. *albus*) in water by flocculation with aluminum sulfate, *Am. J. Hyg.*, 57, 253, 1953.
23. **Carlson, H. J., Ridenour, G. M., and McKhann, C. F.,** Efficacy of standard purification methods in removing poliomyelitis virus from water, *Am. J. Public Health,* 32, 1256, 1942.
24. **Kempf, J. E., Wilson, M. G., Pierce, M. E., and Soule, M. H.,** Effect of aluminum hydroxide sedimentation, sand filtration and chlorination on the virus of poliomyelitis, *Am. J. Public Health,* 32, 1366, 1942.
25. American Water Works Association, *Water Quality and Treatment,* McGraw-Hill, New York, 1971, 89.
26. **Johnson, J. H., Fields, J. E., and Darlington, W. A.,** Removing viruses from water by polyelectrolytes, *Nature (London),* 213, 665, 1967.
27. **Foliguet, J. M., Doncoeur, F., and Block, J. C.,** A contribution to research on the removal of enteroviruses during drinking water treatment, (in French), *J. Fr. Hydrol.*, 14, 11, 1974.
28. **Ives, K. J. and Gregory, J.,** Basic concepts of filtration, *Proc. Soc. Water Treat. Exam.*, 16, 147, 1967.
29. **O'Melia, C. R. and Crapps, D. K.,** Some chemical aspects of rapid sand filtration, *J. Am. Water Works Assoc.*, 56, 1326, 1964.
30. **Craft, T. F.,** Review of rapid sand filtration theory, *J. Am. Water Works Assoc.*, 58, 428, 1966.
31. **Ives, K. J. and Gregory, J.,** Surface forces in filtration, *Proc. Soc. Water Treat. Exam.*, 15, 93, 1966.
32. **Ives, K. J. and Sholji, I.,** Research on variables affecting filtration, *J. Sanit. Eng. Div., Am. Soc. Civ. Eng.*, 91, 1, 1965.
33. **Ison, C. R. and Ives, K. J.,** Removal mechanisms in deep bed filtration, *Chem. Eng. Sci.*, 24, 717, 1969.
34. **Huisman, L. and Wood, W. E.,** *Slow Sand Filtration,* World Health Organization, Geneva, 1974.
35. **Robeck, G. G., Clarke, N. A., and Dostal, K. A.,** Effectiveness of water treatment processes in virus removal, *J. Am. Water Works Assoc.*, 58, 1275, 1972.
36. **Gilcreas, F. W. and Kelly, S. M.,** Relation of coliform-organism test to enteric-virus pollution, *J. Am. Water Works Assoc.*, 47, 683, 1955.
37. **Sriramulu, N. and Chaudhuri, M.,** Dual-media coal sand filter and its virus removal potential, *Indian J. Environ. Health,* 18, 77, 1976.

38. **Poynter, S. F. B. and Slade, J. S.,** The removal of viruses by slow sand filtration *Prog. Water Technol.*, 9, 75, 1977.
39. **Slade, J. S.,** Enteroviruses in slow sand filtered water, *J. Inst. Water Eng. Sci.*, 32, 530, 1978.
40. **Black, H. H. and Spaulding, C. A.,** Diatomite water filtration developed for field troops, *J. Am. Water Works Assoc.*, 36, 1208, 1944.
41. **Bryant, E. A. and Brailey, D.,** Large-scale ozone-DE filtration: an industry first, *J. Am. Water Works Assoc.*, 72, 604, 1980.
42. **Neefe, J. R., Baty, J. B., Reinhold, J. B., and Stokes, J., Jr.,** Inactivation of the virus of infectious hepatitis in drinking water, *Am. J. Public Health*, 37, 365, 1947.
43. **Chaudhuri, M., Amirhor, P., and Engelbrecht, R. S.,** Virus removal by diatomaceous earth filtration, *J. Environ. Eng. Div., Am. Soc. Civ. Eng.*, 100, 937, 1974.
44. **Amirhor, P. and Engelbrecht, R. S.,** Virus removal by polyelectrolyte-aided filtration, *J. Am. Water Works Assoc.*, 67, 187, 1975.
45. **Brown, T. S., Malina, J. F., Jr., and Moore, B. D.,** Virus removal by diatomaceous earth filtration. 1, *J. Am. Water Works Assoc.*, 66, 98, 1974.
46. **Brown, T. S., Malina, J. F., Jr., and Moore, B. D.,** Virus removal by diatomaceous earth filtration. 2, *J. Am. Water Works Assoc.*, 66, 735, 1974.
47. **Berg, G.,** Removal of viruses from sewage, effluents, and waters. I. A review, *Bull. W. H. O.*, 49, 451, 1973.
48. **Wellings, F. M., Lewis, A. L., and Mountain, C. W.,** Demonstration of solids-associated virus in wastewater and sludge, *App. Environ. Microbiol.*, 31, 354, 1976.
49. **Bitton, G.,** Adsorption of viruses onto surfaces in soil and water, *Water Res.*, 9, 473, 1975.
50. **Gerba, C. P., Stagg, C. H., and Abadie, M. G.,** Characterization of sewage solids-associated viruses and behavior in natural waters, *Water Res.*, 12, 805, 1978.
51. **Bitton, G.,** Adsorption of viruses to surfaces: technological and ecological implications, in *Adsorption of Microorganisms to Surfaces*, Bitton, G. and Marshall, K. C., Eds., John Wiley & Sons., New York, 1980, chap. 10.
52. **Schaub, S. A. and Sagik B. P.,** Association of enterioviruses with natural and artificially introduced colloidal solids in water and infectivity of solids-associated virions, *Appl. Microbiol.*, 30, 212, 1975.
53. **Moore, B. E., Sagik, B. P., and Malina, J. F., Jr.,** Viral association with suspended solids, *Water Res.*, 9, 197, 1975.
54. **Ward, R. L. and Ashley, C. S.,** Mode of initiation of cell infection with sludge-associated poliovirus, *Appl. Environ. Microbiol.*, 38, 329, 1979.
55. **Wilson, T., Papahadjopoulos, D., and Taber, R.,** Biological properties of poliovirus encapsulated in lipid vesicles: antibody resistance and infectivity in virus resistant cells, *Proc. Natl. Acad. Sci. U.S.A.*, 74, 3471, 1977.
56. **Murray, J. P. and Parks, G. A.,** Poliovirus adsorption on oxide surfaces. Correspondence with the DLVO-Lifshitz theory of colloid stability, in, *Particulates in Water, Characterization, Fate, Effects, and Removal*, Kavanaugh, M. C. and Leckie, J. O., Eds., American Chemical Society, Adv. Chem. Ser. 189, 1980, chap. 5.
57. **Murray, J. P. and Laband, S. J.,** Degradation of poliovirus by adsorption on inorganic surfaces, *Appl. Environ. Microbiol.*, 37, 480, 1979.
58. National interim primary drinking water regulations, EPA 570/9-76-003, Government Printing Office, Washington, D.C., 1978.
59. **Logsdon, G. S. and Symons, J. M.,** Removal of asbestiform fibers by water filtration, *J. Am. Water Works Assoc.*, 69, 499, 1977.
60. **Logsdon, G. S., Symons, J. M., Hoye, R. L., Jr., and Arozarena, M. M.,** Alternative filtration methods for removal of *Giardia* cysts and cyst models, *J. Am. Water Works Assoc.*, 73, 111, 1981.
61. **Hannah, S. A., Cohen, J. M., and Robeck, G. G.,** Control techniques for coagulation-filtration, *J. Am. Water Works Assoc.*, 59, 1149, 1967.
62. **Kavanaugh, M. C., Tate, C. H., Trussell, A. R., Trussell, R. R., and Treweek, G.,** Use of particle size distribution measurements for selection and control of solid/liquid separation processes, in *Particulates in Water, Characterization, Fate, Effects and Removal*, Kavanaugh, M. C. and Leckie, J. O., Eds., American Chemical Society, Adv. Chem. Ser. 189, 1980, chap. 14.

Chapter 5

INACTIVATION OF VIRUSES IN WATERS AND IN WASTEWATER EFFLUENTS BY DISINFECTION

Gerald Berg

TABLE OF CONTENTS

I. INTRODUCTION

Most treatment processes remove or destroy some viruses in waters and wastewaters, but only disinfection is expected to destroy all viruses in such waters. For many decades, chlorine and some of its compounds have been used to achieve such destruction, although in recent years advancing knowledge has brought about some redirection in the concepts of chlorination. Ozone, moreover, has also become a disinfectant of great interest.

II. THE CHEMISTRY OF WATER DISINFECTANTS

The most important forms of chlorine used for the disinfection of waters and wastewaters are hypochlorous acid (HOCl), the hypochlorite ion (OCl^-), chlorine dioxide (ClO_2), and the chloramines (monochloramine [NH_2Cl] and dichloramine [$NHCl_2$]).

A. Hypochlorous Acid and Hypochlorite Ion

Chlorine (Cl_2) hydrolyzes rapidly in water to form HOCl.[1]

$$Cl_2 + OH^- \rightleftharpoons HOCl + Cl^- \quad (1)$$

At pH levels between 3 and 5, almost all hydrolyzed Cl_2 exists as HOCl. At pH levels above 5, HOCl ionizes to OCl^-.[2]

$$HOCl \rightleftharpoons H^+ + OCl^-$$

$$Ki = 3.3 \times 10^{-8} \text{ at } 20°C \quad (2)$$

At 20°C, the most extensive ionization occurs at pH levels between 7.0 and 8.0. At pH 7, approximately 75% of the chlorine exists as HOCl and 25% occurs as OCl^-. At pH 8, approximately 77% of the chlorine occurs as OCl^- and 23% remains as HOCl. Virtually complete dissociation of HOCl to OCl^- occurs when the pH level reaches 10. At pH levels below 3, hydrolysis of Cl_2 is suppressed and an equilibrium exists between Cl_2 and HOCl. Even at pH 1, however, with chlorine concentrations below 10 mg/ℓ, less than 4% of the chlorine occurs as Cl_2 at 25°C.

Both the hydrolysis and ionization reactions are little affected by temperature.

B. Chloramines

The chloramines derive from reactions of ammonia (NH_3) with HOCl. Monochloramine (NH_2Cl) results as the first step in these reactions:

$$NH_3 + HOCl \rightarrow NH_2Cl + H_2O \quad (3)$$

In the dilute solutions common in water disinfection, the reaction shown in Equation (3) goes to completion at pH levels of 7 to 10 when the ratio of Cl_2 to NH_3 is equimolar (weight ratio of Cl_2 to NH_3 = 5:1). At 25°C, the reaction is fastest at pH 8.3, going to 99% completion in 0.069 sec. At pH 7, the reaction goes to 99% completion in about 0.2 sec. At 0°C, the rate of reaction is reduced considerably; about 10 min are required for 99% completion of the reaction at this pH.[3]

Dichloramine ($NHCl_2$) forms from the reaction between NH_2Cl and HOCl as follows:

$$NH_2Cl + HOCl \rightarrow NHCl_2 + H_2O \quad (4)$$

This reaction is affected by pH and by the concentrations of $NHCl_2$ and NH_3 present. Dichloramine formation is increased by increased concentrations of NH_2Cl or NH_3 and with diminishing pH. Thus at pH 7, when the ratio of Cl_2 to NH_3 is equimolar, the ratio of NH_2Cl to $NHCl_2$ is about 6:4; at pH 8, the ratio is about 8:2, at pH 5, it is about 2:8.[3] At pH 7 to 8, when the weight ratio of Cl_2 to ammonia nitrogen is less than 5:1, almost all of the chloramine present is probably NH_2Cl.[3] As the weight ratio increases and/or as the pH decreases, increasing quantities of $NHCl_2$ form. At pH 7 to 8, the rate for reaction (4) is considerably slower than that for reaction (3).[4] Conversion of 90% of the NH_2Cl to $NHCl_2$ may take an hour at these pH levels. The rates of the reactions that convert HOCl to chloramines are important in disinfection because the existence of the strongly virucidal HOCl for even a short time may achieve a considerable amount of disinfection. Chloramines are much slower virucides than HOCl (see below).

Trichloramine or nitrogen trichloride (NCl_3) results from the reaction of $NHCl_2$ and HOCl, as follows:

$$NHCl_2 + HOCl \rightarrow NCl_3 + H_2O \qquad (5)$$

Trichloramine forms readily in the presence of high chlorine:ammonia nitrogen ratios. It is unstable. Although NCl_3 is a strong oxidizing agent, little is known about its virucidal capabilities in waters. Trichloramine, along with $NHCl_2$, is an important cause of obnoxious taste and odor in chlorinated waters. Trichloramine is also highly corrosive to metal pipes.

C. Chlorine Dioxide

Chlorine dioxide (ClO_2) has been promoted in recent years as a disinfectant alternative to HOCl, OCl^-, and the chloramines. Chlorine dioxide is a rapid virucide that does not form trihalomethanes, which are toxic (see below). Chlorine dioxide can be produced by the reaction of sodium chlorite ($NaClO_2$) and potassium persulfate ($K_2S_2O_8$):

$$1\ ClO_2^- + K_2S_2O_8 \rightarrow 2\ ClO_2 \uparrow + 2K^+ + 2\ SO_4^= \qquad (6)$$

Chlorine dioxide does not react with NH_3 to form chloramines. Chlorine dioxide is highly soluble in water, but it neither hydrolyzes nor dissociates in water.

In the presence of HOCl, ClO_2 oxidizes slowly to form chlorate (ClO_3^-):

$$HOCl + ClO_2 \rightarrow ClO_3^- + HCl \qquad (7)$$

Furthermore, in common water treatment practice, ClO_2 is reduced readily to chlorite (ClO_2^-). Such reduction results from the reaction of ClO_2 with many dissolved materials that occur in water and may completely dissipate the ClO_2 added to water in little more than 30 min.[5] Chlorine dioxide also disproportionates into ClO_3^- and ClO_2^- in water. Thus, ClO_3^- and ClO_2^-, both of which are toxic (see below), commonly result from the introduction of ClO_2 into water. Moreover, waters that possess a demand for HOCl (brought about by species other than ammonia) usually possess an even greater demand for ClO_2.

D. Ozone

Ozone (O_3) is usually produced by exposing dry O_2 to a corona discharge of high voltage electricity. The solubility of O_3 is only 570 mg/ℓ at 20°C; moreover, it is highly volatile and unstable in water, decomposing in minutes. In water, O_3 maintains its oxidizing capacity. Thus, O_3 may not directly be involved in disinfection reactions. When O_3 decomposes, the free radicals HO_2, HO, and possibly H appear to form.[6] These free radicals result also from gamma irradiation of water[7] and may be the chief disinfecting species associated with ozonation. In any event, the short half-life of O_3 disallows the maintenance of disinfecting residuals of the oxidant for extended periods and makes the accurate control of disinfecting residuals for short periods difficult.

Ozone reacts with most oxidizable substances and thus the O_3 demand of impurities in waters is usually high. Among these impurities are the humic acids and other precursors of trihalomethanes. Thus, treatment of waters with O_3 reduces the quantities of trihalomethanes resulting from subsequent chlorination.

III. TOXICITY

A. Hypochlorous Acid and Hypochlorite Ion

In a water system where chlorine is present as HOCl and OCl^-, trihalomethanes form from chlorination of precursors, primarily humic and fulvic acids, much more rapidly at high pH levels than at low pH levels. Thus, in a drinking water dosed at 25°C with 10 mg/ℓ of Cl_2, chloroform ($CHCl_3$) forms twice as quickly at pH 11.5 than at pH 6.5. Other trihalomethanes, especially bromodichloromethane ($CHBrCl_2$) may also form if the proper precursors are present.[8] Chloroform produces central nervous system depression, hepatotoxicity, nephrotoxicity, teratogenicity, and carcinogenicity in mammals. Bromine-containing trihalomethanes are believed to be even more toxic than chloroform. According to the U.S. Environmental Protection Agency (USEPA), studies in man indicate that although

> ". . . causal relationships cannot be proven on the basis of results from epidemiologic studies . . . viewed collectively . . . epidemiologic studies provide sufficient evidence for maintaining the hypothesis that a health risk may be occurring and that the positive relationships may be reflecting a causal association between constituents of drinking water and cancer mortality . . .
>
> "In summary, on the basis of the available toxicological data, chloroform has been shown to be a carcinogen in mice and rats at high dose levels. Because its metabolic pattern in animals is qualitatively similar to that in humans, chloroform should be suspected of being a human carcinogen. Epidemiological studies also suggest a human risk. Although documentation of other trihalomethane toxicity is not so well established, they should be suspected of posing similar risk."[9]

B. Chloramines

Chloramines do not readily form trihalomethanes in water.[10] However, chloramines are toxic to aquatic life, and unless neutralized, produce morbidity and mortality downstream of discharging outfalls even after considerable dilution in receiving waters.[11] This same toxicity is manifested in the hobbyists tropical fish aquarium wherever tap waters contain chloramines. Moreover, chloramines in water may induce hemolytic anemia in patients undergoing kidney dialysis.[12]

C. Chlorine Dioxide

Introduced into water, ClO_2 does not produce trihalomethanes,[10] nor does ClO_2 oxidize trihalomethanes already present in water.[13] Instead, probably by oxidizing humic acids and other precursors, ClO_2 suppresses the formation of trihalomethanes by HOCl and OCl^- added subsequently.[14] Chlorine dioxide, however, disproportionates into ClO_3^- and ClO_2^- in water. Chlorite also results from the oxidation of organic compounds in water by ClO_2. Both ClO_3^- and ClO_2^- are toxic.[10] Because of this toxicity, the USEPA, in its Trihalomethane Regulation, has recommended a total limit of 0.5 mg/ℓ for ClO_2, ClO_2^-, and ClO_3^- in drinking water.[15]

D. Bromine Chloride

Bromine chloride hydrolyzes to form hypobromous acid (HOBr) and Cl^-. Thus, in natural waters, the primary trihalomethane it forms is bromoform ($CHBr_3$). The quantity of trihalomethanes produced in waters by HOBr is greater than that produced by equivalent quantities of HOCl.[10]

E. Ozone

Ozonation of water does not result in the formation of trihalomethanes.[10] Moreover, ozonation appears to oxidize trihalomethane precursors, resulting in a reduction in the quantity of trihalomethanes consequent to subsequent chlorination.[14] However, ozonation does not remove trihalomethanes from waters that already contain them.[13]

F. By-Products of Disinfection

Our knowledge about the toxicities of disinfection by-products is limited. However, ClO_2, O_3, and chloramines do not produce trihalomethanes, and ClO_2 and O_3, at least, destroy trihalomethane precursors. Currently, it would seem that ClO_2, O_3, and chloramines hold some promise as water disinfectants to be used adjunctively or in place of HOCl and OCl^-.

IV. INACTIVATION OF VIRUSES IN WATER

The discovery of trihalomethanes in waters and concern that other toxic substances may result from chlorination of waters stimulated research and debate on alternatives to long standard water disinfection practices. The major questions currently under investigation have sought to explore whether disinfectants such as ClO_2 and O_3 are as effective as HOCl and whether the alternatives to standard chlorination produce trihalomethanes or other toxic substances as by-products of the disinfection process. For the moment at least, other important issues have been relegated to levels of lesser significance. The adequacy of indicator organisms in disinfected waters and the need for comprehensive data on the disinfection of a broad spectrum of viruses, for example, are issues now subjugated. Only the importance of particulates in the disinfection process continues to receive significant attention.

A. Hypochlorous Acid and Hypochlorite Ion

In water disinfection, the equilibrium between HOCl and OCl^- is especially significant, because most potable waters are slightly alkaline and the shift from HOCl to OCl^- is extensive between pH 7 and 8. The difference in the disinfection capabilities of HOCl and OCl^- is important.

HOCl has long been known to be a potent virucide. At 27 to 29°C, 99.6% of a strain of coxsackievirus A2 was inactivated in about 100 sec by 1 mg of HOCl per liter.[16] Hypochlorite is generally a weaker virucide. In parallel experiments, OCl^- in the same concentration inactivated 99.6% of the virus in about 3.5 min. At 3 to 6°C, the same degree of inactivation was achieved by HOCl in about 7 min and by OCl^- in about 30 min.[16] A strain of adenovirus 3 was considerably more sensitive than the coxsackievirus A2 to both HOCl and OCl^-. At 4°C, 0.1 mg of HOCl per liter required only about 20 sec to inactivate 99.8% of the adenovirus and the same concentration of OCl^- required only about 200 sec.[17] At 0°C, 99% of a strain of poliovirus 1 was inactivated in about 1 min by 1 mg of HOCl per liter and in about 10 min by 1 mg of OCl^- per liter.[18] Others reported that under similar conditions, 1 mg of HOCl per liter inactivated 99% of a strain of poliovirus 1 in about 100 sec.[19] Thus, the resistance of the poliovirus 1 was somewhat closer to that of the coxsackievirus A2 than to that of the adenovirus. The germicidal effectiveness of OCl^- is enhanced considerably by the presence of the chloride ion.[19-25] The mechanism for this enhanced activity is unknown. Figure 1 shows the relative resistance of a strain of poliovirus 1 to HOCl and OCl^- in the absence of Cl^-.[26] The virus was inactivated about three times more quickly by HOCl than by OCl^-. In the presence of Cl^-, OCl^- inactivated the poliovirus more rapidly than the HOCl did (not shown).

Although it is understandable that viruses of different size (thus targets of different size) differ in resistance to disinfectants, it is not clear why viruses of similar size and within the same family differ in resistance. It has long been suspected that a real difference exists in

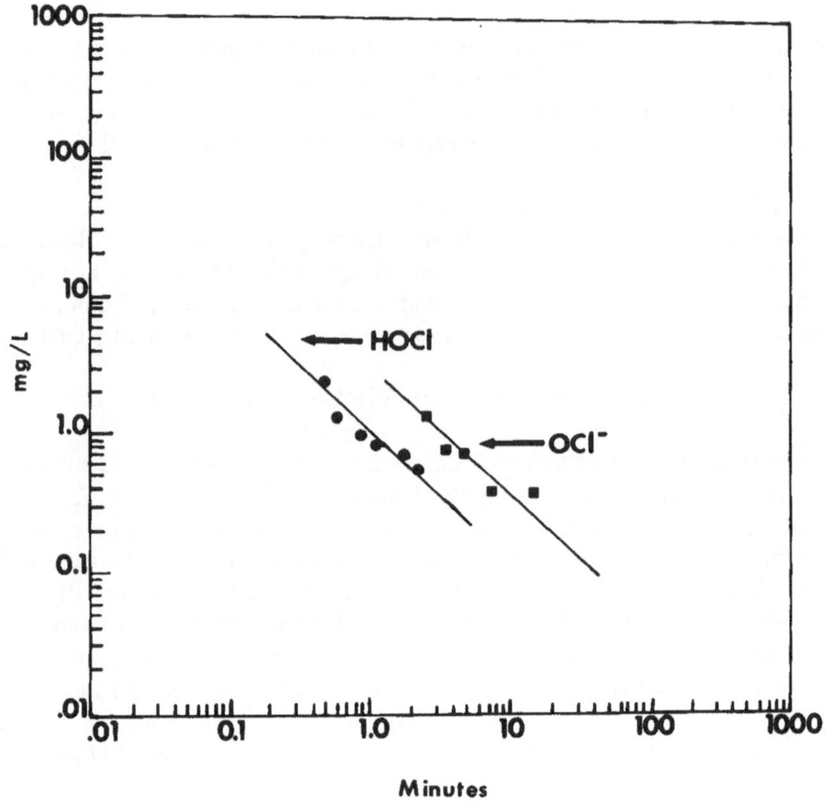

FIGURE 1. Concentration-time relationships for the inactivation of 99% of a strain of poliovirus 1 at 15°C by HOCl and OCl^- in the absence of Cl^- (abstracted). (From Cronier, S., Scarpino, P. V., and Zink, M. L., *Water Chlorination, Environmental Impact and Health Effects*, Vol. 2, Ann Arbor Science, Ann Arbor, Mich., 1978, 651. With permission.)

the resistance to disinfectants of wild strains of a virus and strains of the same virus that have gone through many passages in laboratory host cultures.[27] Moreover, it has been reported that poliovirus 1 (LSc) became progressively more resistant to chlorine during repeated exposures to sublethal doses of chlorine.[28]

In spite of this variability in disinfection resistance and the variability in the resistance to disinfectants that exists among different enteroviruses and among other viruses that are found in water, most viruses including bacterial viruses are more resistant to chlorine than *Escherichia coli* is. And *E. coli,* of course, is the most common of the fecal coliforms.

Most viruses in water are particulate-associated.[29] Theoretically, it has long been apparent that particulates in water must interfere with disinfection. Particulates may protect viruses (and other microorganisms as well) in two ways. If a virion is adsorbed to a particulate, then the virion is shielded on one side from random attack by molecules of a disinfecting species. If a sufficient number of particulates is present, many virions may be shielded on more than one side. The most important protection, however, probably derives from the shielding of cellular debris. This debris is probably comprised of cells in which the virions have multiplied and to which many may still adhere. Many virions may be walled off in such cells and perhaps totally inaccessible to disinfectant molecules, hence the importance in minimizing the numbers of particulates (turbidity) in waters that are to be disinfected. The protection from HOCl and OCl^- of virions associated with cell debris has been demonstrated.[30]

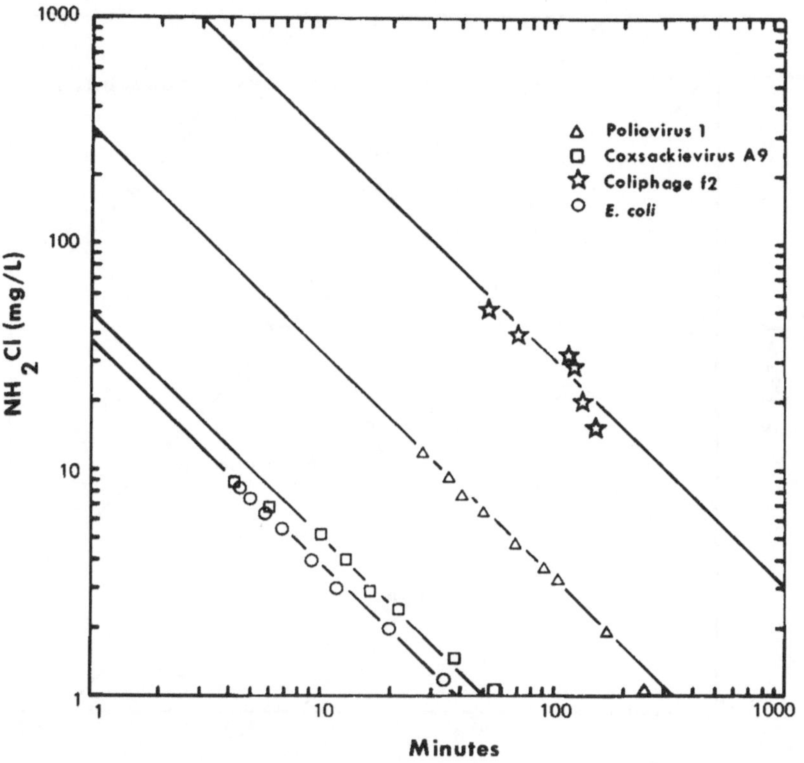

FIGURE 2. Concentration-time relationships for the inactivation of 99% of strains of poliovirus 1, coxsackievirus A9, coliphage f2, and *Escherichia coli* at 25°C by NH_2Cl (at pH 9).

B. Chloramines

Compared with HOCl, the chloramines are slow virucides. However, although the reaction that forms NH_2Cl from NH_3 and HOCl in the pH 7 to 8 range takes less than a second to reach 99% completion at ambient temperatures, the reaction that forms $NHCl_2$ from NH_2Cl and HOCl under these same conditions is relatively slow, requiring perhaps an hour to reach 90% completion (see above). Most natural waters are slightly alkaline; thus, although the addition of HOCl to NH_3-bearing waters usually produces NH_2Cl quickly, some HOCl may remain available for disinfection for at least the hour or more during which $NHCl_2$ is produced. When HOCl (as Cl_2 or OCl^-) is added to NH_3-bearing waters in the 7 to 8 pH range, the rate of inactivation of virions is largely determined by the HOCl (and OCl^- when sufficient Cl^- is present) residual as long as that residual persists. The rate of virion inactivation under these conditions is considerably more rapid than that which occurs when preformed chloramines are added to a water. $NHCl_2$ is usually thought to be more rapidly biocidal than NH_2Cl, but recent findings indicate that in some circumstances at least NH_2Cl is more rapidly virucidal than $NHCl_2$.[31]

The relative rates at which NH_2Cl inactivated strains of poliovirus 1, coxsackievirus A9, coliphage f2, and *E. coli* are shown in Figure 2.[31] The coxsackievirus A9 was inactivated almost as rapidly as the *E. coli* was. The *E. coli* was inactivated 7 to 8 times more rapidly than the poliovirus was and more than 100 times more rapidly than the coliphage.

The *E. coli* was inactivated more than 1000 times more rapidly than the poliovirus was by $NHCl_2$ and almost 100 times more rapidly than the coxsackievirus A9 (Figure 3).[36] The *E. coli* was inactivated by $NHCl_2$ about 10 times more rapidly than coliphage $\Phi \times 174$ was.

The temperature coefficients for a 10°C change (Q_{10}) in the rates of inactivation of the poliovirus by NH_2Cl and $NHCl_2$ were between 2 and 3.

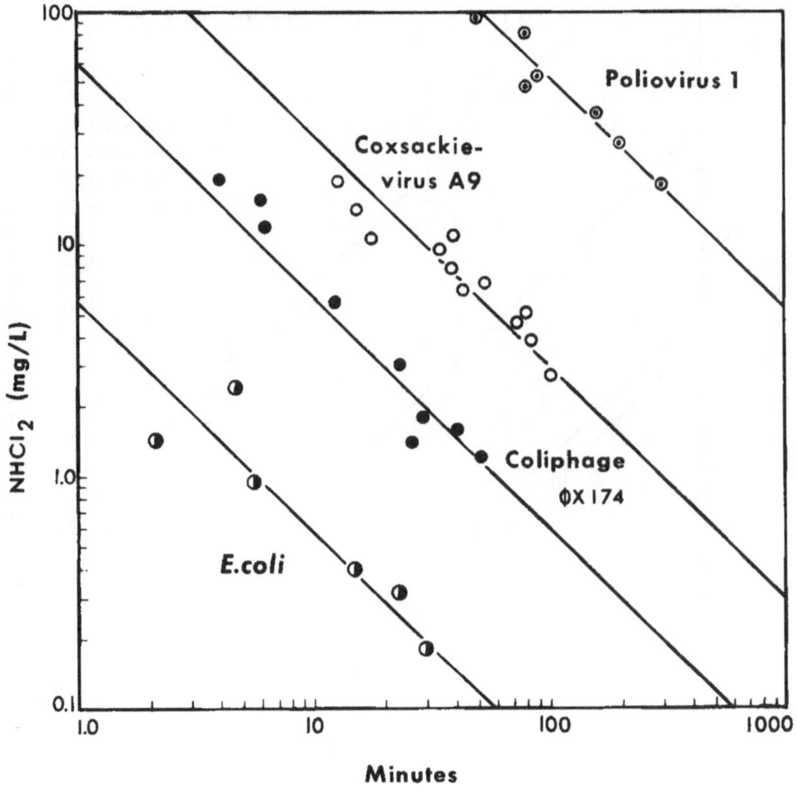

FIGURE 3. Concentration-time relationships for the inactivation of 99% of strains of poliovirus 1, coxsackievirus A9, coliphage ΦX174 and *Escherichia coli* at 15°C by $NHCl_2$ (at pH 4.5).

C. Chlorine Dioxide

Chlorine dioxide is a rapid virucide and bactericide. Although its chemistry seems unaltered over a wide range of pH, its activity as a biocidal agent varies greatly with changing pH. The virucidal capability of ClO_2 increases considerably with increasing pH. At 21°C, 99% of a strain of poliovirus 1 was inactivated by 1 mg of ClO_2 per liter in about 180 sec at pH 4.5, in about 100 sec at pH 7, and in about 20 sec at pH 9 (Figure 4).[26] This same pH effect had been demonstrated with *E. coli*.[33] The reason for this pH effect is not clear. Because ClO_2 apparently remains chemically unaltered over the pH range of these tests, the effect has been attributed to alterations in the virion and bacterial surfaces. Stearic alterations in the ClO_2 molecule may also be a factor.

At pH 7, the rate of inactivation of the poliovirus was a little less than three times greater at 25°C than at 15°C and a little more than three times greater at 15°C than at 5°C (Figure 5).[32] Thus, the temperature coefficient (Q_{10}) for the inactivation reaction was about 3 over the temperature range of the test. These temperature coefficient studies were done with monodispersed virions.

At pH 7, 1 mg of ClO_2 per liter inactivated 99% of a strain of poliovirus 1 in about 90 sec at 15°C. Under similar conditions, 1 mg of ClO_2 per liter inactivated 99% of a strain of coxsackievirus A9 in about 25 sec, and 99% of a strain of *E. coli* in less than 10 sec (Figure 6).[26]

D. Ozone

Ozone is a rapid virucide, probably more rapid than ClO_2 (at pH 7 and less) and HOCl.

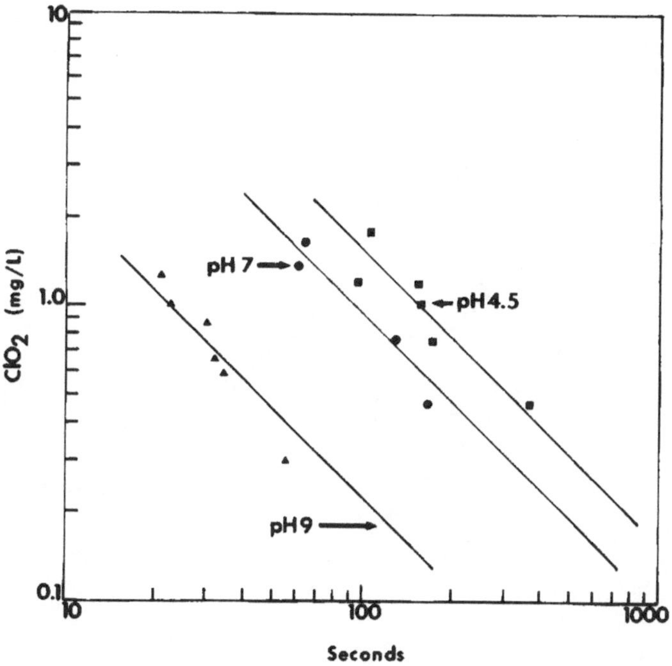

FIGURE 4. Concentration-time relationships for the inactivation at 21°C of 99% of a strain of poliovirus 1 by ClO_2 at pH 4.5, 7, and 9. (From Cronier, S., Scarpino, P. V., and Zink, M. L., *Water Chlorination, Environmental Impact and Health Effects*, Vol. 2, Ann Arbor Science, Ann Arbor, Mich., 1978, 651. With permission.)

Ozone is difficult to quantify, however, at the concentrations at which it is used for water disinfection. Moreover, if O_3-associated disinfection is in reality accomplished by free radicals resulting from ozonation, the dynamics of O_3 disinfection would be difficult to determine. Thus, although data on the inactivation of viruses and bacteria by O_3 exist, rate data that would allow reasonably precise comparisons with other disinfectants are not available.

Some of the earlier data were not temperature related. In distilled water, several tenths of a milligram of O_3 per liter inactivated more than 99% of a strain of poliovirus 1 in 2 to 3 min. In riverwaters, O_3 was almost as effective.[34,35] In water that contained organic material, more than 99% of strains of poliovirus 3 and coxsackievirus B3 were inactivated in 10 min by 0.1 to 0.2 mg of O_3 per liter.[36]

More precise temperature-related data have been reported recently. Inactivation of adenoviruses and a Shigella occurred more rapidly at lower temperatures than at higher temperatures, apparently because of the greater solubility of O_3 at the lower temperatures.[37] At 5°C, more than 99% of a strain of poliovirus 1 was inactivated in distilled water in less than 8 sec by 0.3 mg of O_3 per liter.[38] Almost 1% of the infective units of the poliovirus preparation used in these studies survived for an extended period of time probably reflecting a highly clumped fraction of the virion population. In another study, 99.7% of an attenuated strain of poliovirus 1 was inactivated in buffer in less than 10 sec at 20°C by 0.008 mg of residual O_3 per liter.[39]

E. Comparative Evaluation of Disinfectants

Among the common water virucides, O_3 or its decomposition products appears to be the fastest at neutral pH levels. Hypochlorous acid and ClO_2 are close behind. At elevated pH

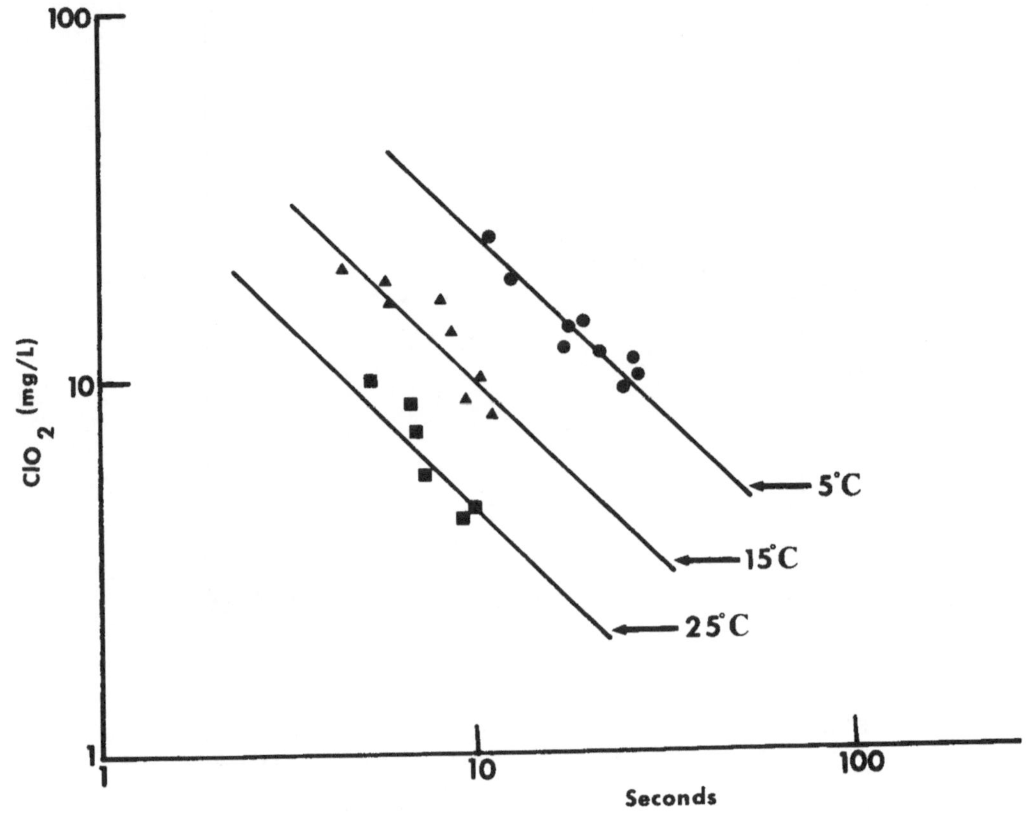

FIGURE 5. Concentration-time relationships for the inactivation of 99% of a strain of poliovirus 1 by ClO_2 (at pH 7) at 5, 15, and 25°C. (From Scarpino, P. V., *Chlorine Dioxide Proc.*, Am. Water Works Assoc., Annu. Conf., Part II, Atlantic City, June 1978, American Water Works Association, Denver, Colo. With permission.)

levels (pH 9), ClO_2 is faster than OCl^- (probably even in the presence of Cl^-) and probably faster than O_3 also. HOCl does not occur in significant quantities at high pH levels. The speed of O_3, HOCl, OCl^- in the presence of Cl^-, and ClO_2 are so great, the small differences among them are probably of little significance. Morever, O_3, HOCl, and ClO_2 at least, are even more effective against bacteria than they are against viruses. Although OCl^- in the presence of Cl^- is as effective as HOCl and ClO_2 (at neutral pH levels) against viruses, it is apparently not as effective against bacteria. The chloramines are slower virucides than HOCl, OCl^- (even in the absence of Cl^-), ClO_2, and O_3. The relative rates for the inactivation of a strain of poliovirus 1 by HOCl, OCl^- (in the absence of Cl^-), ClO_2 (at pH 7), NH_2Cl, and $NHCl_2$ are shown in Figure 7.[26] Comparable quantitative data for ozone are not yet available.

Hypochlorous acid and OCl^- react with humic and fulvic acids to form trihalomethanes which are carcinogens and probable carcinogens. Hypochlorous acid reacts with NH_3 and N-bearing organics in water, also, to form inorganic and organic chloramines. The chloramines are toxic to aquatic life and, in some instances, to man as well. Ozone does not react with NH_3 or with N-bearing organics in water to form chloramines, and it does not form trihalomethanes in water. Ozone decomposes rapidly in water, however; its half-life there is only 4 to 5 min. Moreover, the solubility of O_3 in water is limited. Chlorine dioxide does not form chloramines from NH_3 or N-bearing organics either, nor does it produce trihalomethanes in water. Like O_3 and HOCl, ClO_2 is a strong oxidizing agent and it is not readily maintained in water. Furthermore, two of its major oxidation products, ClO_2^- and ClO_3^-, are toxic to man.

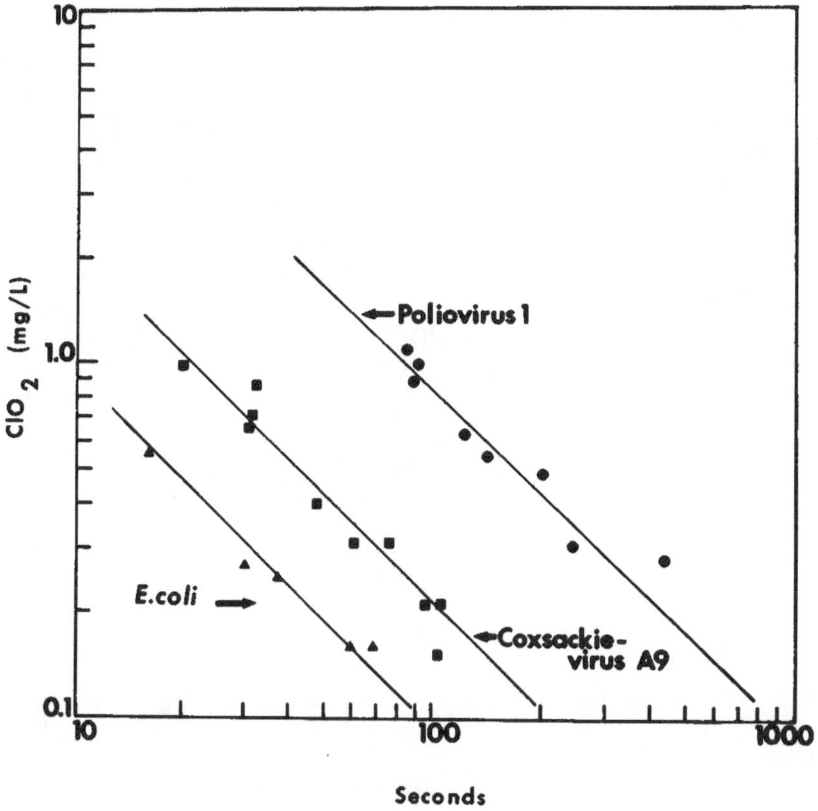

FIGURE 6. Concentration-time relationships for the inactivation of 99% of strains of poliovirus 1, coxsackievirus A9, and *Escherichia coli* at 15°C by ClO_2 at pH 7. (From Cronier, S., Scarpino, P. V., and Zink, M. L., *Water Chlorination, Environmental Impact and Health Effects*, Vol. 2, Ann. Arbor Science, Ann Arbor, Mich., 1978, 651. With permission.)

Ozone oxidizes humic and fulvic acids in water and thereby reduces the formation of trihalomethanes when HOCl or OCl^- is introduced subsequently. Thus, some pretreatment of water with O_3 prior to treatment with HOCl or OCl^- serves to disinfect to some degree and to reduce the subsequent formation of trihalomethanes. Although HOCl is not readily maintained in water, the chloramines that result from the application of HOCl to many waters are relatively stable in distribution systems and continue to disinfect slowly often until waters are consumed. The chloramines are sometimes the only real protection in a water supply against postchlorination contamination, hence the popularity of chloramine-disinfection in some quarters. When chloramines are deliberately used for disinfecting waters, it is better to form them by introducing HOCl into waters into which NH_3 has already been introduced than to add preformed chloramines, because the existence of HOCl in water even for the seconds required to form NH_2Cl is sufficient time for considerable viral (and bacterial) inactivation to take place. When HOCl is introduced into waters before NH_3 is introduced, disinfection occurs at a high rate until the NH_3 is added, but unless the waters have been pretreated to remove or oxidize trihalomethane precursors, trihalomethane formation occurs until the NH_3 is added and converts all of the available HOCl to chloramines. Chlorine dioxide also reacts with the precursors of trihalomethanes in water, and trihalomethane formation is reduced in waters treated with ClO_2 before HOCl is added. Once the toxicities of the oxidation products of ClO_2 are better understood, the rapidly virucidal ClO_2 may be useful for complete disinfection of water supplies or for pretreatment of water supplies subsequently treated with HOCl, OCl^-, or an ammonia chloramine.

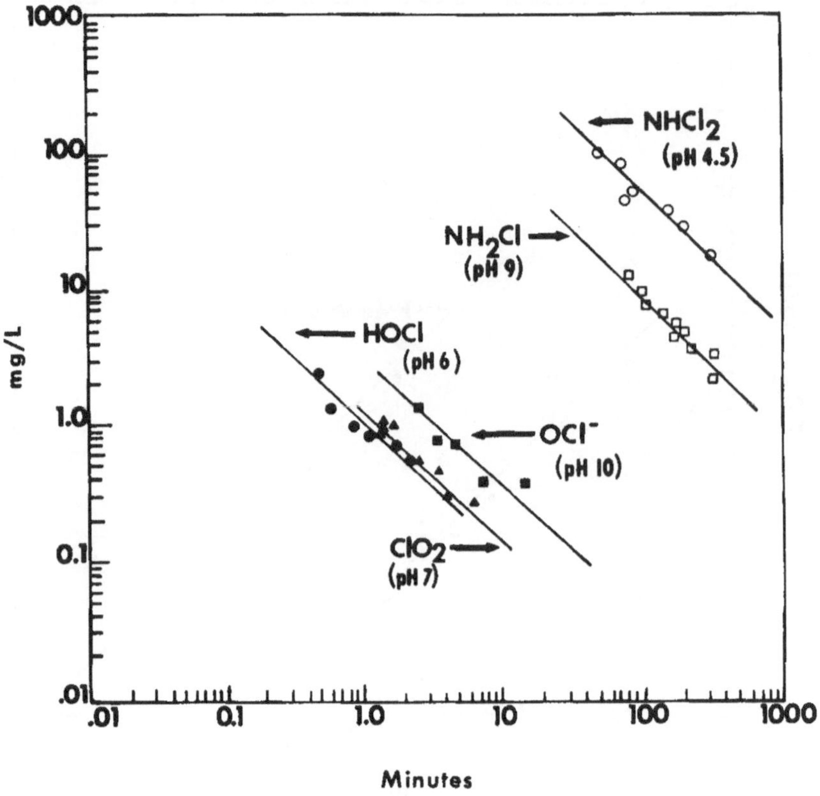

FIGURE 7. Concentration-time relationships for the inactivation of 99% of a strain of poliovirus 1 at 15°C by ClO_2 (pH 7), HOCl, OCl^- in the absence of Cl^-, NH_2Cl, and $NHCl_2$. (From Cronier, S., Scarpino, P. V., and Zink, M. L., *Water Chlorination, Environmental Impact and Health Effects*, Vol. 2, Ann Arbor Science, Ann Arbor, Mich., 1978, 651. With permission.)

Particulates in water present a problem for all chemical disinfectants. When viruses are adsorbed on or within particulates, they receive some measure of protection from all disinfectants.[40]

V. INACTIVATION OF VIRUSES IN SEWAGE

The degree of concern for disinfecting sewage effluents depends upon a number of factors. Trihalomethanes are not produced as rapidly in sewage effluents as in drinking waters, and those that are must often survive relatively long transport before they may become overt hazards to health. The chloramines, which form relatively quickly as the result of chlorination, are of more immediate concern because they are toxic to fish and other aquatic life at the point of discharge and for some distance downstream. When a domestic sewage effluent is discharged into a dry stream bed, into a low-flow stream, or into a stream that quickly impacts a recreational area, the need to disinfect the effluent is usually felt acutely. Thus, with the cost factor always a source of pressure, the circumstance of discharge often determines whether disinfection is applied.

A. Hypochlorous Acid and Hypochlorite Ion

When a domestic sewage effluent (with a pH in the neutral range) is chlorinated with Cl_2 or OCl^-, HOCl forms immediately by hydrolysis and is dissipated in the formation of NH_2Cl

and other chloramines or in other ways almost immediately. In the seconds that pass before all of the HOCl is dissipated, a considerable amount of disinfection occurs. Thus, death rate curves for viruses in a sewage effluent chlorinated with Cl_2 or OCl^- show an immediate rapid inactivation of virions followed by a long-tailing as the HOCl and OCl^- formed initially in the chlorination dissipate in the formation of much less virucidal chloramines and other compounds which may not be virucidal at all.[41] This same effect is seen with indicator bacteria, but the proportion of bacteria inactivated in the short time during which HOCl and OCl^- exist is much greater than the proportion of viruses inactivated.[41] The hypochlorite ion, although it shows considerable virucidal activity in the presence of Cl^-, dissipates at least as quickly as does the HOCl with which it is in equilibrium.

B. Chloramines

In a study now considered classic, Neefe et al.[42] recovered the virus of hepatitis A from the feces of patients ill with the disease and showed that the virus, in fecal filtrates, was destroyed slowly by chlorine. In the early 1940s, when this work was done, chloramines were not measured and are understood only retrospectively to have been the major disinfecting species of chlorine present in the filtrates. Other early studies also showed that chloramines were slow virucides.[43-45] The degree to which HOCl and OCl^- are responsible for the inactivation of viruses and bacteria in effluents undoubtedly varies from effluent to effluent, the dissipation of HOCl and OCl^- occurring more rapidly in some effluents than in others. This variation among effluents probably accounts for the great differences in the time-chlorine concentration coupling required for the same proportion of inactivation of viruses and bacteria in different effluents. Some of the differences, however, may also be accounted for by the formation in different effluents of different organic chloramines, each with a different virucidal and bactericidal capacity. Moreover, in preparing van't Hoff plots, the choice of an arbitrary percent inactivation point on death rate curves that change direction sharply may confound attempts to effectively compare data.

In any event, in an oxidation pond effluent at 20°C, 90% of a strain of poliovirus 1 was inactivated in 6 hr by 1 mg of applied chlorine per liter;[46] 40 mg of applied chlorine per liter inactivated 99.9% of a strain of poliovirus 1 in 10 min. In the same circumstances, 9 mg of applied chlorine per liter inactivated 99.9% of the indigenous coliforms in 10 min.[41] In Haifa sewage effluents, 8 mg of applied chlorine per liter did not decrease the numbers of viruses in an oxidation pond effluent in 1 hr, although some decrease occurred in 2 hr.[47]

In trickling filter effluents at 20°C, 99% of a strain of coliphage T2 was inactivated in 30 min by a residual chloramine concentration of 2.7 mg/ℓ.[48] Coliforms were inactivated even more quickly. In a primary effluent maintained at the same temperature, a chloramine residual of 40 mg/ℓ inactivated 99.99% of a strain of poliovirus 1 in 30 min.[49] In another series of experiments in water, chloramines inactivated strains of poliovirus 1 and coliphage T2 more rapidly than they inactivated a strain of coliphage f2.[50]

In a field study at 22 to 30°C, 5 mg of applied Cl_2 per liter at pH 7.4 to 7.8 inactivated in 3 min only about 50% of the indigenous coliphages in a contact stabilization effluent. In the same time, 99.3% of the indigenous fecal coliforms were inactivated.[51] In a 2-year field study in two primary treatment plants, an average chlorine residual of 0.5 mg/ℓ inactivated 90 to 95% of the viruses in the effluents at ambient temperatures. Sampling in this study was temporally coordinated.[53]

C. Chlorine Dioxide

In field studies at ambient temperatures, ClO_2 inactivated indigenous fecal coliforms and indigenous coliphages in contact stabilization plant effluents more quickly than did the products formed by the addition of Cl_2.[51,53] Chlorine dioxide inactivated the coliphages and fecal coliforms 10 to 100 times faster than Cl_2 (added form) did. Fecal coliforms were inactivated by both forms of chlorine 10 to 100 times more rapidly than the coliphages were.[53]

In microscreened combined wastewater overflow, 12 mg of ClO_2 per liter inactivated 99.99% of a strain of poliovirus 1 in 2 min. In the same time, 16 mg of ClO_2 per liter inactivated 99.99% of the virus. Twice the concentration of Cl_2 was needed to achieve the same results.[54] Coliphages f2 and ΦX174 and coliforms were more sensitive than the poliovirus 1 to the disinfectants, but the fecal streptococci were about as resistant. The application of 8 mg of Cl_2 per liter followed in 15 sec by the application of 2 mg of ClO_2 per liter resulted in the inactivation of the same proportion of coliforms and fecal streptococci as did the application of 12 mg of ClO_2 per liter or 25 mg of Cl_2 per liter alone. These studies were done at ambient temperatures.

D. Ozone

Ozone is used in Europe and in the U.S. for disinfecting wastewaters. Ozone is a powerful virucide and bactericide, but O_3 reacts with many constituents of wastewaters and is rapidly dissipated. Thus, it is desirable usually to reduce the O_3 demand in such waters before ozonation. The reduction of O_3 demand in wastewater by coagulation, sedimentation, and filtration significantly increased the bactericidal effectiveness of applied O_3.[55,56]

In coagulated, sedimented, and filtered wastewater, 10 mg of O_3 per liter inactivated in 18 min more than 99.999% of the viruses indigenous to the wastewater. The same O_3-time coupling in coagulated, filtered wastewater did about as well. The inactivation of indigenous viruses in wastewater that underwent two-stage carbon adsorption before application of 6 mg of O_3 per liter was almost as extensive in 18 min as it was in the coagulated, sedimented, and filtered effluents treated with 10 mg of O_3 per liter. The bactericidal effectiveness of O_3 in all of these experiments, however, was not sufficient to reduce the fecal coliform levels in the effluents to an MPN (most probable number) of 2.2 or less per 100 mℓ.[57]

All of the coliphage f2 seeded into secondary effluent were inactivated in 5 min by 15 mg of applied O_3 per liter. The O_3 residual in the effluent was only 0.15 mg/ℓ.[58]

Continuous ozonation of a buffer inactivated 90% of a seeded strain of poliovirus 1 in 10 sec and 99.999% in less than 1 min at ambient temperatures. The O_3 residual reached about 1 mg/ℓ 1 min after ozonation commenced. In parallel tests with filtered sewage, an O_3 residual was not detected until 90 sec after ozonation commenced, but inactivation of virions appeared to begin about 30 sec after ozonation commenced, and 99% of the virions were inactivated within the subsequent 30 sec. In 2 min, after ozonation of the filtered sewage began, 99.999% of the virions had been inactivated. The O_3 residual at this point was 0.6 mg/ℓ. A maximum O_3 residual of 3 mg/ℓ was reached in both the buffer and filtered sewage about 3 min after ozonation began.[59]

E. Comparative Evaluation of Disinfectants

Neither HOCl nor OCl^- can be maintained in sewage effluents for very long. Rapidly mixing Cl_2 into effluent, by increasing contact of disinfectant molecules with viruses and bacteria before the HOCl (formed rapidly by hydrolysis) and OCl^- (formed by ionization of HOCl) dissipate, increases the rate of inactivation of viruses and bacteria during the first seconds of chlorine contact with the effluent.[60] Even in this circumstance, however, the extent of inactivation is still modest compared to the extensive inactivation of viruses and bacteria that is brought about in the first few minutes of contact by similar quantities of either O_3 or ClO_2. The ammonia chloramines and the organic chloramines are much slower virucides and bactericides than either O_3 or ClO_2. Thus, in a number of secondary effluents that had been subjected to one of several tertiary treatment trains at ambient temperatures, 99.9 to 99.996% of a seeded strain of poliovirus 1 was inactivated in 2 hr by a chloramine residual (at pH levels of 6.7 to 7.6) of 10 mg/ℓ and in 18 min by the same quantity of applied O_3. In effluent from any of the trains, 10 mg of the chloramine residual inactivated in 2 hr the same fraction of influent virus inactivated by 10 mg of applied O_3 in 18 min. In these tertiary effluents, however, the cost for chlorination to a 10 mg/ℓ-2 hr residual and subsequent dechlorination was about equal to the cost for applying 10 mg of O_3 per liter for 18 min.[57]

The need to generate O_3 and ClO_2 on site has long been a point of practical disadvantage that favors chlorination. The need to dechlorinate chlorinated effluents if chloramine-induced toxicity to fish and other aquatic life forms is to be avoided, however, has caused chlorination to be looked upon with less favor in recent years. The application of ClO_2 yields toxic ClO_2^- and ClO_3^- but the sequential application of ClO_2 and Cl_2 may be synergistic and thus may reduce somewhat the total cost of disinfecting effluents and the toxicity imparted to them. The greater demand exhibited by effluents for O_3 than for HOCl can be mitigated somewhat by filtration of effluents prior to disinfection.[55] Ozone is the only important wastewater disinfectant that is not yet known to produce toxic products as a result of the disinfection process.

The presence of large numbers of particulates in wastewaters, of course, protects virions from disinfection and bacteria adsorbed to and within the particulates.[40,61-63] Only coagulation and filtration of effluents can ameliorate this problem.

VI. SUMMARY AND CONCLUSIONS

Ozone, HOCl, OCl^- in the presence of Cl^-, and ClO_2 are the fastest of the virucides used to inactivate viruses in water and wastewater. Ozone appears to be the fastest at neutral and slightly alkaline pH levels, but if so, its slightly greater speed is probably of little practical consequence. The ammonia chloramines are much slower, but given sufficient time, they are as effective as the faster virucides. All of these virucides are even faster bactericides than they are virucides.

Hypochlorous acid and OCl^- react with humic and fulvic acids in natural waters to form trihalomethanes which are carcinogenic and otherwise toxic. Neither O_3, ClO_2, nor the chloramines produce trihalomethanes in natural waters. Moreover, O_3 and ClO_2 react with humic and fulvic acids, major trihalomethane precursors in water and wastewater, and thereby reduce the quantities of trihalomethanes produced when HOCl or OCl^- is subsequently introduced.

Hypochlorous acid reacts with NH_3 and certain N-bearing organics to form chloramines. Neither O_3 nor ClO_2 produce chloramines in similar circumstances. Ozone, ClO_2, and HOCl are strong oxidizing agents and are not readily maintained in water distribution systems and certainly not in wastewater effluents. Ammonia chloramines are relatively stable and can be maintained for some time in water and in wastewater effluents.

Chloramines are toxic to fish and other life forms in the aquatic environment and, under some conditions at least, to man as well. Chlorine dioxide produces ClO_3^- and ClO_2^- as by products of its reactions in water. Both ClO_3^- and ClO_2^- are toxic to man.

Ozone, alone, appears to produce no toxic products in waters and wastewaters. Particulates in waters and wastewaters protect from disinfection virions and bacteria adsorbed to the particulates or embedded within them.

REFERENCES

1. **Morris, J. C.,** The mechanism of the hydrolysis of chlorine, *J. Am. Chem. Soc.*, 68, 1692, 1946.
2. **Fair, G. M., Morris, J. C., Chang, S. L., Weil, I., and Burden, R. P.,** The behavior of chlorine as a water disinfectant, *J. Am. Water Works Assoc.*, 40, 1051, 1948.
3. **Morris, J. C.,** Kinetic reactions between aqueous chlorine and nitrogen compounds, presented at Fourth Rudolphs Research Conference, Rutgers University, New Brunswick, N.J., June 1965.
4. **Morris, J. C., Weil, I., and Culver, R. H.,** Kinetic studies on the breakpoint with ammonia and glycine, 1952, unpublished; cited in White, G. C., *Handbook of Chlorination*, Van Nostrand Reinhold, New York, 1972, 193.
5. **Myhrstad, J. A. and Samdal, J. E.,** Behavior and determination of chlorine dioxide, *J. Am. Water Works Assoc.*, 61, 205, 1969.
6. **Manley, T. C. and Niegowski, S. J.** Ozone, in *Encyclopedia of Chemical Technology*, Vol. 14, 2nd ed., John Wiley & Sons, New York, 1967, 410.
7. **Smith, D. K.,** Disinfection and sterilization of polluted water with ozone, Rep. AM-6704, Ontario Research Foundation, 1969; cited in Venosa, A. D., Ozone as a water disinfectant: a literature review, in *Ozone in Water and Wastewater Treatment*, Evans, F. L., Ed., Ann Arbor Science, Ann Arbor, Mich., 1972, chap. 5.
8. **Symons, J. M., Stevens, A. A., Clark, R. M., Geldreich, E. E., Love, O. T., Jr., and DeMarco, J.,** Treatment techniques for controlling trihalomethanes in drinking water, U.S. Environmental Protection Agency, Cincinnati, Ohio, 1981 (EPA-600/2-81-156), Section III.
9. **Symons, J. M., Stevens, A. A., Clark, R. M., Geldreich, E. E., Love, O. T., Jr., and DeMarco, J.,** Treatment techniques for controlling trihalomethanes in drinking water, U.S. Environmental Protection Agency, Cincinnati, Ohio, 1981 (EPA-600/2-81-156), Section I.
10. **Symons, J. M., Stevens, A. A., Clark, R. M., Geldreich, E. E., Love, O. T., Jr., and DeMarco, J.,** Treatment techniques for controlling trihalomethanes in drinking water, U.S. Environmental Protection Agency, Cincinnati, Ohio, 1981 (EPA-600/2-81-156), Section VIII.
11. **Zillich, J. A.,** Toxicity of combined chlorine residuals to freshwater fish, *J. Water Pollut. Control Fed.*, 44, 212, 1972.
12. **Eaton, J. W., Kolpin, C. F., and Swofford, H. S.,** Chlorinated urban water: a cause of dialysis induced hemolytic anemia, *Science*, 181, 463, 1973.
13. **Symons, J. M., Stevens, A. A., Clark, R. M., Geldreich, E. E., Love, O. T., Jr., and DeMarco, J.,** Treatment techniques for controlling trihalomethanes in drinking water, U.S. Environmental Protection Agency, Cincinnati, Ohio, 1981 (EPA-600/2-81-156), Section VI.
14. **Symons, J. M., Stevens, A. A., Clark, R. M., Geldreich, E. E., Love, O. T., Jr., and DeMarco, J.,** Treatment techniques for controlling trihalomethanes in drinking water, U.S. Environmental Protection Agency, Cincinnati, Ohio, 1981 (EPA-600/2-81-156), Section VII.
15. *Federal Register*, 44, No. 231, 68624-68707 (November 29, 1979); 45, No. 49, 15542-15547 (March 11, 1980).
16. **Clarke, N. A. and Kabler, P. W.,** The inactivation of purified coxsackie virus in water by chlorine, *Am. J. Hyg.*, 59, 119, 1954.
17. **Clarke, N. A., Stevenson, R. E., and Kabler, P. W.,** The inactivation of purified type 3 adenovirus in water by chlorine, *Am. J. Hyg.*, 64, 314, 1956.
18. **Weidenkopf, S. J.,** Inactivation of type 1 poliomyelitis virus with chlorine, *Virology*, 5, 56, 1958.
19. **Scarpino, P. V., Berg, G., Chang, S. L., Dahling, D., and Lucas, M. L.,** A comparative study of the inactivation of viruses in water by chlorine, *Water Res.*, 6, 959, 1972.
20. **Scarpino, P. V.,** Unpublished data, 1973.
21. **Engelbrecht, R. S., Weber, M. J., Salter, B. L., and Schmidt, C. A.,** Comparative inactivation of viruses by chlorine, *Appl. Environ. Microbiol.*, 40, 249, 1980.
22. **Sharp, D. G., Young, D. C., Floyd, R., and Johnson, J. D.,** Effect of ionic environment on the inactivation of poliovirus in water by chlorine, *Appl. Environ. Microbiol.*, 39, 530, 1980.
23. **Sharp, D. G. and Leong, J.,** Inactivation of poliovirus I (Brunhilde), single particles by chlorine in water, *Appl. Environ. Microbiol.*, 40, 381, 1980.
24. **Jensen, H., Thomas, K., and Sharp, D. G.,** Inactivation of coxsackie B3 and B5 viruses in water by chlorine, *Appl. Environ. Microbiol.*, 40, 633, 1980.
25. **Sproul, O. J., Thorup, R. T., Wentworth, D. F., and Atwell, J. S.,** Salt and virus inactivation by chlorine and high pH, *Proc. Natl. Specialty Conf. Disinfection*, July 1970, American Society of Civil Engineers, New York, 1971, 385.
26. **Cronier, S., Scarpino, P. V., and Zink, M. L.,** Chlorine dioxide destruction of viruses and bacteria in water, in *Water Chlorination, Environmental Impact and Health Effects*, Vol. 2, Ann Arbor Science, Ann Arbor, Mich., 1978, 651.

27. **Berg, G., Chang, S. L., and Harris, E. K.,** Devitalization of microorganisms by iodine. 1. Dynamics of the devitalization of enteroviruses by elemental iodine, *Virology,* 22, 469, 1964.
28. **Shaffer, P. T. B., Metcalf, T. G., and Sproul, O. J.,** Chlorine resistance of poliovirus isolants recovered from drinking water, *Appl. Environ. Microbiol.,* 40, 1115, 1980.
29. **Berg, G.,** Discussion of the fate of viruses in a marine environment by Yehuda Kott, in *Advances in Water Pollution Research,* Proc. 4th Int. Conf., Int. Assoc. of Water Pollut. Res., Prague, Jenkins, S. H., Ed., Pergamon Press, Oxford, 1969.
30. **Hoff, J. C.,** The relationship of turbidity to disinfection of potable water, Evaluation of the Microbiology Standards for Drinking Water, Hendricks, C. H., Ed., U.S. Environmental Protection Agency, Cincinnati, (EPA-570/9-78-OOC), 1978.
31. **Scarpino, P. V., Esposito, M. P., Dorn, J. M., and Siders, D. L.,** Unpublished data, 1982.
32. **Scarpino, P. V.,** Virucidal effectiveness of disinfecting processes, *Chlorine Dioxide Proc.,* Am. Water Works Assoc. Annu. Conf., Part II, Atlantic City, June 1978, American Water Works Association, Denver, Colo.
33. **Benarde, M. A., Israel, B. M., Olivieri, V. P., and Granstrom, M. L.,** Efficiency of chlorine dioxide as a bactericide, *Appl. Microbiol.,* 13, 776, 1965.
34. **Coin, L., Hannoun, C., and Gomella, C.,** Inactivation par l'ozone du virus de la poliomyelite present dans les eaux, *Presse Med.,* 72, 2153, 1964.
35. **Coin, L., Gomella, C., Hannoun, C., and Trimoreau, J.-C.,** Inactivation par l'ozone du virus de la poliomyelite present dans les eaux (Nouvelle contribution), *Presse Med.,* 75, 1883, 1967.
36. **Suchkov, B. P.,** Studies of the ozonation of drinking water containing pathogenic bacteria and viruses, *Hyg. Sanit.,* 29, 24, 1964.
37. **Yakovleva, G. P. and Il'nitskii, A. P.,** Ozonation of water containing drug-resistant forms of Shigella and adenoviruses, *Hyg. Sanit.,* 32, 330, 1967.
38. **Shuval, H. I. and Katzenelson, E.,** Detection and Inactivation of Enteric Viruses in Wastewater, U.S. Environmental Protection Agency, Cincinnati, Ohio, 1977 (EPA-600/2-77-095,), 232.
39. **Foster, D. M., Emerson, M. A., Buck, C. E., Walsh, D. S., and Sproul, O. J.,** Ozone inactivation of cell- and fecal-associated viruses and bacteria, *J. Water Pollut. Control Fed.,* 52, 2174, 1980.
40. **Hoff, J. C. and Geldreich, E. E.,** Comparison of the biocidal efficiency of alternative disinfectants, *J. Am. Water Works Assoc.,* 73, 40, 1981.
41. **Shuval, H. I.,** Detection and control of enteroviruses in the water environment, in *Developments in Water Quality Research,* Shuval, H. I., Ed., Ann Arbor-Humphrey Science, Ann Arbor, Mich., 1970, 47.
42. **Neefe, J. R., Stokes, J., Jr., Baty, J. B., and Reinhold, J. G.,** Disinfection of water containing causative agent of infectious (epidemic) hepatitis, *JAMA,* 128, 1076, 1945.
43. **Kelly, S. and Sanderson, W. W.,** The effect of sewage treatment on viruses, *Sewage Ind. Wastes,* 31, 683, 1959.
44. **Kelly, S., Sanderson, W. W., and Neidl, C.,** Removal of enteroviruses from sewage by activated sludge, *J. Water Pollut. Control Fed.,* 33, 1056, 1961.
45. **Mack, W. N., Frey, J. R., Riegle, B. J., and Mallmann, W. L.,** Enterovirus removal by activated sludge treatment, *J. Water Pollut. Control Fed.,* 34, 1133, 1962.
46. **Shuval, H. I., Cymbalista, S., Wachs, A., Zohar, Y., and Goldblum, N.,** The inactivation of enteroviruses in sewage by chlorination, in *Advances in Water Pollution Research,* 1965, Pergamon Press, Elmsford, N.Y., 1966, 37.
47. **Lindeman, S. and Kott, Y.,** The effect of chlorination on enteroviruses in the effluents of the Haifa sewage treatment plant, *Isr. J. Med. Sci.,* 7, 1111, 1972.
48. **Burns, R. W. and Sproul, O. J.,** Virucidal effects of chlorine in wastewater, *J. Water Pollut. Control Fed.,* 39, 1834, 1967.
49. **Lothrop, T. L. and Sproul, O. J.,** High-level inactivation of viruses in wastewater by chlorination, *J. Water Pollut. Control Fed.,* 41, 567, 1969.
50. **Shah, P. C. and McCarmish, J.,** Relative chlorine resistance of poliovirus 1 and coliphages f2 and T2 in water, *Appl. Microbiol.,* 24, 658, 1972.
51. **Longley, K. E., Moore, B. E., and Sorber, C. A.,** Comparison of chlorine and chlorine dioxide as disinfectants, *J. Water Pollut. Control Fed.,* 52, 2098, 1980.
52. **Sattar, S. A. and Westwood, J. C. N.,** Viral pollution of surface waters due to chlorinated primary effluents, *Appl. Environ. Microbiol.,* 36, 427, 1978.
53. **Longley, K. E., Moore, B. E., and Sorber, C. A.,** Disinfection efficiencies of chlorine and chlorine dioxide in a gravity flow contactor, *J. Water Pollut. Control Fed.,* 54, 140, 1982.
54. **Tifft, E. C., Moffa, P. E., Richardson, S. L., and Field, R. I.,** Enhancement of high rate disinfection by sequential addition of chlorine and chlorine dioxide, *J. Water Pollut. Control Fed.,* 49, 1652, 1977.
55. **Venosa, A. D., Meckes, M. C., Opatken, E. J., and Evans, J. W.,** Disinfection of filtered and unfiltered secondary effluent in two ozone contactors, presented at 52nd Annu. Water Pollut. Control Conf., Houston, 1979.

56. **Venosa, A. D.**, Comparative disinfection of wastewater effluent with chlorine, bromine chloride, and ozone, presented at Forum on Disinfection with Ozone, Chicago, 1976.
57. **Selna, M. W., Miele, R. P., and Baird, R. B.**, Disinfection for water reuse, presented at Disinfection Seminar at the Annu. Am. Water Works Assoc. Conf., Anaheim, Calif., 1977.
58. **Pavoni, J. L., Tittlebaum, M. E., Spencer, H. T., Fleischman, M., Nebel, C., and Gottschling, R.**, Virus removal from wastewater using ozone, *Water Sewage Works*, 119, 59, 1972.
59. **Shuval, H. I. and Katzenelson, E.**, Detection and Inactivation of Enteric Viruses in Wastewater, U.S. Environmental Protection Agency, Cincinnati, Ohio, 1977 (EPA-600/2-77-095), 262.
60. **Kruse, C. W. and Longley, K. E.**, Improvement in terminal disinfection of sewage effluent, U.S. Army Research and Development Command Report, Ft. Detrick, Frederick, Md., 1974.
61. **Howser, D. M., Emerson, M. A., Buke, C. E., Walsh, D. S., and Sproul, O. J.**, Ozone inactivation of cell and fecal associated viruses and bacteria, presented at 51st Annu. Water Pollut. Control Fed. Conf., Anaheim, Calif., 1978.
62. **Brigano, F. A. O., Scarpino, P. V., Cronier, S., and Zink, M. L.**, Effect of particles on inactivation of enteroviruses in water by chlorine dioxide, in Progress in Water Disinfection Technology, Venosa, A. D., Ed., U.S. Environmental Protection Agency, Cincinnati, Ohio, (EPA-600/9-79-018), 1979.
63. **Scarpino, P. V., Cronier, S., Zink, M. L., and Brigano, F. A. O.**, Effect of particulates on disinfection of enteroviruses and coliform bacteria in water by chlorine dioxide, Am. Water Works Assoc. Water Quality Tech. Conf., Kansas City, Mo., 1977.

Chapter 6

DESTRUCTION OF VIRUSES IN SLUDGES BY TREATMENT PROCESSES

Richard L. Ward

TABLE OF CONTENTS

I. INTRODUCTION

Human and animal excreta contain valuable and recoverable resources which have been used throughout recorded history. However, the potential spread of human disease due to the presence of pathogenic microorganisms in human excreta has been recognized by modern man. This, coupled with altered aesthetic values and various economic considerations, had greatly decreased its use. This is especially true in developed countries where human excreta is typically disposed of in the fastest, most-confidential manner available. During the past few years, it has been recognized that this practice is not only wasteful, but also causes the pollution of our land, water, and air. Therefore, laws have been enacted to regulate the disposal of human waste and to promote its utilization.

The only natural components of human excreta present in sufficient concentration to constitute a health hazard are pathogenic microorganisms. When human excreta becomes a part of sewage, as usually occurs in developed countries, it often becomes mixed with toxic chemicals from industrial wastes. However, much of the sewage produced throughout the world contains few industrial contaminants. Furthermore, the amount of toxic chemicals in sewage that does contain high concentrations of industrial waste can often be considerably reduced or eliminated by industrial source control. For these reasons, safe utilization of most sewage or sewage products should require only the removal or inactivation of pathogenic microorganisms.

The main function of most sewage treatment plants is to remove the dissolved and suspended solids from wastewater and produce a biologically and chemically clean effluent. As a result, the products of such plants should be clean, useable water fractions and solids fractions referred to as sludge. Recent laws enacted by the U.S. Congress have mandated a reduction in sewage contamination of waterways. This has caused increases in the number of treatment plants placed in operation and in the amount of treatment given in existing plants which, in turn, has caused the production of much greater volumes of sludge.

What is to be done with the mountains of sludge that are expected to be produced in the years ahead? Ocean dumping on the continental shelf is being phased out, trenching is often unrealistic because of the unavailability of land, and incineration is not only extremely energy-intensive but also causes air pollution. For these reasons, disposal is often no longer a viable sludge management option. Furthermore, it is estimated that nearly 50% of the costs of sewage treatment plants is already directed towards sludge management.[1] It is impossible to recover even a portion of these costs if sludge is merely buried in a landfill or dumped in the ocean.

Efficient sludge management requires an end product that can be used. This requires that the treatment processes involved in the production and handling of sludge are optimized towards this end. If the toxic chemicals in wastewater that can ultimately become concentrated in sludge are eliminated by source control, the only hazard generally associated with sludge utilization would be its load of biological pathogens. Most standard treatment processes reduce the numbers of these pathogens. It is possible that slight changes in many of these processes could be made with minimal expense and vastly improve the level of pathogen reduction.

One of the major groups of human pathogens found in wastewater and its sludges is viruses. By far the greatest number of viruses found in sewage are those shed in fecal material and referred to as enteric viruses. This chapter will review present knowledge on the inactivation of human enteric viruses during plant operations commonly used to produce and treat sludges, and the effects of more elaborate sludge disinfection processes on this group of human pathogens.

II. TRANSFER OF VIRUSES FROM SEWAGE INTO SLUDGE

It has been repeatedly demonstrated that enteric viruses can be excreted in concentrations of more than 10^6/g of feces, and more than $10^5/\ell$ have been detected in raw sewage.[2] Enteric viruses are naturally embedded in fecal materials when they are released into the environment or into the sewage system. Numerous reports strongly indicate that even after sewage solids are broken up in some fashion, viruses are likely to remain solids-associated.[3-14] Therefore, quantitative methods for virus detection in sewage must account for viruses that are bound to particulates. Attempts to develop methods which will detect solids-associated viruses in sewage have only begun in earnest in the past few years. Also, many enteric viruses such as hepatitis A virus are not readily detectable by present laboratory techniques. For these reasons, it is generally agreed that many, if not most, infectious viruses in sewage escape detection.

The established tendency of enteric viruses to remain or become bound to the particulate matter in sewage indicates that a large percentage will eventually become ingredients of sludge during treatment plant operations. During primary sedimentation, which is the first operation after grit removal in most treatment plants, about 60% of the sewage solids settle to become a primary sludge.[1] Although it is difficult to prove because viruses enclosed in sludge are not easily extracted and because a number of investigators have reported little or no removal of "detectable" viruses during primary sedimentation (see Reference 15 for review) it is expected that at least 60% of the enteric viruses will accompany these settled solids. Data presented recently from studies conducted in India support this expectation.[16] There it was also shown that virus removal during primary sedimentation increased with longer retention times and with higher concentrations of suspended solids in the raw sewage.

Secondary and tertiary treatments of wastewater are designed to remove most of the remaining particulates and dissolved solids in sewage. Passage of wastewater through a trickling filter is a typical secondary treatment. This operation alone has been reported to have little effect on the number of detectable enteric viruses in sewage.[15,17] However, the solids that settle as a sludge after this secondary treatment of sewage should be accompanied by many bound enteric viruses.

Activated sludge treatment is another common sewage plant operation. Inactivation and removal of enteric viruses from sewage during this process has been the subject of a large number of investigations carried out over a period of more than 20 years. In general, it is agreed that this operation accompanied by a sedimentation step can remove in excess of 99% of the detectable enteric viruses remaining in sewage. Most early reports suggested that removal was due primarily to association of viruses with solids which end up as components of sludge.[5,17-23] These reports also suggest the possibility that some fraction of virus loss is due to inactivation. More recent studies have uncovered further evidence that viruses are truly inactivated during activated sludge treatment.[24-26]

Because the retention time of sewage during activated sludge treatment can be relatively long and because some of the microorganisms in sewage cause inactivation of enteric viruses,[27,28] it is expected that at least some virus inactivation should occur during activated sludge treatment. It would be satisfying if it could be stated with certainty that this is the case and if the amount of inactivation could be accurately quantified. However, the experimental methodologies used in these studies show inadequacies that limit the interpretation of the results. The main problem associated with all these studies is that viruses embedded in sewage or sludge solids would not have been detected and the observed decreases in viral titer could have been due only to masking effects.

It has been reported that solids-associated viruses retain their infectivities.[8,9] However, the infectivity of a virus can only be detected if it can infect a cell. The process whereby a virus enters a cell normally requires a specific binding interaction between the virus and receptor on the surface of the cell.[29] This entry mechanism can be by-passed and the possibility existed that sewage solids could act as a vehicle to carry viruses into cells. This route of viral infection was investigated. However, cells lacking proper receptors for poliovirus could not be infected by this virus even when associated with sludge.[30] Therefore, it is assumed that infectious viruses associated with sewage solids are detectable only if they are attached to the surfaces of the particulates. Viruses totally embedded in sewage solids should remain undetectable unless they are exposed.

Other treatment methods that significantly reduce the number of detectable enteric viruses in sewage are also used during plant operations. One of the more common is coagulation or flocculation with chemicals such as ferric chloride, aluminum sulfate, or lime. Studies indicate that the extent of removal of enteric viruses from wastewater by chemical treatment is quite variable but efficiencies of greater than 99.9% have been found for some viruses.[31-42] Many viruses removed from wastewater by this treatment can be recovered from the settled floc or chemical sludge.

The rates of viral inactivation in chemical sludges are not well-defined but clearly depend on such factors as the type and concentration of flocculent and on the temperature, as shown by Sattar and co-workers.[41,43] Because chemical sludges are essentially unusable, they must normally be disposed. Although it is not a foregone conclusion, their disposal should be performed in a manner which will present no health hazard to man, animals, or plants, either from toxic materials or from pathogens.

A number of methods are used to disinfect wastewater before it leaves the treatment plant. However, the processes discussed above are the main ones used that result in the production of sludge. Although enteric viruses can still be detected in the wastewater effluents of some treatment plants, even after chlorination,[40,41-51] it is clear that the great majority are normally removed with the different sludge fractions. Therefore, if infectious viruses are to be eliminated during treatment plant operations, most must be inactivated during the processing of sludge.

III. VIRUS INACTIVATION IN SLUDGE

The sludge fractions recovered from wastewater after primary sedimentation and trickling filtration, or activated sludge treatment are often combined for further treatment unless an individual fraction is used in a special way. For example, activated sludge is used by the City of Milwaukee to make Milorganite®, a commercial fertilizer. Methods commonly employed to treat combined sludges typically have been directed toward sludge stabilization and methane generation.[1] Effects of these treatments on sludge pathogens have had little if any importance, especially when the sludge was disposed of rather than utilized. It is serendipitous that many of these processes also reduce the levels of sludge pathogens including viruses. It is now possible to utilize these processes, along with other treatments specifically directed toward pathogen reduction, to produce usable sludges that are free of pathogens.

A. Anaerobic Digestion

Probably the most common method of sludge stabilization in a typical sewage treatment plant in the U.S. is anaerobic digestion. During this process, large organic molecules are biodegraded into much smaller molecules, a large fraction of which are gases if the process is carried to completion. The process itself does not generate heat and the organisms involved in biodegradation grow much better above ambient temperatures. Therefore, a portion of the methane generated by the process is normally burned to heat the sludge. The usual temperature increases are to about 35°C where mesophilic bacteria have optimal growth rates, or to about 50°C where thermophilic bacteria metabolize more rapidly. Most digesters are operated in the mesophilic range because of the lesser energy requirements.

Table 1
RECOVERY OF RADIOACTIVELY LABELED POLIOVIRUSES FROM RAW AND ANEROBICALLY DIGESTED SLUDGES

Incubation time/ temperature	Percentage recovery of PFU[a]		
	No sludge	Raw sludge	Digested sludge
15 min/20°C	100	100	74
5 days/4°C	100	100	3.8
5 days/20°C	90	87	0.003

[a] Plaque-forming units.

Data from Ward, R. L. and Ashley, C. S., *Appl. Environ. Microbiol.*, 31, 921, 1976.

Changes in the numbers of different pathogens in sludge during anaerobic digestion depend on several factors such as chemical environment and biological activity. Probably the most important, however, is temperature. Because enteric viruses cannot multiply outside of host cells, their numbers can only remain constant or decrease during sludge treatment. The rate of decrease of infectious viruses in sludge during anaerobic digestion is temperature-dependent and is much more rapid at thermophilic than at mesophilic temperatures.[52,53]

Several studies have been performed to determine the effects of mesophilic anaerobic digestion of sludge on both indigenous and seeded viruses.[5,52-58] Although the number of recoverable viruses was consistently reduced about one order of magnitude or more during digestion, it was unclear whether they were irreversibly inactivated and, if so, what caused this effect. In studies designed to answer these questions purified, radioactively-labeled polioviruses (enteroviruses that belong to the Picornaviridae family) were much more rapidly inactivated in anaerobically digested than in raw sludge at the same temperature (Table 1). These typical enteric viruses were partially broken down in digested sludge and their ribonucleic acid (RNA) genomes were degraded (Figure 1).[59,60] Therefore, poliovirus can be irreversibly inactivated in digested sludge.

Poliovirus inactivation was significantly more rapid at the temperatures of these experiments with labeled viruses than expected on the basis of the results of other investigators who measured virus inactivation in digesting sludge rather than in previously digested sludge.[56-58] The difference in rates was due primarily to the pH of the digested sludge which is generally higher than that of raw sludge and increases during storage following digestion.[61] As a result of this increase in pH, indigenous ammonium ion is partially converted into aqueous ammonia which inactivates polioviruses and all other enteroviruses examined (Table 2).

The mechanism of poliovirus inactivation by ammonia is cleavage of the viral RNA within physically intact poliovirus particles.[62] Neither the isoelectric point (Figure 2) nor the binding potential of inactivated virions to cells (Table 3) was altered; thus, the viral proteins were apparently unmodified by ammonia. Poliovirus inactivation by ammonia is greatly enhanced by increases in temperature.[60,63] Therefore, temperature, ammonia concentration, and pH are at least three factors that affect the rates of viral inactivation during anaerobic digestion of sludge.

Chemicals with structures similar to ammonia were also virucidal for polioviruses.[64] In fact, ethylamine was significantly more active than ammonia (Figure 3). However, slightly larger amines such as butylamine and diethylamine were inactive which suggests that the specific effect of compounds like ammonia and ethylamine on polioviruses are limited to a few compounds with similar structures.

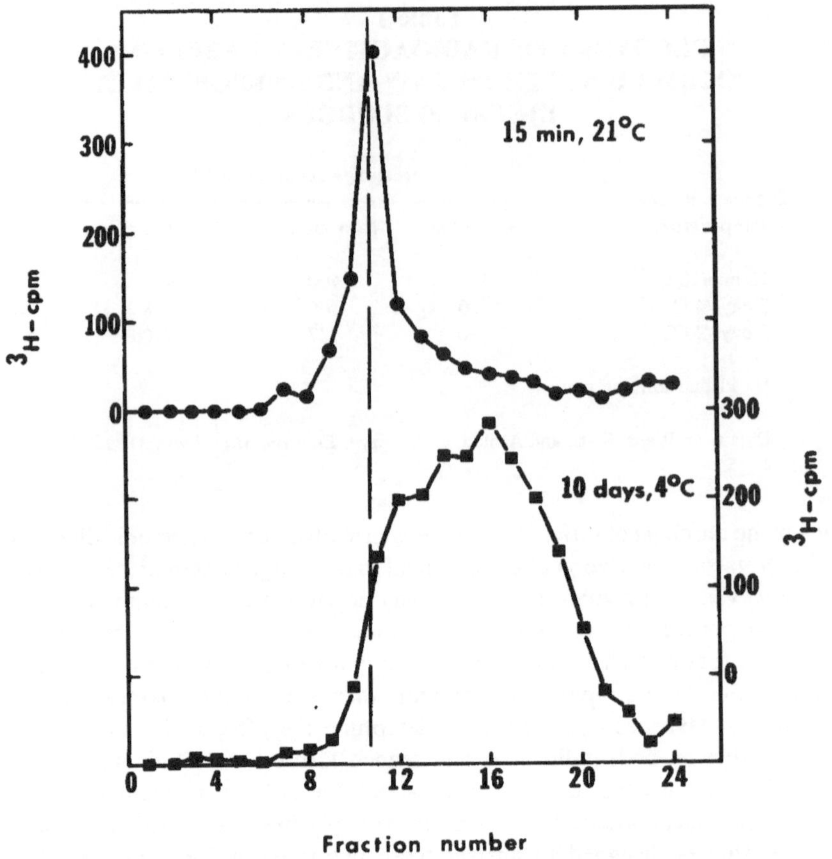

FIGURE 1. Sedimentation profile of RNA extracted from radioactively-labeled polioviruses after different treatments in anaerobically digested sludge. Sedimentation is from right to left. (From Ward, R. L. and Ashley, C. S., *Appl. Environ. Microbiol.*, 31, 921, 1976. With permission.)

Table 2
EFFECT OF EXPOSURE TO AMMONIA[a] FOR 3 HR AT 21°C ON INFECTIVITIES OF SEVERAL STRAINS OF ENTEROVIRUSES

	Percentage recovery of PFU[b]	
Virus	**pH 9.5, − NH_3**	**pH 9.5, + NH_3**
Poliovirus 1 (CHAT)	63	<0.000035
Poliovirus 1 (Mahoney)	100	<0.000014
Poliovirus 2 (712)	60	<0.000044
Coxsackievirus A13	28	<0.00012
Coxsackievirus B1	100	<0.000046
Echovirus 11	9.6	<0.00015

[a] 0.5 *M* NH_4Cl.
[b] Plaque-forming units.

From Ward, R. L. and Ashley, C. S., *Appl. Environ. Microbiol.*, 33, 860, 1977. With permission.

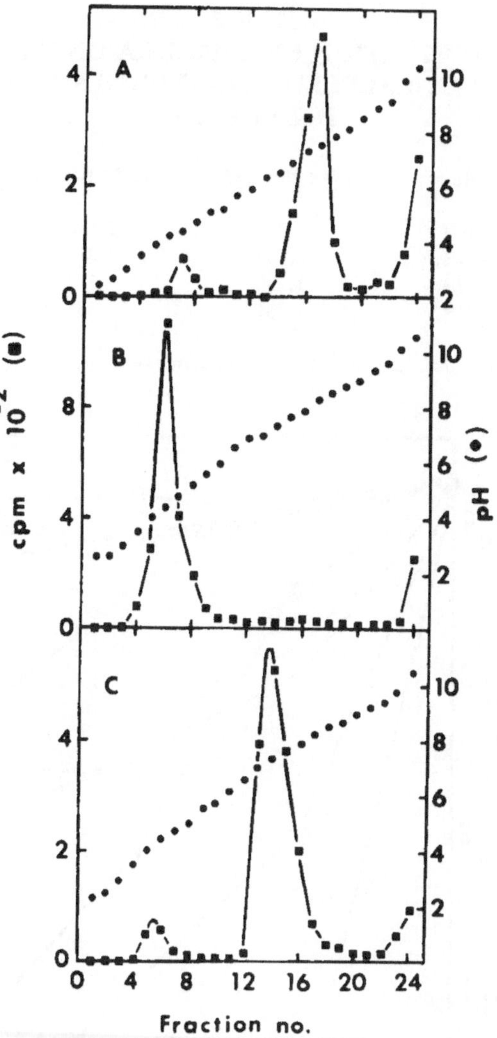

FIGURE 2. Isoelectric point determinations of labeled polioviruses without treatment (A) or after inactivation by either heat (B) or ammonia (C). (From Ward, R. L., *J. Virol.*, 26, 299, 1978. With permission.)

Sludge also contains agents that protect viruses against the chemical and physical stresses which occur during digestion. For example, raw sludge protects polioviruses and other enteroviruses against inactivation by heat, and even anaerobically digested sludge is protective under certain conditions.[63] Certain ionic detergents that occur in sludge, such as sodium dodecyl sulfate (SDS), have this property.[65] The stabilizing effect of SDS on enteroviruses, however, is pH-dependent;[66] i.e., SDS protected at neutral and alkaline pHs but was virucidal at pH values below 5. The latter effect was also noted by Mandel.[67-69]

Another group of enteric viruses found in sludge belong to the Reoviridae family. *Reovirus* is the prototype of this family, but rotaviruses, probably the major causative agents of infantile gastroenteritis, are also classified as members of this family.[70-72] Reoviruses were resistant to ammonia[61] and both reoviruses and rotaviruses were relatively heat resistant. From these results, it was expected that these viruses would be much more heat resistant in raw and digested sludges than enteroviruses. Studies showed, however, that these viruses are destabilized in both types of sludge (Table 4).

Table 3
EFFECT OF INACTIVATION BY AMMONIA ON THE ABILITY OF RADIOACTIVELY-LABELED POLIOVIRUSES TO ADSORB TO HeLa CELLS

Label/species	Treatment	Total CPM added	CPM bound	% CPM bound
^3H/RNA	$-NH_3$	15,220	2,488	16.3
	$+NH_3$	14,252	2,033	14.3
^{14}C/protein	$-NH_3$	1,188	317	26.7
	$+NH_3$	1,731	382	22.1

From Ward, R. L. , *J. Virol.*, 26, 299, 1978. With permission.

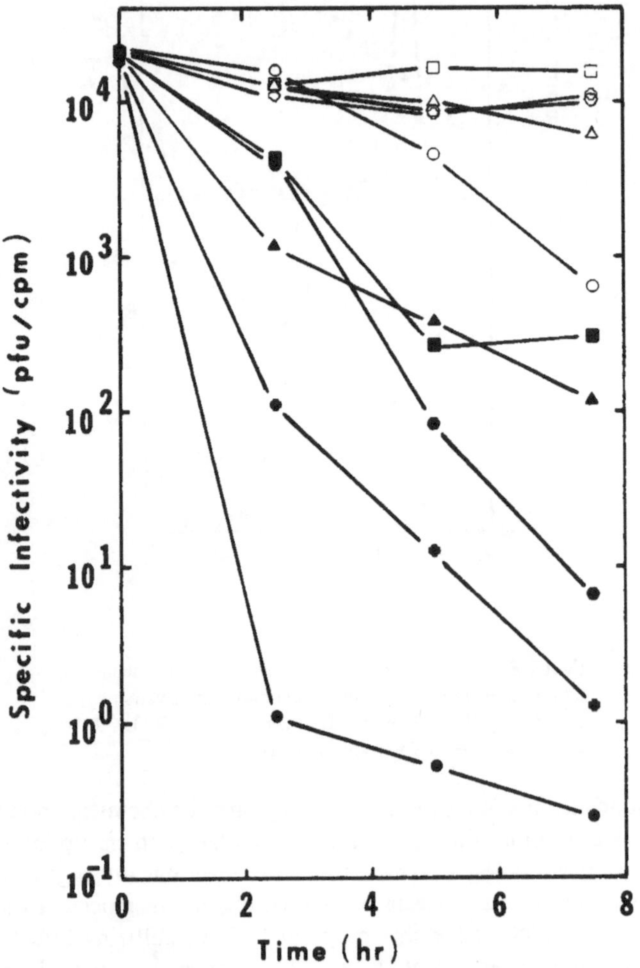

FIGURE 3. Comparative effects of ammonia and related compounds on polioviruses (21°C, pH 9.5). Symbols: ●, ethylamine; ⬢, ammonium chloride; ⬣, propylamine; ▲, methylamine; ■, dimethylamine; ○, 2-methoxyethylamine; △, 2-aminoethanol; ⬡, trimethylamine; ⬠, butylamine; □, diethylamine. (From Ward, R. L. and Ashley, C. S., *Appl. Environ Microbiol.*, 36, 198, 1978. With permission.)

Table 4
HEAT INACTIVATION OF REOVIRUSES AND ROTAVIRUSES IN ANAEROBICALLY DIGESTED SLUDGE

Virus	Temperature (°C)	Percentage survival after heat treatment	
		−Sludge	+Sludge
Reovirus	45(20 min)	103	0.01
	50	100	0.01
	55	24	0.002
	60	0.02	0.0008
Rotavirus	40(15 min)	100	65
	45	110	50
	50	40	0.03
	55	<0.02	<0.02

From Ward, R. L. and Ashley, C. S., *Appl. Environ. Microbiol.*, 34, 681, 1977; *Appl. Environ. Microbiol.*, 40, 1154, 1980. With permission.

The agents in sludge responsible for destabilization of reoviruses and rotaviruses have been identified as ionic detergents,[65,76] the same agents that stabilize enteroviruses against heat inactivation at neutral and alkaline pHs. Inactivation of reoviruses in sludge resulted in a breakdown of the viral particles,[74] and inactivation of rotaviruses by SDS decreased the sedimentation values of viral particles and prevented their attachment to host cells.[77] SDS-inactivated rotavirions lost at least one viral protein (Figure 4) which probably caused the observed effects on these viral particles. In contrast, EDTA, which is known to alter capsid structure, caused loss of two or more rotavirus proteins (see Figure 4). Heat inactivation of both reoviruses and rotaviruses was opposed in sludge by the presence of agents that countered the destabilizing effects of ionic detergents. Nonionic detergents were probably at least one group of agents responsible for this effect because they can stabilize rotaviruses against the potent destabilizing effects of SDS (Table 5).

Another property of sludge that modifies the inactivation rates of reoviruses and rotaviruses is pH. Maximum stability of reovirus in digested sludge occurs near neutrality (Figure 5). A similar result for both reoviruses and rotaviruses occurs in the presence of SDS[66,76] which supports the conclusion that ionic detergents are the sludge agents primarily responsible for destabilization of these viruses in sludge.

From these results it is clear that a number of competitive factors influence the inactivation rates of enteric viruses during sludge handling and treatment. Furthermore, these factors can often influence the rates of inactivation of different groups of viruses in different and, sometimes, opposite ways. The most important of these factors that have been identified in connection with anaerobic digestion are temperature, pH, ammonia, and detergents. Undoubtedly other physical and chemical factors are also involved and will eventually be identified.

B. Lime Stabilization

Lime and other chemicals are used to remove solids from wastewater by coagulation and sedimentation, and these same chemicals can be used to stabilize sludge. Removal of viruses from wastewater by chemical coagulation can be quite effective, as previously discussed, but the stabilities of enteric viruses in chemical sludges or in chemically-stabilized sludges have not been extensively studied.

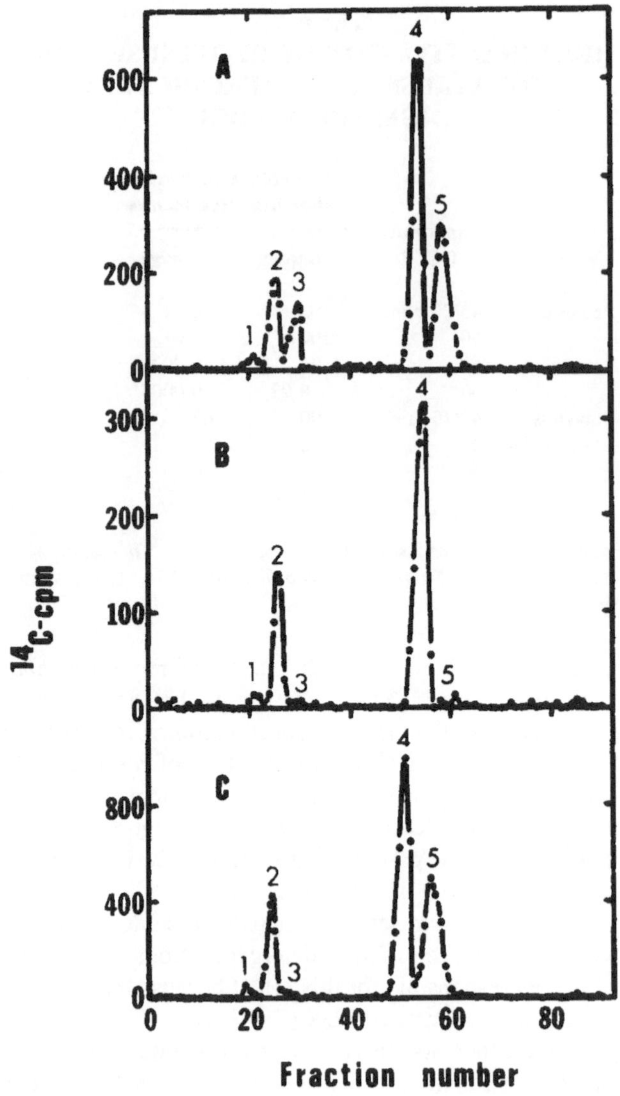

FIGURE 4. Electrophoretic patterns of rotavirus proteins obtained from untreated virus (A), EDTA-treated virus (B), and SDS-treated rotavirus particles (C). (From Ward, R. L. and Ashley, C. S., *Appl. Environ. Microbiol.*, 40, 1148, 1980. With permission.)

Although most chemical coagulants used to treat wastewater and sludge would not be expected to significantly affect the rates of viral inactivation, lime should have a large effect if used in sufficient concentration. The reason is that lime increases the pH of the sludge. In general, it appears that rapid inactivation of enteric viruses in wastewater requires pH values of greater than 11.[36,41,42] However, polioviruses can be easily inactivated at pH 10 in buffer at 40°C.[78,79]

There appears to be no reports on the effect of lime treatment per se on the inactivation of enteric viruses in sludge. However, polioviruses and reoviruses are greatly destabilized in liquid sludges at alkaline pHs, especially those greater than pH 9.[61,66] The effect on polioviruses is apparently due primarily to ammonia and the effect on reoviruses is at least partially caused by ionic detergents. Because the concentration of both chemicals are much

Table 5
STABILIZATION OF ROTAVIRUSES AGAINST THE IONIC DETERGENT SDS BY THE NONIONIC DETERGENT IGEPAL CO-630

Treatment medium	% Recovery of PFU[a] after 20 min, 40°C
0.1 *M* Tris, pH 7	100
0.1 *M* Tris, pH 7 + 0.001% SDS	100
0.1 *M* Tris, pH 7 + 0.01% SDS	0.88
0.1 *M* Tris, pH 7 + 0.1% SDS	0.03
0.1 *M* Tris, pH 7 + 0.1% Igepal Co-630	130
0.1 *M* Tris, pH 7 + 0.1% Igepal + 0.001% SDS	110
0.1 *M* Tris, pH 7 + 0.1% Igepal + 0.01% SDS	130
0.1 *M* Tris, pH 7 + 0.1% Igepal + 0.1% SDS	0.02

[a] Plaque-forming units.

From Ward, R. L. and Ashley, C. S., *Appl. Environ. Microbiol.*, 40, 1154, 1980. With permission.

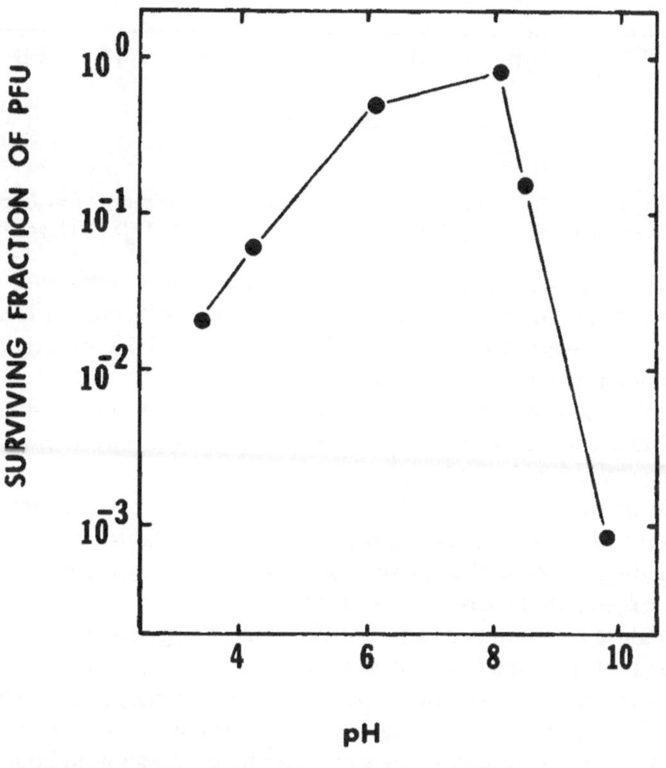

FIGURE 5. Effect of pH on heat inactivation (45°C, 20 min) of reoviruses in anaerobically digested sludge. (From Ward, R. L. and Ashley, C. S., *Appl. Environ. Microbiol.*, 38, 314, 1979. With permission.)

greater in sludge than in wastewater, it is expected that enteric virus inactivation will occur more rapidly in sludge at the same pH and temperature. However, this suggestion has not yet been verified by either laboratory or field experimentation.

C. Drying

Dewatering of sludge to facilitate handling and to reduce treatment or disposal costs is a

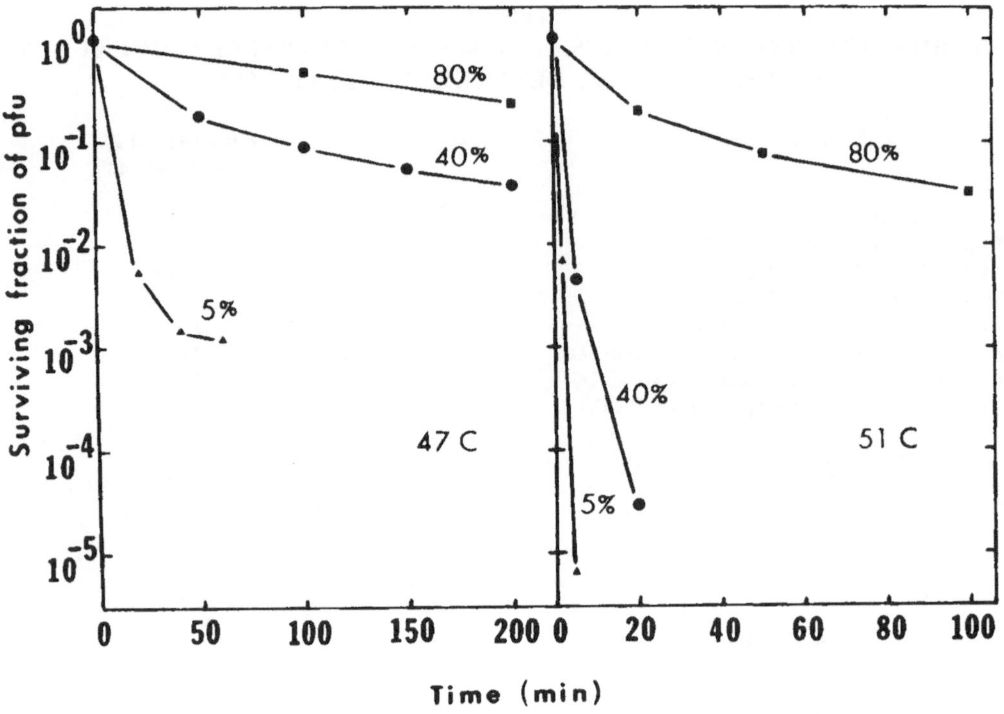

FIGURE 6. Heat inactivation of poliovirus in raw sludge as a function of moisture content (percentages of solids). (From Ward, R. L. and Ashley, C. S., *Appl. Environ. Microbiol.*, 36, 898, 1978. With permission.)

common practice in many treatment facilities. Because viruses are mainly associated with sludge solids, few should be removed with the water during treatments such as centrifugation or filter-pressing. Likewise, these kinds of processes are neither expected nor known to cause a significant amount of viral inactivation.

Virus inactivation in sludge by factors such as heat, on the other hand, can be greatly affected by moisture content. For example, the rate of heat inactivation of polioviruses in seeded raw sludge was considerably reduced at low moisture levels (Figure 6). Also, the destabilizing effects of liquid raw sludge on reoviruses were partially reversed when the sludge was dewatered.[80] These results are probably due to the increased concentration of protective agents in dewatered sludges such as ionic detergents in the case of polioviruses and nonionic detergents in the case of reoviruses.

A method to reduce moisture that can, in itself, cause a large amount of virus inactivation in sludge is natural evaporation. Loss of water by this mechanism decreased the infectivity of enteric viruses in seeded sludge more than four orders of magnitude (Table 6).[81] Similar effects have been noted by other investigators both in sludge[82] and in soil.[55, 83-85]

Inactivation of polioviruses through loss of water by evaporation in both soil and sludge systems was irreversible because viral particles released their RNA genomes which were degraded.[81,86] The cause of inactivation appears to be somehow related to the evaporation process itself because similar effects occurred in distilled water when the virus particles were kept continuously in solution.[81] Perhaps this mechanism is similar to that observed in aerosols where virions are degraded, presumably at the air-surface interface, and release naked RNA.[87-90]

Inactivation of viruses through loss of water by evaporation may have broad applicability because a number of different viruses are destroyed by evaporation in both sludge and soil systems.[81,85] Furthermore, the ova of *Ascaris lumbricoides*, which are generally quite resistant

Table 6
RECOVERY OF VIABLE ENTERIC VIRUSES FROM SLUDGE AFTER DEWATERING BY EVAPORATION

Water content of sludge (%)	Recovery (%)		
	Poliovirus	Coxsackievirus	Reovirus
95	100	100	100
70	40	65	65
40	22	ND[a]	31
20	0.01	ND	ND
5	0.004	0.006	<0.009

[a] Not determined.

From Ward, R. L. and Ashley, C. S., *Appl. Environ. Microbiol.*, 34, 564, 1977. With permission.

to environmental stresses, are inactivated in sludge drying beds that contain less than about 20% moisture.[91] In a study just recently completed, enteric bacteria in sludge were generally reduced in number only about one half to one order of magnitude during drying.[92] However, the die-off rates of these same bacterial species were rapid during long-term storage at 21°C in all but the dryest dewatered sludges, i.e., sludges containing less than 10% water.[93] If the results observed in these studies, which are primarily from laboratory experimentation, are consistently found in field studies, drying by evaporation followed by long-term storage may be one of the most efficient and economical methods for disinfecting sludge.

D. Composting

Composting is one of the most effective methods for biodegrading and stabilizing sludge and other organic wastes. Because this process is performed under aerobic conditions, the biochemical reactions that occur produce large quantities of heat. As a result, efficient composting of sludge results in a highly disinfected product that has the texture and odor of a rich soil.

There are several methods for composting sludge, but the ones most commonly used in the U.S. are the aerated-pile method as developed by the U.S. Department of Agriculture Laboratory in Beltsville, Md. and the windrow method as used in the Los Angeles County Sanitation District and elsewhere. Sludge properly composted by the pile method reaches temperatures in excess of 60°C for periods of nearly a week or longer.[94] At those temperatures, inactivation of enteric viruses occurs rapidly in either raw or composted sludges.[80] Therefore, efficient virus inactivation is expected during composting by the pile method, an expectation that is in agreement with results found with the coliphage f2 in seeded compost piles.[95]

Temperatures reached and maintained during sludge composting by the windrow method are generally much lower than those attained during pile composting. Although virus inactivation during this procedure should be quite effective over an extended period of time, the rate of virus destruction is expected to be significantly slower than attained during pile composting. Again, this expectation was verified by studies conducted with coliphage f2.[95,96] A comparison of the rates of coliphage f2 inactivation during aerated-pile and windrow composting is shown in Table 7.

Factors such as the pH, chemical composition, and moisture content of sludge also effect virus inactivation rates during composting. Because sludge is sometimes treated with lime prior to composting, its pH may remain somewhat above neutrality for a period of time after composting has begun. The presence of ammonia, coupled with high pH and above ambient

Table 7
RATES OF COLIPHAGE f2 INACTIVATION DURING AERATED-PILE AND WINDROW COMPOSTING OF SLUDGE

Composting method	Time required for 90% reduction in PFU[a]
Windrow (favorable weather)	3.2 days
Windrow (unfavorable weather)	11 days
Aerated-Pile	1.8 days

[a] Plaque-forming units.

Data from Burge, W. D., Marsh, P. B., and Millner, P. D., *Natl. Conf. Composting of Municipal Residues and Sludges*, Rockville, Information Transfer, Rockville, Md., 1977, 128.

temperatures, should cause rapid inactivation of enteroviruses.[61] Many other interactions of chemical and physical factors are expected to affect the rates of virus inactivation during composting, but the main contributions of these factors is probably in the modification of the effects of heat.[80] For this reason, the degree of virus inactivation during these operations should be directly proportional to the temperature maintained which, in turn, depends on the efficiency of composting.

E. Other Methods of Disinfection

The U.S. Environmental Protection Agency (USEPA) recently published regulations on solid waste disposal practices.[120] This document specifies, among other things, the types of treatments required for wastewater sludge before it can be utilized. Certain uses require treatment with a "Process to Significantly Reduce Pathogens (PSRP)" and others require this treatment plus a "Process to Further Reduce Pathogens (PFRP)." The types of sludge treatments described above have been PSRPs with the exception of aerated-pile composting which can serve as both a PSRP and PFRP. The remainder of the sludge treatments that will be discussed are all PFRPs.

1. Heat Treatment

Numerous reports exist on heat inactivation of viruses and the chemical agents that modify the effects of heat. For example, divalent cations,[73,97] detergents,[65] cystine,[98] and 2-thiouracil[99] all stabilize enteric viruses against heat inactivation. The effects of sludge on heat inactivation of these viruses at moderate temperatures (35°C to about 60°C) during anaerobic digestion and composting have been discussed in some detail in previous sections.

Other sludge treatment processes designed primarily for pathogen reduction utilize higher temperatures. For example, the minimum temperature of pasteurization specified as a PFRP in the recent USEPA guidelines is 70°C for 30 min. At temperatures of this magnitude, enteric viruses are inactivated quite rapidly in sludge.[100] Other PFRPs, such as heat drying, require even higher temperatures. Because no enteric viruses are known to be stable at temperatures above 70°C for any extended period of time, it is assumed that proper heat treatment of sludge will destroy all infectious enteric viruses.

2. Ionizing Radiation Treatment

Sludge may also be disinfected with ionizing radiation. Although such irradiation can alter the physical and chemical properties of sludge, the dose of 1 mrad specified by the most recent USEPA guideline as a PFRP should have much smaller effects on these sludge properties than the other treatment processes that have been discussed. Therefore, irradiation of sludge is considered, almost exclusively, as a means for disinfection rather than stabilization.

Sludge disinfection with ionizing radiation can be accomplished with either beta rays (high-energy electrons emitted from an accelerator) or gamma rays (photons of energy released during the decay of unstable isotopes such as cobalt-60 or cesium-137). Because the energy of gamma rays is primarily deposited on electrons, causing them to be accelerated and released from their orbitals,[101] disinfection is initiated in both cases by activation of an electron. Therefore, comparable absorbed doses of beta and gamma rays should produce comparable amounts of disinfection. However, because a gamma ray can travel relatively large distances in an aqueous environment before its energy is dissipated, its ability to penetrate sludge samples is much greater than that of a beta ray.

The rates of inactivation of biota by ionizing radiation depend upon the nature of the medium surrounding the biota.[102] In dilute aqueous environments most damage is attributed to indirect effects (toxic radiation products of water or other materials in the medium), but in the presence of high concentrations of organic substances, inactivation is thought to be caused primarily by direct effects (direct interactions of radiation with molecules).[103-106] Therefore, pathogen inactivation in sludge by ionizing radiation should be almost solely due to direct effects.

Because of their smaller sizes, viruses are generally much more radiation resistant than other enteric pathogen under direct effect conditions. Doses of 150 to 500 krad are normally required to inactivate 90% of viable viruses (D_{10}-value) under these conditions.[107-113] This is about ten times the dose generally required to inactivate enteric bacteria under similar conditions.[92,114-119]

It is clear that inactivation of bacterial pathogens by irradiation of sludge should be much more effective than inactivation of enteric viruses. In fact, the 1 mrad dose specified by the USEPA as a PFRP is expected to reduce the virus population only about three orders of magnitude. Although this is a very substantial reduction, it may not be sufficient in itself to insure that the sludge product can be utilized with essentially no health risks of virally-induced disease. However, coupling irradiation with a PSRP, as also required by the USEPA regulation, should be an effective way to reduce or eliminate these risks.

IV. SUMMARY AND CONCLUSIONS

Most enteric viruses in wastewater are bound to solids and become components of sludge during normal operations at a treatment facility. Anaerobic digestion of sludge causes viral inactivation but the amount depends on the temperature and time of digestion as well as on certain properties of the sludge such as ammonia concentration and pH. Enteric viruses are readily inactivated by lime treatment if the pH is held above 11 for a sufficient period of time and by drying of sludge through evaporation.

Heat is one of the most effective methods for inactivating viruses. Temperatures attained during efficient composting should cause thorough viral inactivation even though certain ingredients of sludge such as detergents can stabilize some enteric viruses against heat. The exposure time required for virus inactivation by heat drops rapidly at temperatures above 60°C. Irradiation of sludge with either beta or gamma rays also causes viral inactivation but the doses required are greater than needed to inactivate larger enteric pathogens.

It is quite likely that future experimentation will demonstrate that alternative treatments not discussed in this review will be even more effective and less costly than methods available today. It is hoped that these methods will help make it possible to utilize sludge, a renewable resource with a variety of uses, to its maximum potential.

REFERENCES

1. **Vesilind, P. A.,** Treatment and Disposal of Wastewater Sludges, *Ann Arbor Science,* Ann Arbor, Mich., 1974.
2. **Gerba, C. P., Wallis, C., and Melnick, J. L.,** Viruses in water: the problem, some solutions, *Environ. Sci. Technol.,* 9, 1122, 1975.
3. **Carlson, G. F., Jr., Woodard, F. E., Wentworth, D. F., and Sproul, O. J.,** Virus inactivation on clay particles in natural waters, *J. Water Pollut. Control Fed.,* 40, R89, 1968.
4. **Lund, E.,** Observations on the virus binding capacity of sludge, in *Advances in Water Pollution Research,* Proc. 5th Int. Conf., Jenkins, S. H., Ed., Pergamon Press, Elmsford, N.Y., 1970.
5. **Lund, E. and Ronne, V.,** On the isolation of virus from sewage treatment plant sludges, *Water Res.,* 7, 863, 1973.
6. **Lund, E.,** The effect of pretreatments on the virus contents of sewage samples, *Water Res.,* 7, 873, 1973.
7. **Schaub, S. A., Sorber, C. A., and Taylor, G. W.,** The association of enteric viruses with natural turbidity in the aquatic environment, in *Virus Survival in Water and Wastewater Systems,* Malina, J. F., Jr., and Sagik, B. P., Eds., Center for Research in Water Resources, University of Texas at Austin, 1974, 71.
8. **Moore, B. E., Sagik, B. P., and Malina, J. F., Jr.,** Viral association with suspended solids, *Water Res.,* 9, 197, 1975.
9. **Schaub, S. A. and Sagik, B. P.,** Association of enteroviruses with natural and artificially introduced colloidal solids in water and infectivity of solids-associated virions, *Appl. Microbiol.,* 30, 212, 1975.
10. **Cliver, D. O.,** Virus association with wastewater solids, *Environ. Lett.,* 10, 215, 1975.
11. **Wellings, F. M., Lewis, A. L., and Mountain, C. W.,** Demonstration of solids-associated virus in wastewater and sludge, *Appl. Environ. Microbiol.,* 31, 354, 1976.
12. **Boardman, G. D. and Sproul, O. J.,** Protection of viruses during disinfection by adsorption to particulate matter, *J. Water Pollut. Control Fed.,* 49, 1857, 1977.
13. **Sattar, S. A. and Westwood, J. C. N.,** Comparison of four elements in the recovery of indigenous viruses from raw sludge, *Can. J. Microbiol.,* 22, 1586, 1976.
14. **Subramanyan, T. P.,** Persistence of enteroviruses in sewage sludge, *Bull. W.H.O.,* 55, 431, 1977.
15. **Berg, G.,** Removal of viruses from sewage, effluents, and waters. 1. A review, *Bull. W.H.O.,* 49, 451, 1973.
16. **Rao, V. C., Lakhe, S. B., Wagmare, S. V. and Ramon, V.,** Virus removal during primary sedimentation of raw sewage, presented at 179th Natl. Meet. American Chemical Society, Houston, Tex., March 1980.
17. **Isherwood, J. D.,** The fate of coxsackie virus A-13 in water reclamation, *Am. J. Public Health,* 55, 1945, 1965.
18. **Kelly, S., Sanderson, W. W., and Neidl, C.,** Removal of enteroviruses from sewage by activated sludge, *J. Water Pollut. Control Fed.* , 33, 1056, 1961.
19. **Clarke, N. A., Stevenson, R. E., Chang, S. L., and Kabler, P. W.,** Removal of enteric viruses from sewage by activated sludge treatment, *Am. J. Public Health,* 51, 1118, 1961.
20. **Mack, W. N., Frey, J. R., Riegle, B. J., and Mallman, W. L.,** Enterovirus removal by activated sludge treatment, *J. Water Pollut. Control Fed.,* 34, 1133, 1962.
21. **Lund, E., Hedstrom, C. E., and Jantzen, N.,** Occurrence of enteric viruses in wastewater after activated sludge treatment, *J. Water Pollut. Control Fed.,* 41, 169, 1969.
22. **Safferman, R. S. and Morris, M. E.,** Assessment of virus removal by a multi-stage activated sludge process, *Water Res.,* 10, 413, 1976.
23. **Berg, G.,** Virus transmission by the water vehicle. II. Virus removal by sewage treatment procedures, *Health Lab. Sci.,* 3, 90, 1966.
24. **Malina, J. F., Jr., Ranganathan, K. R., Moore, B. E., and Sagik, B. P.,** Poliovirus inactivation by activated sludge, in *Virus Survival in Water and Wastewater Systems,* Malina, J. F., Jr. and Sagik, B. P., Eds., Center for Research in Water Resources, University of Texas at Austin, 1974, 95.
25. **Malina, J. F., Jr., Ranganathan, K. R., Sagik, B. P., and Moore, B. E.,** Poliovirus inactivation by activated sludge *J. Water Pollut. Control Fed.,* 47, 2178, 1975.
26. **Glass, J. S. and O'Brien, R. T.,** Enterovirus and coliphage inactivation during activated sludge treatment, *Water Res.,* 14, 877, 1980.
27. **Cliver, D. O. and Herrmann, J. E.,** Proteolytic and microbial inactivation of enteroviruses, *Water Res.,* 6, 797, 1972.
28. **Herrmann, J. E. and Cliver, D. O.,** Degradation of coxsackie virus type A9 by proteolytic enzymes, *Infect. Immun.,* 7, 513, 1973.
29. **McLaren, L. C., Holland, J. J., and Syverton, J. T.,** The mammalian cell-virus relationship. I. Attachment of poliovirus to cultivated cells of primate and non-primate origin, *J. Exp. Med.,* 109, 475, 1959.
30. **Ward, R. L. and Ashley, C. S.,** Mode of initiation of cell infection with sludge-associated poliovirus, *Appl. Environ. Microbiol.,* 38, 329, 1979.

31. **Chang, S. L., Stevenson, R. E., Bryant, A. R., Woodward, R. L., and Kabler, P. W.,** Removal of coxsackie and bacterial viruses in water by flocculation. I. Removal of coxsackie and bacterial viruses in water of known chemical content by flocculation with aluminum sulfate or ferric chloride under various testing conditions, *Am. J. Public Health,* 48, 51, 1958.
32. **Chang, S. L., Stevenson, R. E., Bryant, A. R., Woodward, R. L., and Kabler, P. W.,** Removal of coxsackie and bacterial viruses in water by flocculation. II. Removal of coxsackie and bacterial viruses and the native bacteria in raw Ohio River water by flocculation with aluminum sulfate and ferric chloride, *Am. J. Public Health,* 48, 159, 1958.
33. **Robeck, G. G., Clarke, N. A., and Dostal, K. A.,** Effectiveness of water treatment processes in virus removal, *J. Am. Water Works Assoc.,* 54, 1275, 1962.
34. **Wentworth, D. F., Thorup, R. T., and Sproul, O. J.,** Poliovirus inactivation in water-softening precipitation processes, *J. Am. Water Works Assoc.,* 60, 939, 1968.
35. **Sproul, O. J.,** Virus removal by adsorption in treatment processes, *Water Res.,* 2, 74, 1968.
36. **Berg, G., Dean, R. B., and Dahling, D. R.,** Removal of poliovirus from secondary effluents by lime flocculation and rapid sand filtration, *J. Am. Water Works Assoc.,* 60, 193, 1968.
37. **Farrah, S. R., Goyal, S. M., Gerba, C. P., Conklin, R. H., and Smith, E. M.,** Comparison between adsorption of poliovirus and rotavirus by aluminum hydroxide and activated sludge flocs, *Appl. Environ. Microbiol.,* 35, 360, 1978.
38. **Thorup, R. J., Nixon, F. P., Wentworth, D. F., and Sproul, O. J.,** Virus removal by coagulation with polyelectrolytes, *J. Am. Water Works Assoc.,* 62, 97, 1970.
39. **Manwaring, J. F., Chaudhuri, M., and Engelbrecht, R. S.,** Removal of viruses by coagulation and flocculation, *J. Am. Water Works Assoc.,* 63, 298, 1971.
40. **York, D. W. and Drewry, W. A.,** Virus removal by chemical coagulation, *J. Am. Water Works Assoc.,* 66, 711, 1974.
41. **Sattar, S. A., Ramia, S., and Westwood, J. C. N.,** Calcium hydroxide (lime) and the elimination of human pathogenic viruses from sewage: studies with experimentally contaminated (poliovirus type 1, Sabin) and pilot plant samples, *Can. J. Public Health,* 67, 221, 1976.
42. **Grabow, W. O. K., Middendorff, I. G., and Basson, N. C.,** Role of lime treatment in the removal of bacteria, enteric viruses, and coliphages in a wastewater reclamation plant, *Appl. Environ. Microbiol.,* 35, 663, 1978.
43. **Sattar, S. A. and Ramia, S.,** Viruses in sewage: effect of phosphate removal with calcium hydroxide (lime), *Can. J. Microbiol.,* 24, 1004, 1978.
44. **Parkhurst, J. D.,** Pomona Virus Study Final Report, Sanitation Districts of Los Angeles County, 1977.
45. **Sattar, S. A. and Westwood, J. C. N.,** Viral pollution of surface waters due to chlorinated primary effluents, *Appl. Environ. Microbiol.,* 36, 427, 1978.
46. **Grabow, W. O.,** The virology of wastewater treatment, *Water Res.,* 2, 675, 1968.
47. **Lothrop, T. L. and Sproul, O. J.,** High-level inactivation of viruses in wastewater by chlorination, *J. Water Pollut. Control Fed.,* 41, 569, 1969.
48. **Varma, M. M., Christian, B. A., and McKinstry, D. W.,** Inactivation of Sabin oral poliomyelitis type 1 virus, *J. Water Pollut. Control Fed.,* 46, 987, 1974.
49. **Cramer, W. N., Kawata, K., and Krusé, D. W.,** Chlorination and iodination of poliovirus and f2, *J. Water Pollut. Control Fed.,* 48, 61, 1976.
50. **Young, D. C. and Sharp, D. G.,** Partial reactivation of chlorine-treated echovirus, *Appl. Environ. Microbiol.,* 37, 766, 1979.
51. **Fujioka, R. S. and Loh, P. C.,** Recycling of water for irrigation: persistence of enteroviruses in sewage effluent and natural waters receiving the effluent, *Water Soil Air Pollut.,* 9, 213, 1978.
52. **Eisenhardt, A., Lund, E., and Nissen, B.,** The effect of sludge digestion on virus infectivity, *Water Res.,* 11, 579, 1977.
53. **Berg, G. and Berman, D.,** Destruction by anaerobic mesophilic and thermophilic digestion of viruses and indicator bacteria indigenous to domestic sludges, *Appl. Environ. Microbiol.,* 39, 361, 1980.
54. **Palfi, A.,** Survival of enteroviruses during anaerobic sludge digestion, in *Advances in Water Pollution Research,* Proc. 6th Int. Conf., Jerusalem, Jenkins, S. H., Ed., Permagon Press, Elmsford, N.Y., 1973, 99.
55. **Moore, B. E., Sagik, B. P., and Sorber, C. A.,** An assessment of potential health risks associated with land disposal of residual sludges, in *Proc. 3rd Natl. Conf. Sludge Manage., Disposal, Utilization,* Miami, Fla., Information Transfer, Rockville, Md., 1976, 108.
56. **Sanders, D. A., Malina, J. F., Jr., Moore, B. E., Sagik, B. P., and Sorber, C. A.,** Fate of poliovirus during anaerobic digestion, *J. Water Pollut. Control Fed.,* 51, 333, 1979.
57. **Bertucci, J. J., Lue-Hing, C., Zenz, D., and Sedita, S. J.,** Inactivation of viruses during anaerobic sludge digestion, *J. Water Pollut. Control Fed.,* 49, 1642, 1977.

58. **Moore, B. E., Sagik, B. P., and Sorber, C. A.,** Land application of sludges: minimizing the impact of viruses on water resources, in *Proc. Conf. Risk Assessment Health Effects Land Appl. Munic. Wastewater Sludges,* Sagik, B. P. and Sorber, C. A., Eds., Center for Applied Research and Technology, The University of Texas at San Antonio, 1977, 154.
59. **Ward, R. L. and Ashley, C. S.,** Inactivation of poliovirus in digested sludge, *Appl. Environ. Microbiol.,* 31, 921, 1976.
60. **Ward, R. L.,** Inactivation of enteric viruses in wastewater sludge, in *Proc. 3rd Natl. Conf. Sludge Manage., Disposal, Utilization,* Miami, Fla., Information Transfer, Rockville, Md., 1976, 138.
61. **Ward, R. L. and Ashley, C. S.** Identification of the virucidal agent in wastewater sludge, *Appl. Environ. Microbiol.,* 33, 860, 1977.
62. **Ward, R. L.,** Mechanism of poliovirus inactivation by ammonia, *J. Virol.,* 26, 299, 1978.
63. **Ward, R. L., Ashley, C. S., and Moseley, R. H.,** Heat inactivation of poliovirus in wastewater sludge, *Appl. Environ. Microbiol.,* 32, 339, 1976.
64. **Ward, R. L. and Ashley, C. S.,** Comparative effects of ammonia and related compounds on poliovirus, *Appl. Environ. Microbiol.,* 36, 198, 1978.
65. **Ward, R. L. and Ashley, C. S.,** Identification of detergents as components of wastewater sludge that modify the thermal stability of reovirus and enteroviruses, *Appl. Environ. Microbiol.,* 36, 889, 1978.
66. **Ward, R. L. and Ashley, C. S.,** pH Modification of the effects of detergents on the stability of enteric viruses, *Appl. Environ. Microbiol.,* 38, 314, 1979.
67. **Mandel, B.,** The use of sodium dodecyl sulfate in studies on the interaction of poliovirus and HeLa cells, *Virology,* 17, 288, 1962.
68. **Mandel, B.,** Early stages of virus-cell interaction as studied by using antibody, *Cold Spring Harbor Symp. Quant. Biol.,* 27, 123, 1962.
69. **Mandel, B.,** The extraction of ribonucleic acid from poliovirus by treatment with sodium dodecyl sulfate, *Virology,* 22, 360, 1964.
70. **Flewett, T. H., Bryan, A. S., Davies, H., Woode, G. N., Bridger, J. C., and Derrick, J. M.,** Relation between viruses from acute gastroenteritis of children and newborn calves, *Lancet,* ii, 61, 1974.
71. **Davidson, G. P., Bishop, R. F., Townley, R. R. W., Holmes, I. H., and Ruck, B. J.,** Importance of a new virus in acute sporadic enteritis in children, *Lancet,* i, 242, 1975.
72. **McNulty, M. S.,** Rotaviruses, *J. Gen. Virol.,* 40, 1, 1978.
73. **Wallis, C., Melnick, J. L., and Rapp, F.,** Different effects of $MgCl_2$ and $MgSO_4$ on the thermostability of viruses, *Virology,* 26, 694, 1965.
74. **Ward, R. L. and Ashley, C. S.,** Discovery of an agent in wastewater sludge that reduces the heat required to inactivate reovirus, *Appl. Environ. Microbiol.,* 34, 681, 1977.
75. **Van Donsel, D. J. and Larkin, E. P.,** Thermal inactivation studies on bovine and simian rotaviruses, Abstr. Ann. Mett. Am. Soc. Microbiol., 1980, 209.
76. **Ward, R. L. and Ashley, C. S.,** Effects of wastewater sludge and its detergents on the stability of rotavirus, *Appl. Environ. Microbiol.,* 40, 1154, 1980.
77. **Ward, R. L. and Ashley, C. S.,** Comparative study on the mechanisms of rotavirus inactivation by sodium dodecyl sulfate and ethylenediaminetetraacetate, *Appl. Environ. Microbiol.,* 40, 1148, 1980.
78. **Boeyé, A. and Van Elsen, A.,** Alkaline disruption of poliovirus: kinetics and purification of RNA-free particles, *Virology,* 33, 335, 1967.
79. **Katagiri, S., Aikawa, S., and Hinuma, Y.,** Stepwise degradation of poliovirus capsid by alkaline treatment, *J. Gen. Virol.,* 13, 101, 1971.
80. **Ward, R. L. and Ashley, C. S.,** Heat inactivation of enteric viruses in dewatered wastewater sludge, *Appl. Environ. Microbiol.,* 36, 898, 1978.
81. **Ward, R. L. and Ashley, C. S.,** Inactivation of enteric viruses in wastewater sludge through dewatering by evaporation, *Appl. Environ. Microbiol.,* 34, 564, 1977.
82. **Pancorbo, O. C., Bitton, G., Farrah, S. R., Davidson, J. M., and Overman, A. R.,** Survival and transport of viruses following sludge appliation to soil columns exposed to natural conditions, Abstr. Ann. Meet. Am. Soc. for Microbiol., Miami, 1980, 211.
83. **Bagdasáryan, G. A.,** Survival of viruses of the enterovirus group (poliomyelitis, ECHO, coxsackie) in soil and on vegetables, *J. Hyg. Epidemiol. Microbiol. Immunol.,* 7, 497, 1965.
84. **Lance, J. C., Gerba, C. P., and Melnick, J. L.,** Virus movement in soil columns flooded with secondary sewage effluent, *Appl. Environ. Microbiol.,* 32, 520, 1976.
85. **Yeager, J. G. and O'Brien, R. T.,** Enterovirus inactivation in soil, *Appl. Environ. Microbiol.,* 38, 694, 1979.
86. **Yeager, J. G. and O'Brien, R. T.,** Structural changes associated with poliovirus inactivation in soil, *Appl. Environ. Microbiol.,* 38, 702, 1979.
87. **Trouwborst, T., de Jong, J. C., and Winkler, K. C.,** Mechanism of inactivation in aerosols of bacteriophage T1, *J. Gen. Virol.,* 15, 235, 1972.

88. **de Jong, J. C., Harmsen, M., and Trouwborst, T.,** The infectivity of the nucleic acid of aerosol-inactivated poliovirus, *J. Gen. Virol.*, 18, 83, 1973.
89. **de Jong, J. C., Harmsen, M., Trouwborst, T., and Winkler, K. C.,** Inactivation of encephalomyocarditis virus in aerosols: fate of virus protein and ribonucleic acid, *Appl. Microbiol.*, 27, 59, 1974.
90. **Trouwborst, T., Kuyper S., de Jong, J. C., and Plantinga, A. D.,** Inactivation of some bacterial and animal viruses by exposure to liquid-air interfaces, *J. Gen. Virol.*, 24, 155, 1974.
91. **Reimers, R. S., Little, M. D., Leftwich, D. B., Bowman, D. D., Englande, A. J., and Wilkinson, D. F.,** Investigation of parasites in Southern sludges and disinfection by standard sludge treatment processes, Draft final report, U.S. Environmental Protection Agency, Cincinnati, Ohio, 1980.
92. **Ward, R. L., Yeager, J. G., and Ashley, C. S.,** Response of bacteria in wastewater sludge to moisture loss by evaporation and the effects of moisture content on bacterial inactivation by ionizing radiation, *Appl. Environ. Microbiol.*, 41, 1123, 1981.
93. **Yeager, J. G. and Ward, R. L.,** Effects of moisture content on the long-term survival and regrowth of bacteria in wastewater sludge, *Appl. Environ. Microbiol.*, 41, 1117, 1981.
94. **Willson, G. B., Epstein, E., and Parr, J. R.,** Recent advances in compost technology, in *Proc. 3rd Natl. Conf. Sludge Manage., Disposal, Utilization,* Miami, Fla., Information Transfer, Rockville, Md., 1976, 167.
95. **Burge, W. D., Marsh, P. B., and Millner, P. D.,** Occurrence of pathogens and microbial allergens in the sewage sludge composting environment, in *Natl. Conf. Composting of Munic. Residues and Sludges,* Rockville, Information Transfer, Rockville, Md., 1977, 128.
96. **Kawata, K., Cramer, W. N., and Burge, W. D.,** Composting destroys pathogens in sewage solids, *Water Sewage Works,* 124, 76, 1977.
97. **Wallis, C. and Melnick, J. L.,** Cationic stabilization — a new property of enteroviruses, *Virology,* 16, 504, 1962.
98. **Pohjanpelto, P.,** Stabilization of poliovirus by cystine, *Virology,* 6, 472, 1958.
99. **Steele, F. M. and Black, F. L.,** Inactivation and heat stabilization of poliovirus by 2-thiouracil, *J. Virol.*, 1, 653, 1967.
100. **Foliquet, J. M. and Doncoeur, F.,** Inactivation assays of enteroviruses and Salmonella in fresh and digested wastewater sludges by pasteurization, *Water Res.*, 6, 1399, 1972.
101. **Dertinger, H. and Jung, H.,** Primary processes of energy adsorption, in *Molecular Radiation Biology,* Springer-Verlag, New York, 1970, 33.
102. **Luria, S. E. and Exner, F. M.,** The inactivation of bacteriophages by x-rays. Influence of the medium, *Proc. Natl. Acad. Sci. U.S.A.*, 27, 370, 1941.
103. **Watson, J. D.,** The properties of x-ray-inactivated bacteriophage. I. Inactivation by direct effects, *J. Bacteriol.*, 60, 697, 1950.
104. **Watson, J. D.,** The properties of x-ray-inactivated bacteriophage. II. Inactivation by indirect effects, *J. Bacteriol.*, 63, 473, 1952.
105. **Michaels, H. B. and Hunt, J. W.,** A model for radiation damage in cells by direct effect and by indirect effect: a radiation chemistry approach, *Radiat. Res.*, 74, 23, 1978.
106. **Ward, R. L.,** Mechanisms of poliovirus inactivation by the direct and indirect effects of ionizing radiation, *Radiat. Res.*, 83, 330, 1980.
107. **McCrea, J. F.,** Ionizing radiation and its effects on animal viruses, *Ann. N.Y. Acad. Sci.*, 83, 692, 1960.
108. **Sullivan, R., Fassolitis, A. C., Larkin, E. P., Read, R. B., Jr., and Peeler, J. T.,** Inactivation of thirty viruses by gamma radiation, *Appl. Microbiol.*, 22, 61, 1971.
109. **Sullivan, R., Scarpino, P. V., Fassolitis, A. C., Larkin, E. P., and Peeler, J. T.,** Gamma radiation inactivation of coxsackievirus B-2, *Appl. Microbiol.*, 26, 14, 1973.
110. **Sinskey, A. J., Shah, D., and Metcalf, T. J.,** Biological effects of irradiation with high energy electrons, in *Proc. 3rd Natl. Conf. Sludge Manage., Disposal, Utilization,* Miami, Fla., Information Transfer, Rockville, Md., 1976, 160.
111. **Ward, R. L.,** Inactivation of poliovirus in wastewater sludge with radiation and thermoradiation, *Appl. Environ. Microbiol.*, 33, 1218, 1977.
112. **Ward, R. L.,** Pathological aspects of inactivation of enteric viruses in dewatered wastewater sludge, in *Natl. Conf. Design of Munic. Sludge Compost Facilities,* Chicago, Information Transfer, Rockville, Md., 1978, 42.
113. **Metcalf, T. G.,** Control of virus pathogens in municipal wastewater and residuals by irradiation with high energy electrons, U.S. National Tech. Inf. Serv. Rep. No. PB-272347, 1977.
114. **Lowe, H. N., Jr., Lacy, W. J., Surkiewicz, B. F., and Jaeger, R. F.,** Destruction of microorganisms in water, sewage, and sewage sludge by ionizing radiations, *J. Am. Water Works Assoc.*, 48, 1363, 1956.
115. **Ridenour, G. M. and Armbruster, E. H.,** Effect of high-level gamma radiation on disinfection of water and sewage, *J. Am. Water Works Assoc.*, 48, 671, 1956.

116. **Etzel, J. E., Born, G. S., Stein, J., Helbing, T. J., and Baney, G.,** Sewage sludge conditioning and disinfection by gamma irradiation, *Am. J. Public Health,* 59, 2067, 1969.
117. **Touhill, C. J., Martin, E. C., Fujihara, M. P., Olesen, D. E., Stein, J. E., and McDonnell, G.,** The effects of radiation on Chicago Metropolitan Sanitary District municipal and industrial wastewater, *J. Water Pollut. Control Fed.*, 41, R44, 1969.
118. **Wizigmann, I. and Würsching, E.,** Experience with a pilot plant for the irradiation of sewage sludge. Bacterial and parasitological studies after irradiation, in *Proc. IAEA Symp. Radiat. Clean Environ.*, Munich, International Atomic Energy Agency, Vienna, 1975, 477.
119. **Brandon, J. R. and Langley, S.,** Inactivation of bacteria in sewage sludge by ionizing radiation, heat, and thermoradiation, SAND75-0168, Sandia Laboratories, Albuquerque, N.M., 1976.
120. *Federal Register,* 40 CFR Part 257, September 13, 1979.

Occurrence and Survival of Viruses in the Environment

Chapter 7

VIRUSES IN ENVIRONMENTAL WATERS

Jean-Claude Block

TABLE OF CONTENTS

I. INTRODUCTION

Many problems are still posed by viral contamination of environmental water because of the risks of human contamination through swimming, consumption of contaminated sea food, and deficient treatment of tapwater. As these risks are not well-known, they have to be better evaluated in order to define a new health policy or environmental protection policy. In spite of the practice of vaccination, wild strains of poliovirus are still recovered from water[1,2] and numerous other pathogenic strains of enteric viruses also flow in our immediate environment. However, it is still very difficult to evaluate the viral contamination of environmental waters. Results obtained from human viral analyses are limited for at least three reasons. First, compared to animal viruses (cattle, pigs, cats, dogs, etc.), the human viruses only represent a small proportion of those viruses carried through the water route. Animal viruses are worthy of more attention and are beginning to be more thoroughly studied.[3-5] Second, the methods for concentrating viruses from water are far from 100% effective (see below) and underestimate the viral flow in water. Third, the use of specific cell cultures selects certain recoverable strains[6,7] and restricts the range of recovered viruses. Thus Denis'[8] bibliographical review shows that the investigators are primarily interested in enteroviruses; from 4600 indexed viral strains, approximately 75% were enteroviruses and among these 37% were polioviruses. Today, these observations are mostly identical even if real efforts are made to study rotaviruses, coronaviruses, hepatitis viruses, etc. in water. Even in the best cases, our analyses only give a partial idea of the viral contamination of water.

Contamination of environmental waters consists of two contradictory phenomena: (1) the discharge of feces containing virus into our environment via wastewater, and (2) the dispersion and the inactivation of viruses, they being obligate parasites which cannot multiply in water.

A priori field investigations always give a particular view of viral pollution because of the multiplicity of factors which influence viral dispersion and survival. Consequently it is most important to know whether it is possible to draw up a general law on the fate of viruses from all of these particular observations.

II. OCCURRENCE OF VIRUSES IN ENVIRONMENTAL WATERS

Viruses are widely dispersed in our environment and today one cannot claim that any type of water is virus-free. Thus groundwater, generally considered to be well-protected from contamination by soil can contain enteric viruses, especially after land disposal of wastewater effluents.[9-13] In the same way, enteroviruses have been recovered from constantly chlorinated swimming pool water[14-16] even in the absence of fecal indicators. Furthermore, tapwater has been proved to carry viruses (poliovirus 1, coxsackievirus B5, echovirus 7, reoviruses) in different concentrations: 1 plaque-forming unit (PFU)/60 ℓ to 1 PFU/ℓ.[17-24] The principal source of water contamination by viruses is, of course, sewage. Numerous recent studies have estimated types and the quantities of viruses in water. Theoretically, this approach should allow us to make a more precise appraisal of the significance of surface water contamination. The viral concentrations found in wastewater after biological treatment (secondary effluent) are relatively high because wastewater treatment plants were never intended to eliminate viral micropopulation (see Chapter 3). In addition, the high level of organic matter and suspended solids limits the efficiency of chlorination, and in 56 to 58% of cases, viruses were recovered from chlorinated effluents.[25,26] The results from viral analysis of secondary effluents are shown in Table 1. Most of the results give an average number between 1 and 100 PFU or most probable number of cytopathic units (MPNCU) per liter. Nevertheless, for no obvious reasons, some numbers are much higher.[25,27,41] Such differences can be explained by different methods of virus concentration and by other parameters such

Table 1
SELECTIVE QUANTITATIVE DATA FROM VIRAL ANALYSES OF SECONDARY WASTEWATERS

Geographic location	Number of viruses recovered	Ref.
Haifa (Israel)	3000 to 28,000 PFU/ℓ	27
Bombay (India)	5 to 15 PFU/ℓ	28
Los Angeles (U.S.)	About 5.2 PFU/ℓ	29
Hawaii (U.S.)	7 to 5222 PFU/ℓ	25
Haifa (Israel)	18 to 65 PFU/ℓ	30
Kiel (German)	About 55 PFU/ℓ	31
Fairbanks (Alaska)	27 to 65 PFU/ℓ	32
Los Angeles (U.S.)	5.3 to 13.2 PFU/ℓ	33
Cincinnati (U.S.)	1 to 12 PFU/380/ℓ	34
Cagnes sur mer (France)	<1 to 41.8 MPNCU/ℓ	35—37
Laval (Quebec)	0 to 30 PFU/ℓ	38
Nancy (France)	0 to 9.6 MPNCU/ℓ	39
Texas (U.S.)	<10 to 140 PFU/ℓ	40
Pretoria (South Africa)	About 4100 DI_{50}/ℓ	41
Nancy (France)	2.7 to 29 MPNCU/ℓ	42

as viral adsorption, which prevents the recovery of viruses on cell cultures.[43] In addition, as Berg[44] has stated, the effluent sampling time is a decisive factor. Classically, not only have seasonal, qualitative, and quantitative variations in virus concentrations been observed[27,45-47] but also hourly variations.[28,34,48,49] A recent study conducted by Rolland[49] into the secondary effluent of a Nancy wastewater treatment plant (Nancy, France) shows a minimal flow of enteric viruses at 9 a.m. (with apparently zero PFU/20 ℓ) and a maximal flow at 6 p.m. (with 75 PFU/ℓ). In a 24-hr study, it was shown that after activated sludge treatment at the Nancy plant serving 350,000 inhabitants; 150,000 m^3/day) approximately 360 × 10^6 PFU were discharged every day into the Meurthe River.[49] Because of the low efficiency of the virus detection method (concentration at pH 3.5; adsorption onto BGM [Buffalo Green Monkey] cells), these results should be considered as a minimum; they underestimate, by at least a factor of 10, the quantity of enteric viruses discharged into receiving waters (stream, river, lake, ocean, etc.) The receiving waters can be easily contaminated and despite the dilution (from 2- to 1000-fold), enteric viruses are still often found in such environmental waters. Tables 2 and 3 show quantitative results from virological analyses of surface water and seawater. In extremely polluted seawater, a relatively smaller level of contamination occurs: with 28 DI 50/ℓ,[63] 22.4 PFU/ℓ,[35] and 12.6 PFU/ℓ.[67]

No constant ratio of fecal indicator bacteria to enteric viruses can be demonstrated and therefore it is not possible to make a correlation between the concentration of bacteria and viruses. For example, Berg[50] found in the Missouri River 1 and 4 PFU of viruses per 50 gal together with concurrent levels of fecal coliforms of 60 and 10,000/100 mℓ. Similarly, Sarrette et al.[51] found at two points of the river Seine 233 and 350 total coliforms per 100 mℓ and concurrent concentration of enteric viruses of 9 and 1906/10 ℓ. However in some situations, it has been possible to find similitudes in variations of concentrations of both viruses and other microorganisms.[53] The same observations have been made in seawater; 8 PFU of viruses per gallon were recovered both in Penatoguit Creek and in the Great South Bay; however, the levels of fecal coliforms were 460/100 and 4/100 mℓ. Katzenelson[66] found enteric viruses at a fecal coliform level of 80/100 mℓ.

The viral contamination of receiving waters (river and sea) is characterized by four factors common to bacterial contamination:

1. The sources of contamination are numerous and diversified in origin; in other words we talk of diffuse pollution.

Table 2
SELECTED QUANTITATIVE DATA FROM VIRAL ANALYSES OF FRESH SURFACE WATERS (RIVERS AND LAKE)

Geographic location	Number of viruses recovered	Ref.
Missouri River (U.S.)	1 to 19 PFU/190 ℓ	50
Seine River (France)	0 to 14 PFU/10 ℓ (one sample contained 1906 PFU)	51
Thames River (U.K.)	4 to 22 PFU/ℓ	52
Lee River (U.K.)	0 to 11.8 PFU/ℓ	
Rhine River (France)	0 to 283 PFU/ℓ	54
Moselle River (France)	2 MPNCU/10 ℓ	53
Tanana River (Alaska)	0 to 22 PFU/380 ℓ	32
Rio Besos (Spain)	23 to 55 MPNCU/ℓ	55
Rio Lhobregat (Spain)	0.5 to 5 MPNCU/ℓ	
Lake Ronkonkoma (N.Y., U.S.A.)	0 to 6.5 PFU/4.7 ℓ	56
Seine River (France)	0.3 to 173 PFU/ℓ (when 500 1 were analyzed)	57
— (Germany)	0 to 16 MPNCU/ℓ	58
Wear River (U.K.)	0 to 30 PFU/5 ℓ	59
Avon and Sowe Rivers (U.K.)	2 to 620 PFU/ℓ	60

Table 3
SELECTED QUANTITATIVE DATA FROM VIRAL ANALYSES OF SEAWATER

Number of viruses recovered	Ref.
1 to 17 DI 50/ℓ	61
0.3 to 3.3 PFU/ℓ	62
1 to 28 DI 50/ℓ	63
0 to 0.2 PFU/ℓ	64
0.2 to 1.6 PFU/ℓ	65
0 to 0.06 PFU/ℓ	66
0.5 to 12.6 PFU/ℓ	67
0 to 0.1 PFU/ℓ	68
4 to 167 PFU/380 ℓ	69
1 to 22.4 PFU/ℓ	35
0.12 to 0.92 MPNCU/ℓ	55
0 to 25 PFU/3.8 ℓ	56
About 1100 DI 50/ℓ	41
1 to 40 PFU/60 ℓ	70
0 to 20 PFU/10 ℓ	71

2. Virus carriage can be effective at long distances.
3. Most of the enteric viruses in water are associated with solid particles. Consequently viruses may settle with suspended solids to the river bottom.
4. Most enteric viruses are aggregated in clumps which increase the aquatic medium heterogeneity.

A. Multiplicity of Contamination Sources

On the coast as in rivers, many effluents discharge diluted feces into environmental waters. These effluents are of varying sizes and characteristics according to their origin, but the principal source of pollution is not really known; secondary effluent from urban areas, septic tank effluent,[72] or raw wastewater coming from farms, small villages, or isolated houses, storm water runoffs. During a study on the Moselle River, the viral contamination upstream from all the big cities was approximately 2 MPNCU/10 ℓ (50,000 and 500,000 times lower than the quantities of *Streptococcus faecalis* and *Escherichia coli*, respectively). No correlations have been shown to exist between the viruses in the water and the density of the population living along the river but, in spite of a dispersed habitat, enteric viruses were detected at a frequency of 20 to 60%. These observations put into question our capacity to control even the smaller sources of viral contamination. In the same way, disinfection of wastewater will be applied preferentially in treating the sewage of large cities where the effluents will have been well-treated and where the urban community pays for such treatment. Consequently, it has not been shown that, if only the effluents of great cities are virus-free, that the level of viruses in surface water significantly decreases.

B. Virus Carriage by Water Flow

The transportation of viruses by water flow, even for long distances, is well-known.[73] Examination of the Houston ship channel led to the detection of enteroviruses as far away as 13 km from its source.[62] Hugues et al.[35] detected viruses in the Baie des Anges near Cagnes-sur-Mer, more than 200 m from the coast and in concentrations often equivalent to those observed near the beaches (from 0.1 to 27 MPNCU/ℓ). Simkova and Wallnerova[74] recovered enteric viruses in a river 200 m below a sewage outfall. Recently, Dahling and Safferman[32] gave one of the best demonstrations of this phenomenon. Their study was conducted on the Tanana River (Alaska) in a zone where it was possible to follow virus transport for 317 km without encountering other apparent sources of contamination (no other villages or cities, no storm water runoffs because of the winter period). They observed that at least 30% of the enteric viruses presented at the source of domestic pollution were present 300 km downstream after 7.1 days of transport.

Such results point out that no source of pollution is insignificant and that from one point of contamination, viruses can spread throughout a large geographical zone and then increase the hazard of virus contact with susceptible hosts.

C. Association of Viruses with Solids in Water

Virus association with solids has been well-studied and now appears to be of the utmost importance in viral ecology. Viruses, negatively charged biocolloids, have a great potential for surface adsorption, whatever the solids (biologic or nonbiologic). Viral adsorption occurs in the digestive tract with feces and with other solids carried by water. After adding poliovirus 1 to an activated sludge mixture, Block et al.[76] observed that viral adsorption onto the floc was instantaneous for 90% of the viruses and increased with time: 99.9% after 5 min, 99.99% after 30 min, 99.997% after 60 min. In the secondary effluent of the wastewater treatment plant, most microorganisms were adsorbed and after centrifugation, numerous viral particles could be recovered from the pellets.[76-78] Gerba et al.[79] and Wellings et al.,[80] attempting to evaluate the proportion of solids-associated viruses, found 3 to 100% and 16 to 100% of the viruses bound to suspended matter, respectively. According to our observations on solids-associated bacteria (Figure 1), one can expect the proportion of "free virus" to be low. During wastewater treatment or during discharge into the receiving water solids-associated viruses follow the fate of suspended matter. Rao et al.[28] showed that primary treatment of sewage removes 24 to 83% of viruses. Rolland[49] demonstrated that the strong correlation between viral concentration and suspended matter in raw wastewater totally disappeared after primary and secondary treatment. Even if virus desorption occurs in receiving water, adsorption phenomena are certainly preponderant on clays,[81] bacteria,[82] silicates, algal cells,[75] etc. These phenomena give rise to a typical situation; on the one hand

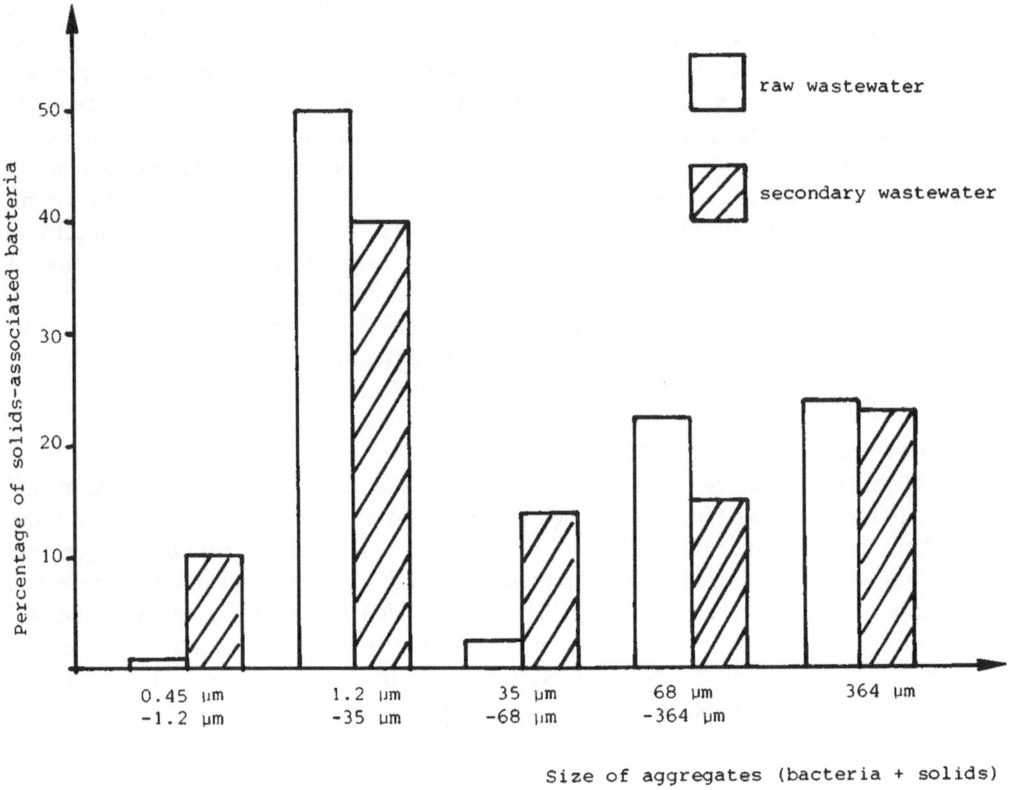

FIGURE 1. Percentage of solids-associated bacteria according to the size of the aggregates in raw and secondary wastewater. (Adapted from Block, J. C. and Dollard, M. A., unpublished data).

Table 4
RECOVERY OF VIRUSES FROM WATER AND SEDIMENTS (GALVESTON BAY, TEXAS)

Sample number	1	2	3	4
Number of viruses in 20 ℓ of seawater (PFU)	6	528	5	30
Number of viruses in 20 ℓ of sediment (PFU)	200	140	50	480

Adapted from Labelle, R. L., Gerba, C. P., Goyal, S. M., Melnick, J. L., Cech, I., and Bogdan, G., *Appl. Environ. Microbiol.*, 39, 588, 1980.

solid-associated viruses settle if they are bound onto "settleable" solids, on the other hand free viruses may spontaneously adsorb onto river or marine sediments.[83]

In both cases, the quantities of viruses on sediments are always greater, on a unit volume basis, than they are in overlying water.[63,84,85] The same relationship holds for bacteria. Labelle et al.,[85] recovering viruses both in seawater and sediment, found the latter to be at least 10 times more contaminated by viruses (Table 4).

The effect of viral adsorption on their infectivity has been studied. Bacteriophages or enveloped viruses may be inactivated when bound to solids; however enteric viruses do not seem so easily inactivated by such adsorption.[85-87] But, in contrast, adsorbed viruses are protected from inactivation by environmental factors.[63,82,88-90] According to Labelle and Gerba,[89] adsorption of poliovirus 1 onto sediments decreased the inactivation rate for the virus by a factor of 4 in an unpolluted zone and by a factor of 96 in a polluted zone. In this respect, the presence in high concentration of virus particles which remain viable for a long time on solids poses some problems in representative sampling because often only water is analyzed. Moreover, rains or melting snows increase the flow rate of riverwater, bring about a resuspension of the upper layers of sediments, and increase virus concentration in the water column. The main consequence of this is a modification of apparent viral contamination of the surface water.

D. Viral Clumps in Environmental Waters

In parallel with adsorption phenomena, viruses can spontaneously and naturally aggregate. During the synthesis of viruses in cells, viruses accumulate in the cytoplasmic area or in the nucleus and form pseudo-crystalline aggregates which are easily observed under ultramicroscope. When cell lysis occurs, these aggregates are not totally disunited and can be found in feces. Moreover, on some occasions, viruses form clumps and cause an apparent loss in the viral numbers in water[92-94] which can be interpreted as an inactivation. Galasso and Sharp[95] seemed to be among the first to describe clumps of viral suspensions in water. It is very difficult to estimate both the number of clumps in a water sample and the number of viruses per clump. After the treatment of samples by freon, Floyd et al.[96] showed that 5% of a viral population was in clump form. In these aggregates, the number of viruses seemed to be between 2 and 10.[96,97] However, the clumped state is not irreversible and is dependent on pH and salinity. Consequently, it can be supposed that in receiving waters, aggregation of viruses and disaggregation of clumps occurs constantly.[93] The presence of aggregated viruses has significant and diversified consequences. The impossibility of inactivating viruses clumped in cells, by chlorine for example, has been described by an extensive literature. A fraction of the viruses present always escapes treatment and causes aberrations in the curves of inactivation. As far as environmental questions are concerned, the state of the viruses (aggregated or dispersed) is being constantly modified, especially as viruses flow from wastewater to surface water to tapwater.[100] The result is a modification of the viral population's resistance to environmental factors. On the other hand, the presence of clumps contributes to the increase in the heterogeneity of the aquatic media. As shown in Table 5, consecutive samples collected at 1-min intervals from the same spot in the river contained 0 to 283 PFU/ℓ. Once more the representativeness of sampling is questionable. According to Vilagines and colleagues,[57] larger volume samples such as 500 ℓ could partly resolve the problem.

III. SURVIVAL OF VIRUSES IN ENVIRONMENTAL WATERS

Survival of viruses in environmental waters has been extensively reviewed.[102-106] Human enteric viruses cannot multiply in the environment or in shellfish;[107] consequently, virus populations in such waters decrease with time. Dilution, aggregation and sedimentation are decreasing factors for viruses in our environment. Moreover, inactivation, actual loss of infectivity, will result from capsid or nucleic acid damage which prevent virus attachment to cells or virus transcription in cells, respectively. Hermann et al.[108] pointed out that retention of labeled proteins on two 20 nm filters results from microbial utilization of ^{14}C leucine from the virus coat protein. But they did not determine whether degradation occurred as or after the virus lost infectivity.

Cords et al.[109] recently reported that coxsackievirus A13 could be rendered noninfectious by dissociation of the capsid protein. O'Brien and Newman[110] suggested that the inactivation of poliovirus 1 and coxsackievirus B1 was due to the cleavage of the capsids which exposed

Table 5
CONCENTRATIONS OF ENTERIC VIRUSES IN DIP SAMPLES OF SURFACE WATERS COLLECTED AT LESS THAN 2-min INTERVALS AT THE SAME SPOT.

Number of consecutive samples	Geographic location		
	Moselle River+	Rhine River+	Seine River+
1	0	56	1
2	0	0	0
3	210	59	0
4	0	62	0.2
5	22	0	0.1
6	283	N.D.	N.D.

Note: + Virus concentrations expressed as PFU/ℓ (20 ℓ were analyzed); N.D.: not determined.

Adapted from Block, J. C., Joret, J. C., Morlot, M., and Foliguet, J. M., *Tech. Sci. Munic. — L'eau*, 73, 181, 1978.

the RNA to inactivating factors and determined the loss of its infectivity in less than 24 hr. The mechanisms of inactivation may be different in freshwater and in marine water.[111] In freshwater, hydrolysis of RNA may be a primary inactivating mechanism, while in marine water the initial inactivation reaction may be capsid associated and degradation of the viral RNA may be a secondary, dependent on the capsid changes. Considering the observations of Dawe and Penrose[112] on the debilitation of coliforms in contrast to inactivation, the question could be asked concerning viruses: debilitation or inactivation? Moreover, recent investigations[113] demonstrated that chlorine-treated echoviruses could be at least partially reactivated, by a multiplicity reactivation-related phenomenon which occurred when the host cell was infected with several damaged virions. Such observations induce us to be very careful in the interpretation of virus inactivation. In our aquatic environment, viruses come under the ecological pressures of the biotop. The number of factors which may affect the survival of enteric viruses in water is very high. The principal ones are listed in Table 6 and can be divided into three groups: physical and chemical factors (i.e., abiotic factors) and biological factors. The majority of the influencing parameters are very complex and explain, in part, why survival studies are so difficult to conduct. As is shown in Figure 2, many of the parameters have a direct reciprocal implication. For example, bacterial and algal activity are directly involved in virus survival, but they are affected by protozoan density and predation, pH, ammonia, dissolved oxygen, light, and temperature. Also, these last two parameters can be regarded as key factors because of their implication in microfauna activity and so on. In these conditions, only very well-standardized experiments will help to gain a better knowledge of the effects of environmental factors.

The majority of these studies were conducted in laboratories; moreover seawater has been studied more often than freshwater. In spite of interrelations between environmental factors that may lead to virus inactivation, in the interest of clarity, each will be discussed separately.

A. Physical Factors Affecting Virus Survival

1. Light

Studies on the effect of light on virus inactivation have been carried out only recently. The effect of natural solar light may be direct upon the viral particle or indirect through microfauna stimulation.

Table 6
FACTORS AFFECTING VIRUS SURVIVAL IN ENVIRONMENTAL WATERS

Physical	Chemical	Biological
Light	pH (natural or from industrial effluent)	Virus types
Temperature	Ionicity	Bacterial and algal activity (enzymes, polypeptides, polysaccharides, fatty acids production)
Hydrostatic pressure	Cations (Ca^{2+}, Mg^{2+}, etc.)	Predation by protozoa
Adsorption	Heavy metals (Hg, etc.)	
Aggregation	Organic chemicals (ammonia, detergent, petroleum molecules, etc.)	
	Dissolved oxygen	

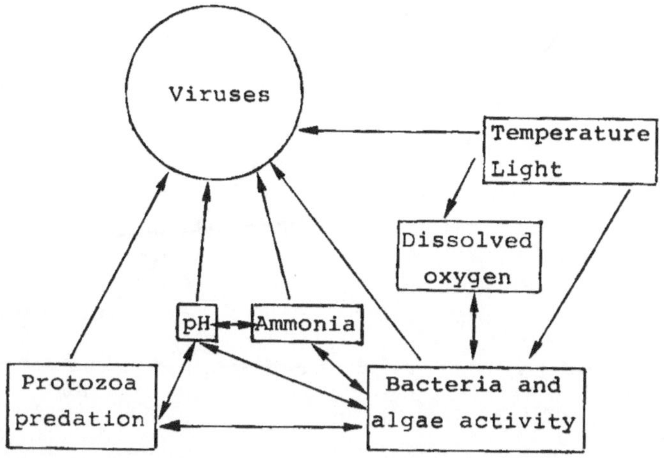

FIGURE 2. Schematic representation of relationships between some environmental factors affecting virus survival. →, direct implication; ↔, reciprocal implication.

The direct effect of light, probably at wavelengths lower than 370 nm, determines virus inactivation.[92,114-117] This radiation is absorbed by proteins and nucleic acids which leads to changes in the configuration of proteins and to the breaking down of the nucleic acids.[103] Some data support a photoinactivation theory, because a number of reactive oxidants can be formed photochemically in seawater.[118,119] Moreover such inactivation can be stimulated by natural molecules, which can act as photosensitizers (lignins, humic acids)[105] or by photoreactive dyes (methylene blue).[120] Hill et al.[115, 116] inactivated 2 logs of poliovirus 1 in unstirred estuarine water exposed to artificial UV radiations (116 ergs/mm^2/sec). The same authors found significant differences in results in static and dynamic study; for example, the exponential rate of devitalization was significantly increased in a flowing-seawater system. Working with seawater and ground water, respectively, Attree-Pietri and Breittmayer[114] and Bitton et al.[117] showed under laboratory conditions that the T 90 of phage T4 and poliovirus 1 was approximately 3 hr (Figure 3). During the same period, the viruses were stable in the dark. Poliovirus 1 inactivation was obtained with a low turbidity groundwater (1.7 JTU) subjected to a relatively high light intensity (0.646 cal/cm^2/min^1). Higher turbidity and greater depths limited the virucidal activity[92,117,121] by adsorption of the light and attenuation of the inactivating energy. However, according to Cubbage et al.,[92] the mechanisms by which turbidity affects the rate of loss of poliovirus infectivity needs to be further defined.

The indirect effect of light on viruses can be related to microfauna activation (bacteria,

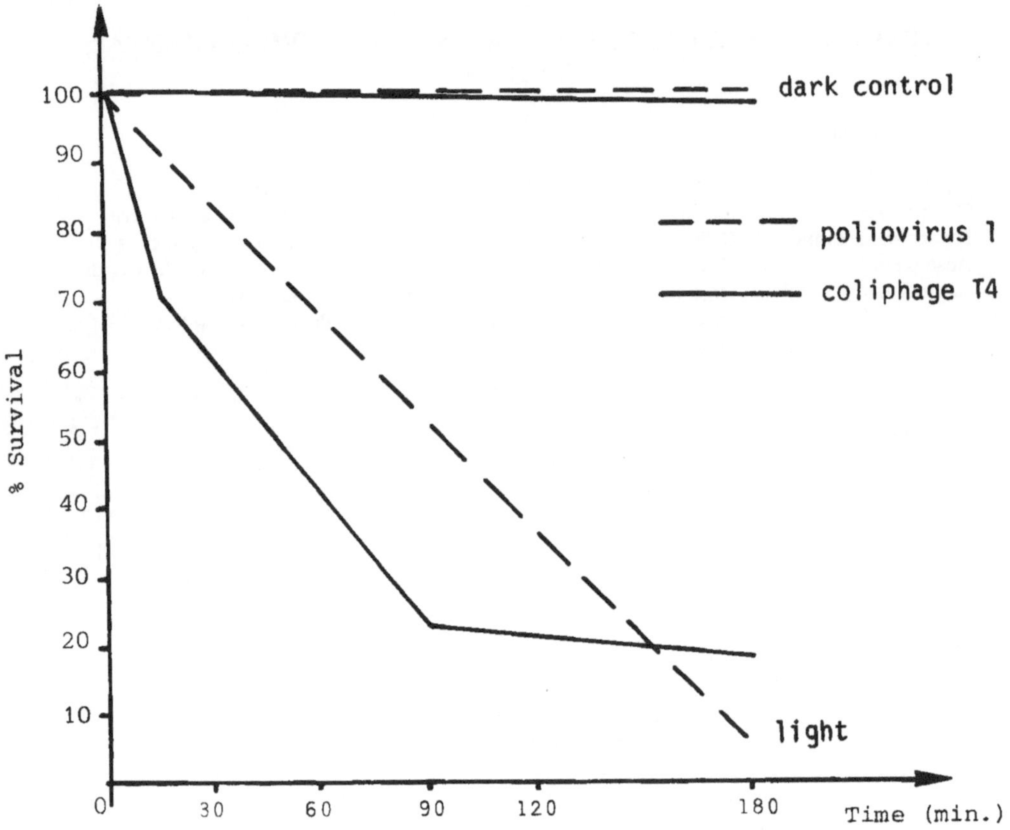

FIGURE 3. Effect of solar radiations on the survival of poliovirus type 1 in groundwater and phage T 4 in seawater. (Adapted from Bitton, G., Fraxedas, R., and Gifford, G. E., *Water Res.*, 13, 225, 1979; Attree-Pietri, C. and Breittmayer, J. P., *J. Fr. Hydrol.*, 10, 103, 1970.

algae, etc.). Malherbe and Strickland-Cholmley[122] observed better inactivation in stabilization ponds under light exposure. Fukada et al.[123] demonstrated that *Chlorella* extracts can photodynamically inactivate VSV virus. However, algal cells probably provide a shading effect against light inactivation also. Experiments with *Anabaena* sp. and *Microcystis* sp. showed that these alga protected viruses incubated under light.[117]

2. Temperature

Temperature is certainly one of the most important factors influencing virus survival, whatever the type of water. This parameter has been frequently studied.[124-125] According to Katzenelson,[104] inactivation of viruses in water is associated with the movement of water molecules. It seems that increasing temperature intensifies the bombardment of the virions by these molecules and decreases survival time. Freezing allows long survival of enteric virus strains; however Block et al.[146] observed that freezing to −26°C followed by thawing to laboratory temperature inactivated 40% of poliovirus 1 in water. At temperatures up to 0°C, inactivation mechanisms are quite complex. Gard and Maale[125] demonstrated that the effect of temperature on survival times is related to the quantity of oxygen dissolved in the water, which may inactivate viruses. Cystine added to a poliovirus 1 suspension increased the thermal stability of the virus by preventing oxidation.[129] The type of inactivation is temperature dependent and 37°C appears to be a critical level.[131-138] At temperatures between 37 and 50°C, oxidation appeared to be the most important inactivating factor.[147] At 60°C, viral protein denaturation occurs much more rapidly than does infectious RNA inactivation.[148] At 37°C, the loss of the minor viral polypeptide V P-4 has been observed for several group A coxsackieviruses incubated at 37°C in a hypotonic buffer at neutral pH.[109] Dimmock[134] demonstrated differences between heat inactivation of poliovirus at temperatures of 56°C and below 37°C. He observed that at the high temperature, rapid denaturation of viral proteins occurred while at the low temperature, crucial RNA alteration occurred. The viral particles inactivated at the low temperatures did not show the C-antigenicity characteristics of virions inactivated at 56°C. Temperature also acts indirectly on viral inactivation by influencing the activity of the aquatic flora or fauna and their products (e.g., enzymes) involved in the virus inactivation.[149,150] On the other hand, adsorption onto solids can protect microorganisms.[117,151] Working with enteric viruses in surface water and wastewater samples, Clarke et al.[132] investigated the effect of temperature on virus survival.

For poliovirus 1, an increase in temperature from 4 to 28°C caused a decrease in the time necessary for 99.9% (T 99.9) destruction of microorganisms by a factor of two in Miami river water and by a factor of six in wastewater. Poynter[135] added 200 PFU/mℓ of poliovirus 3 to Lee River water and found a T 99.9 as high as 9 weeks at 5 to 6°C. With an increase in temperature to 22°C, the T 99.9 occurred in 11 days. Won and Ross[152] observed decreasing titers of echovirus 6 in 1 log and 5 logs at 3 to 5°C and 22°C, respectively, in seawater. Schwartzbrod et al.[140] attempted to determine the survival time of poliovirus 1 and 2 in sterile, chemically defined water. The T 90 was 60 days and up to 600 days in water at +18 to 24°C and +4°C, respectively. During these experiments the reproductive capacity temperature (RCT) 40 marker did not change and the inactivation was pseudo first order suggesting a single site of inactivation (Figure 4). In the laboratory, 3 logs of poliovirus 1 added to water (salinity of 10 ‰) were inactivated within 3 weeks at 4°C and in only 5 days at 25°C. Kokina et al.[145] demonstrated that in wellwater, the T 100 of coxsackievirus B3 and poliovirus 1 increased from 66 to 113 days and from 33 to 55 days, respectively, when the temperature decreased from 20 to 10°C. Denis et al.[139,144] pointed out that water temperature plays a considerable role in the survival of infectious particles and on the stability of the hemagglutinins. The time for inactivating 99% (T 99) of poliovirus 2 in autoclaved seawater is divided by 15 when the temperature increases from 4 to 22°C (Figure 5).

3. *Hydrostatic Pressure*

As pointed out by Bitton,[103] there are no available data on enteric virus survival under high pressure following deep-sea disposal of human wastes. The structure of enteric viruses suggests that viral particles survive the effect of pressure. Field studies are difficult to carry out on the variations of parameters such as pressure, temperature, microfauna density, etc.

4. *Aggregation and Adsorption*

In water, most viral particles are in an aggregated and adsorbed state. Aggregation and adsorption generally increase the survival time of viruses. The mechanisms are not well-known, and they totally differ from those concerning the survival of adsorbed bacteria. Sörensen[153,154] showed that montmorillonite fixed enzyme proteins and protected them from degradation. During at least the early part of the decomposition period, amino acid metabolites produced in the decomposition process subsequently were stabilized for a time by the silt-clay fraction. The degree of adsorption of enteric virus surfaces was type and strain dependent.[155] Viruses associated with suspended solids are not inactivated and their infectivity for humans or cell cultures can be masked. Protection against inactivation can be solid type-dependent.[157,158] Babich and Stotzky,[158] studying the inactivation rate of *E. coli* and *Staphylococcus aureus* bacteriophages, found that montmorillonite and kaolinite were less protective than attapulgite and vermiculite. The effects of suspended solids and colloids in water can be related both to the adsorption of viral particles which are then protected and to

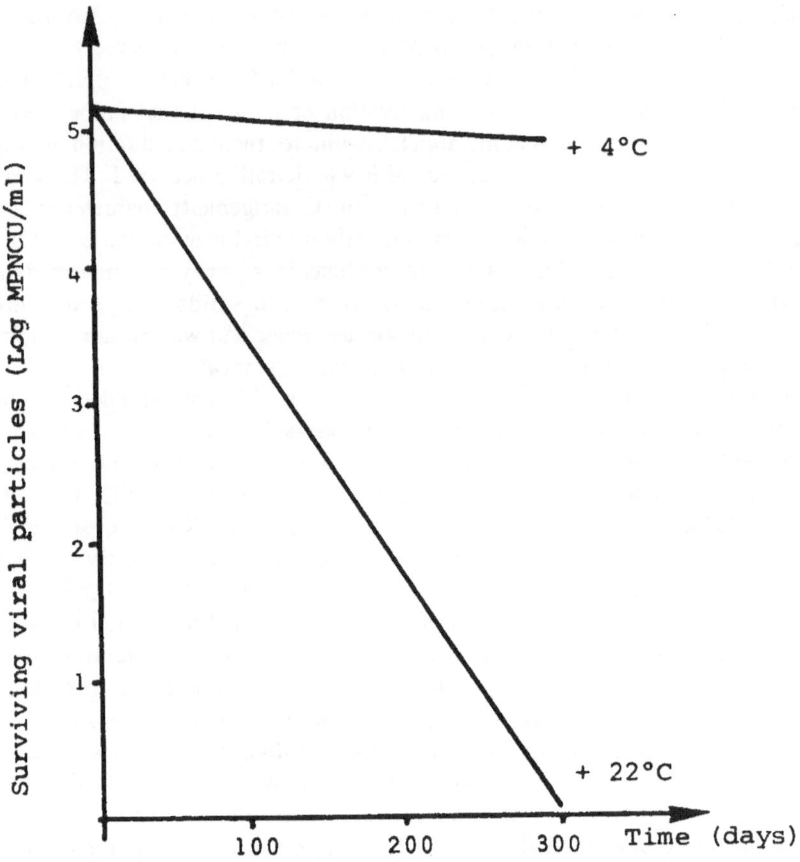

FIGURE 4. Survival of poliovirus 1 in sterile, synthetic water. (Adapted from Schwartzbrod, J., Dixneuf, P., Schwartzbrod, L., Brochet, J. C., and Foliguet, J. M., *Rev. Epidemiol. Med. Soc. Santé Publique*, 23, 235, 1975.

adsorption of viral inactivating factors.[158] According to Liew and Gerba,[151] adsorption of the virus particle may stabilize the virus against disruption by steric effects or the apparent increase in stability of the virus may be due to the adsorption of released nucleic acid by sediment particles. DeFlora et al.[63] showed, however, that the interstitial water of marine sediments was less virucidal than the water from the overlying water column. Whatever the mechanisms implicated in the protection of viruses in water, several experiments have shown such protection.[156] Mitchell and Jannasch[157] found that the inactivation of the bacteriophage ΦX174 in natural seawater was decreased by the presence of microbial cells killed by autoclaving or by UV radiation. Bitton and Mitchell[82] found that bacteriophage T7 adsorbed to montmorillonite or *E. coli K* cells in seawater was relatively resistant to inactivation. The addition of 750 μg/mℓ of montmorillonite significantly increased the bacteriophage survival time. After 60 days of incubation, the viral titer in seawater that contained the clay was more than 10^4/mℓ of seawater and less than 10/mℓ in the absence of clay. A linear relationship was observed between virus particle survival and the clay concentration in the concentration range of 0 to 200 μg/mℓ. The protective effect exerted by montmorillonite was similar to that of *E. coli* cells. Similar results have been obtained with bacteriophage T2 and kaolinite[88] and with enteric viruses (echovirus 1, coxsackieviruses B3 and A9, and poliovirus 1) associated with marine sediments.[90] According to Bitton et al.,[117] clay offers the virus some protection against photoinactivation under a mean light intensity of 0.647 cal/cm^2/min. The

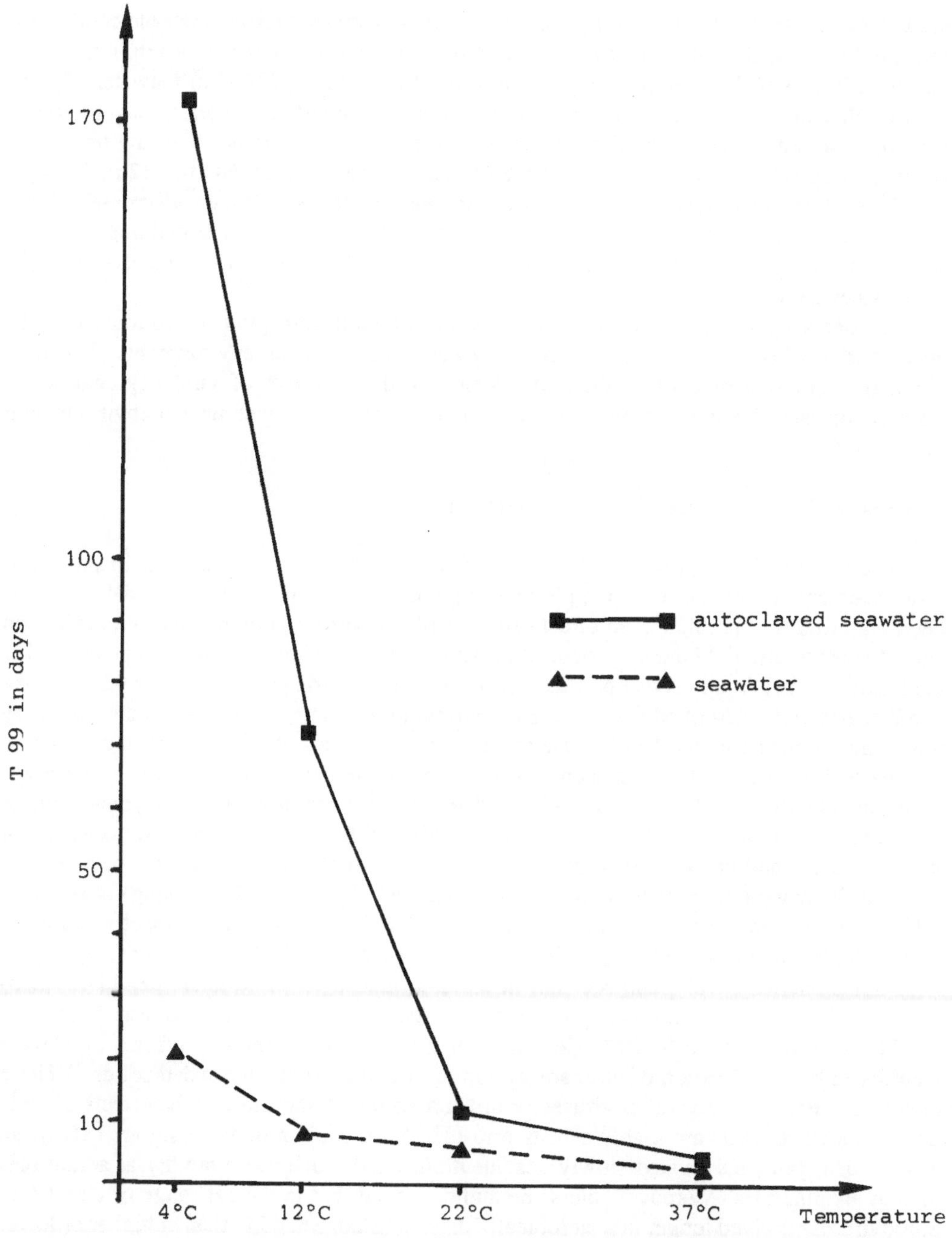

FIGURE 5. T 99 values of poliovirus 2 according to the temperature of the water. (Adapted from Denis, F., Brisou, J. F., and Dupuis, T., *C. R. Acad. Sci.*, 281, 471, 1975; *J. Fr. Hydrol.*, 8, 25, 1977.

T 90 value in the tank under light exposure was 75 min in groundwater and 163 min in groundwater amended with clay. Protection against the effect of direct light by turbid water was not the only factor in the phenomenon because virus survival was highest in the groundwater sample amended with nontronite and incubated in the dark.[117] Labelle and Gerba[89] found that the time required to inactivate 99% of a poliovirus population increased from 1.4 days in seawater along to 6.0 days for virus adsorbed to sediment at a relatively nonpolluted site. In parallel, an echovirus 1 T 90 was less than 1 day in seawater and up to 3 days in seawater that contained sediment. Moreover, they demonstrated that under sterile conditions, sediment was capable of protecting viruses from inactivating factors other than microorganisms, factors such as temperature and salinity. With coliphage MS2 in seawater, Tyler[71] showed that the protective effect of sediment was only significant after a long period of storage in seawater. They found no difference in the rate of inactivation in raw or filtered seawater after 0 to 12 days of storage (95% of inactivation at 10°C) but after 12 to 32 days, the slope of the inactivation curve for seawater that contained sediment decreased. Liew and Gerba[151] noted that the level of infectivity for viruses in seawater containing sediment was at least 1 log higher on each sampling date than that for viruses incubated at 24 or 37°C in seawater alone.

The effects upon the rate of virus inactivation of viral adsorption to colloids may be analagous to effects resulting from virus aggregation in aqueous environments. However although there is a diversified literature dealing with the effect of viral aggregation on disinfection rate,[100] almost nothing can be related to protective effect against environmental factors.

B. Chemical Factors Affecting Virus Survival

1. pH

The effect of pH on virus infectivity was first studied in water treatment. In particular lime treatment, which increases the pH to as high as 11 or 12, is rapidly virucidal. This has been reviewed by Sproul.[159] The direct effect of pH on virus can be related to its effect on capsid configuration. Mandel[160] found that capsid proteins show pH dependent reversible configurational alternations that produce virions with isoelectric points of 7.0 or 4.5. Fujioka and Ackermann[161] described four different configurational states of poliovirus controlled by ionic environment which alters the inactivation characteristics of temperature and exposure to urea and guanidine-HCl. Apparently these configurational modifications can affect enterovirus infectivity[109,160-162] and sensitivity of these viruses to proteolytic enzymes[143] under inactivating conditions. According to Salo and Cliver[143] the capsid is sensitive to RNase at less than pH 7 and becomes susceptible to chymotrypsin at pH 3. Moreover, between pH 3 and 6, hemagglutinins of echovirus 7 are destroyed. Poliovirus 2 (Dansing) is stable at 4°C for 24 hr at levels ranging from 1.5 to 10.5.[163] Poliovirus 1 (LSc) is inactivated more rapidly at pH 4 to 5 than at pH 7.2 at elevated temperatures (50°C).[164] Experiments conducted with industrial wastewater showed that in alkaline industrial wastewater (pH 11 or 13.5) and in acid industrial wastewater, poliovirus 2, coxsackievirus B3 and echovirus 7 survived for 11 to 18 days at 16 to 20°C (less than at neutral pH).[165] Indirect effects of pH are manifested by modification of viral adsorption to or elution from suspended solids.[75] These effects influence the survival of viruses. Ward and Ashley[166] showed that the effects of ionic detergents on reovirus are highly sensitive to pH. Also, sodium dodecyl sulfate (SDS), an anionic detergent, inactivated slowly near neutrality and much more rapidly at acidic pHs than at alkaline pHs. Moreover, the same authors found that in the pH range of about 6 to 8, reoviruses survived longer in anaerobically digested sludge at 45°C than at higher or lower pH levels.

2. Inorganic Ions

The ionic composition of water has been reported as a factor in virus survival. However, all inorganic ions have not been tested, and only three major groups are described: di- or bivalent cations, heavy metals, salinity. Thermostabilization by divalent cations (Mg^{2+} or Ca^{2+}) has been reported for RNA unenveloped viruses while other viral types are quickly inactivated at 50°C in 1 *M* Mg^{2+} solution.[167,168] Moreover, at 37°C divalent cations protect viruses from loss of infectivity when a 2 *M* solution of Na^+ increases the rate of inactivation. Di- or trivalent cations also increase viral adsorption onto clay. In parallel, Skujins et al.[169] showed the importance of ionic conditions in adsorption and activity of enzymes.

Heavy metals may be virucidal. Thus, Jones[170] observed that organic chelating agents could be substituted for autoclaving in reducing bactericidal activity in filtered seawater and suggested that metals were biocidal only in their uncomplexed state. Babich and Stotzky[171] showed that mixtures of anionic Hg Cl_3^-/Hg Cl_4^{2-} complexes were less toxic to *S. aureus* and *E. coli* bacteriophages than were equivalent concentrations of Hg as cationic Hg^{2+}. This can be explained both by differing affinities of heavy metals for organics and by the influence of abiotic factors, including the Cl^- concentration. A decrease in the inactivating capacity of autoclaved seawater may be related to the formation of a minor precipitate of aragonite which scavenges many trace metals from solutions.[172] Wegrzyn[173] showed that metal ions such as Al^{3+}, Ca^{2+}, Co^{2+}, Fe^{3+}, Mg^{2+}, Mn^{2+}, Pb^{2+}, and Sn^{3+} had no capacity to inactivate RNA, whereas Cu^{2+} was inactivating. L-Histidine in quantities directly proportional to the amount of Cu^{2+} present protected RNA. According to some authors,[104,171,175] salinity does not influence enteric virus inactivation. However, Cords et al.[109] inactivated coxsackievirus A13 in solutions of low ionic strength at 37°C although the virus was stable in phosphate-buffered saline. In contrast, Dimmock[134] and Salo and Cliver[143] found that viral inactivation occurred more rapidly in saline solution than in distilled water. On the other hand, Akin et al.[138] observed, depending upon salinity, from 3.0 to 4.5 logs of poliovirus 1 inactivation within 11 days. In artificial seawater, more inactivation occurred at a salinity of 10 g/kg than at 1 or 20 g/kg, suggesting an optimal salinity for poliovirus inactivation of approximately 10 g/kg (Figure 6). In the experiments conducted by Lo et al.,[142] the salinity of the water seemed to have only minor effects on virus survival, except with poliovirus 1. For this virus, an increase in salinity from 10 to 20 ‰ decreased the T 90 from 7 to 1 days at 4 or 15°C. At 25°C, the virus appeared to be slightly more stable in the low salinity water. It is reasonable to assume that salinity of seawater differs from one geographical location to another and that such differences influence virus survival. Fujioka et al.[176] found a poor correlation between the stability of poliovirus 1 in natural seawater and the water source.

3. Organic Molecules

Organic pollution has been reported by some investigators to increase survival of viruses in estuarine water or in the sediments.[89,90,124,149] Such an effect may derive from coating of viruses with organic molecules or from adsorption of inactivating factors.[170,177] Detergents are known to alter the lipid envelopes of some viruses, but more recently these compounds and ammonia have been reported to inactivate enteroviruses and rotaviruses.[178-181] Ionic detergents apparently alter the stability of nonenveloped viruses through their interactions with capsid proteins. These interactions may be described as a combination of the charged portion of a detergent molecule with oppositely charged residues of a protein molecule, thus leading to protein denaturation.[166] Treatment of rotaviruses with 0.1% SDS for 60 min at 21°C caused a reduction in infectivity by a loss of the ability to adsorb to CV1 cells.[180] However, in another study conducted with wastewater sludge, SDS and dodecyltrimethylammonium chloride stabilized enteroviruses against heat inactivation, a very different result than that found with either reoviruses or rotaviruses.[180] Ammonia causes a slight reduction in the sedimentation coefficients of poliovirus particles. These virions still attach to cells but are unable to depress host protein synthesis or bring about the replication of detectable amounts of viral ribonucleic acid. The RNA genome appears to be the component of polioviruses, which is significantly affected by ammonia. Field studies are very rare and it is difficult to assess the impact of viruses of river waters contaminated by 0.3 mg/ℓ of ammonia or 0.1 mg/ℓ of anionic detergents. Observations collected by Bagdasaryan et al.[182] showed that enteric viruses were unaffected in a stream polluted by petroleum wastes and by anion-active SPAV.

4. Dissolved Oxygen

Inactivation of viruses may be due to accelerated oxidation of virions in water,[138] although

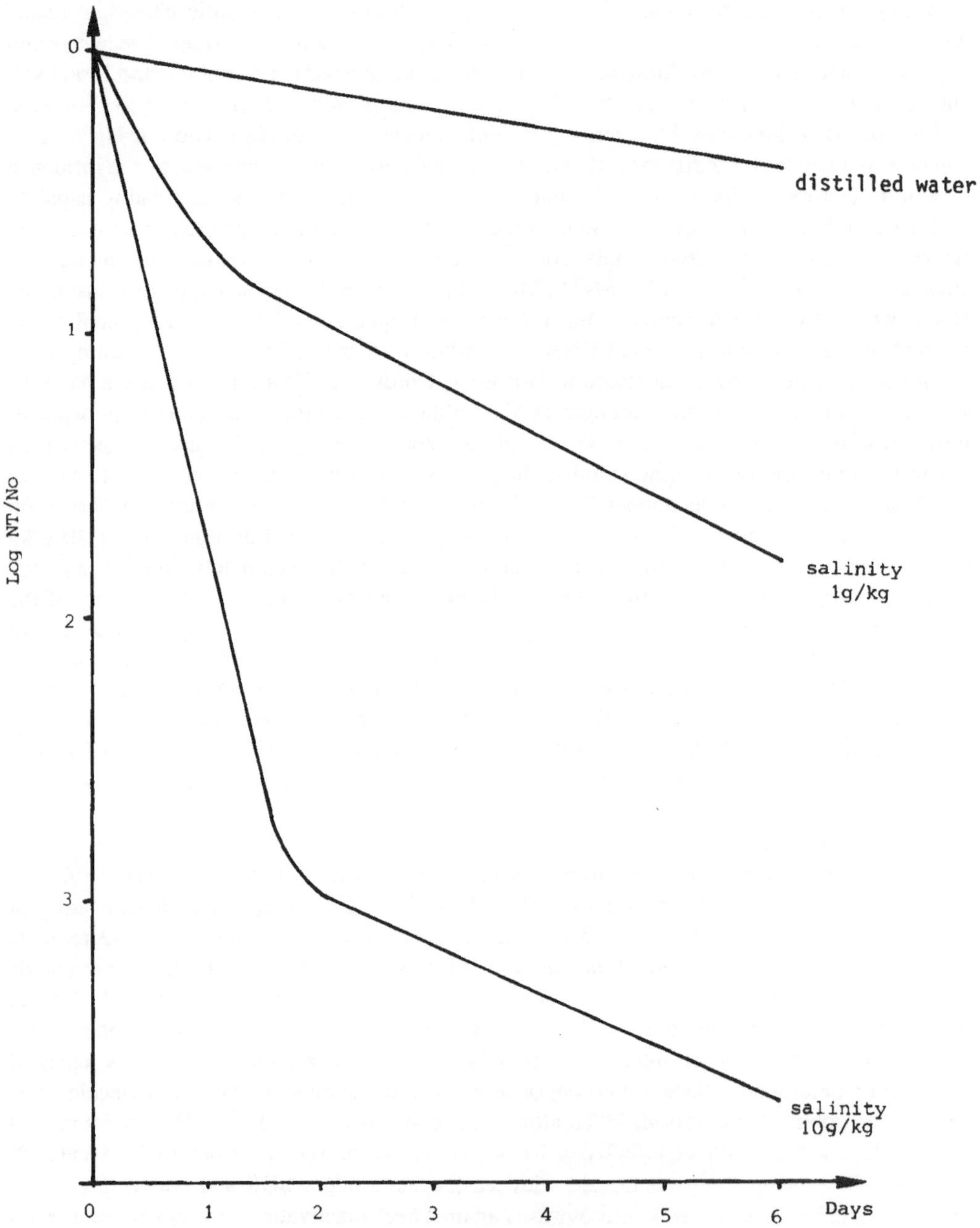

FIGURE 6. Inactivation of poliovirus 1 in distilled water and artificial seawater at salinity of 1 and 10 g/kg. (Adapted from Akin, E. W., Hill, W. F., Jr., and Clarke, N. A., presented at Int. Symp. Discharge of Sewage from Sea Outfalls, London, August 1974.

certain molecules in water stabilize virions by preventing oxidation. Zepp et al.[183] have presented evidence of rapid photochemical generation of singlet oxygen in coastal waters. However, we have no evidence of the effect of such compounds on viruses in water. Moreover, the concentration of oxygen dissolved in water both reflects and helps microflora activity.

C. Biological Factors Affecting Virus Survival

1. Virus Type

Virus type is a key parameter which influences the survival time of these microorganisms. In studying survival times, most investigators have tested only 2 to 5 strains. The relative resistance of enteric viruses is shown in Table 7; it is quite difficult to draw a coherent picture. According to Metcalf and Stiles[149] and Lo et al.,[142] echovirus 6 survives longer than poliovirus 1 in water, but according to Denis et al.[144] the reverse occurs. According to Metcalf and Stiles[149] working at 0 to 1.5°C and Kokina et al.[145] working in the dark, coxsackievirus B3 was more resistant than poliovirus 1. In contrast, Denis et al.,[144] Smith et al.,[90] Laveran et al.,[189] Hurst and Gerba,[190] working in the light, reported the reverse. Akin et al.[138] also reported these discrepancies. Temperature of incubation,[149] presence of light,[189] and sterilization of water[139] can all reverse the apparent scale in virus sensitivity. Moreover, as pointed out by Denis et al.,[139] significant differences are observed not only among serotypes but from strain to strain and even between vaccinal and wild strains.[189]

In conclusion, it is clear that viral sensitivity to inactivation depends on virus type and strain, and that unless viruses are standardized, no relevant information will be obtained in the future.

2. Biological Activities

Biological inactivation is one of the most important factors in virus survival; it is also the most controversial one. Numerous experiments have been carried out in raw, sterilized, and filtered freshwater and seawater. Leogrande and Iadolo[192] compared the survival of ten strains of poliovirus 1 in autoclaved seawater and in raw seawater and did not find any differences in survival time. Their findings were confirmed by other experiments in which autoclaving or filtering water did not significantly prevent viral loss.[138] Metcalf and Stiles[149] found inactivation rates for coxsackievirus B3, echovirus 6, and poliovirus 1 in autoclaved seawater to be the same as in natural seawater. However, most other investigators have found otherwise. Heating seawater to 45°C for 1 hr or filter sterilization affected the viral inactivating capacity of the water for poliovirus.[174,177] Matossian and Garabedian[175] showed that water boiled for 5 min or stored at room temperature for 2 weeks prior to the addition of virus required about twice as long to achieve a 3-log reduction, when compared to fresh untreated seawater. Shuval et al.[150] observed that heating, ether treatment, or filtration through a 0.22-μm porosity membrane filter eliminated marine antiviral activity (MAVA). Herrmann et al.[108] in one of the rare studies in freshwater showed that the inactivation of poliovirus 1 and coxsackievirus A9 was more rapid in natural lake water than in sterilized lake water at 21 to 23°C (Figure 7). Denis et al.[139] by autoclaving Atlantic Ocean water increased the T_{90} of poliovirus 2 from 13 to 30 days at 4°C (Figure 5). Pietri and Breittmayer[193] found that poliovirus 1 survives longer in Mediteranean seawater filtered through a 0.22 μm porosity filter or autoclaved at 115°C for 20 min than in untreated seawater. O'Brien and Newman[110] observed that freshwater of the Rio Grande lost inactivating capacity after the water was autoclaved but not after it was filtered. The rate of inactivation of coliphage MS2 in seawater decreased in water filtered or stored before inoculation.[71] According to Fujioka et al.,[176] antiviral activity was lost when Hawaiian ocean water was boiled, autoclaved, or filtered through a 0.22 or 0.45 μm membrane filter. Filtration through a 1.0 μm filter did not affect infectivity. Most investigators have found differences in virus survival in different waters. Distilled water, tapwater, and synthetic sterile seawater are less inactivating than sewage, surface water, or seawater.[89,142,149,150,194-202] In consequence of this, and because the viral inactivating factors are thermolabile and particulate bacteria and algae have been incriminated as responsible factors. Gunderson et al.[203] isolated from seawater a marine bacterium which, grown in pure culture and suspended in heat treated seawater, imparted the viral inactivating capacity of the water. Thus lytic marine bacterium was

Table 7
RELATIVE RESISTANCE OF VIRUSES IN WATER (LABORATORY EXPERIMENTS)

Type of water	Relative resistance of viruses	Ref.
Wellwater	Coxsackievirus B3 > Coxsackievirus B1 > Coxsackievirus B5	184
Riverwater	Poliovirus 1 > Echovirus 7 > Echovirus 12 > Coxsackievirus A9	132
Wastewater	Poliovirus 1 (strain Brunhilde) > Poliovirus 2 (strain M.E.F.) > Poliovirus 3 (strain Leon)	133
Seawater	Coxsackievirus A8 > Poliovirus 1	185
Seawater	Coxsackievirus B3 (strain Nancy) > Echovirus 6 (strain D'Amori) > Poliovirus 1 (strain Chat)	149
Riverwater	Poliovirus 2 = Coxsackievirus B3 = Echovirus 7	165
Riverwater	Echovirus 6 = Echovirus 11 = Echovirus 30 = Echovirus 33	186
Lakewater	Poliovirus 1 (strain Chat) > Coxsackievirus A9 (strain Bozek)	108
Seawater	Poliovirus 1 > Adenovirus 5 > Vaccinia virus	187
Seawater	Coxsackievirus B5 (strain Faulkner) > Echovirus 6 (strain D'Amori) > Poliovirus 1 (strain Mahoney)	142
Sterilized seawater	Poxvirus >Echovirus 30 > Coxsackieviruses B4, B5, B6, A9 = Poliovirus 1,2 > Echovirus 6 > Herpesvirus 1 > Coxsackievirus B3 = Poliovirus 3 > Herpesvirus 2	139, 144
Wellwater	Echovirus 7 = Coxsackievirus B3 > Poliovirus 1 (strain LSC)	145
Riverwater	Poliovirus 1 = Coxsackievirus B1 = Poliovirus 3 > Coxsackievirus A13	110
Riverwater	Poliovirus 1 > Adenovirus	188
Seawater	Poliovirus 1 > Echovirus 1 > Coxsackievirus B3 > Coxsackievirus A9	90
Seawater	Echovirus 1 (strain Farouk or lab'isolated) > Poliovirus 1 (strain LSC)	89, 151
River- and lakewater	Poliovirus 2 (strain MEF) > Poliovirus 2 (strain Sabin) > Coxsackievirus B3 > Coxsackievirus B1	189
Unpolluted freshwater	Rotavirus S A 11 > Poliovirus 1 (strain LSC) > Coxsackievirus B3 (strain Nancy) > Echovirus 7 (strain Wallace)	190

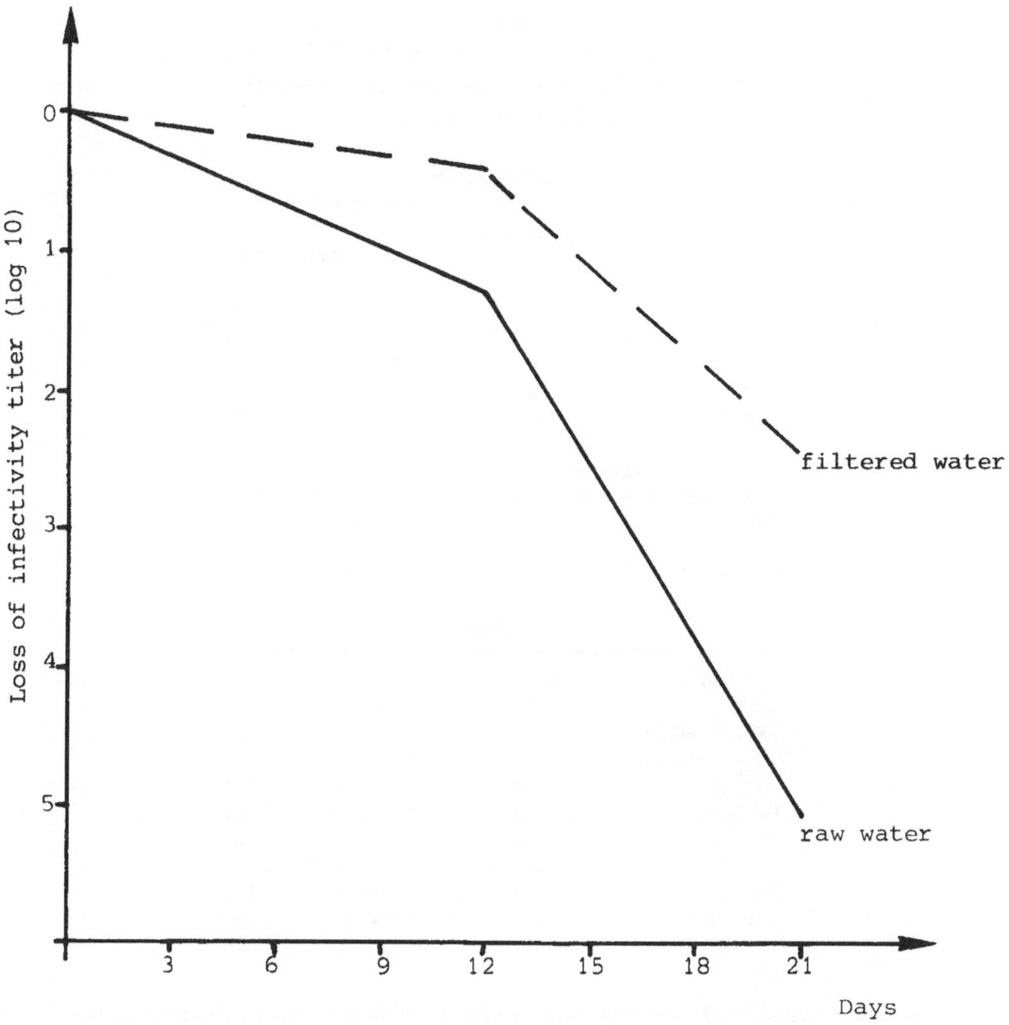

FIGURE 7. Inactivation of poliovirus 1 in filtered or raw lake water. (Adpated from Hermann, J. E., Kostenbader, K. D., Jr., and Cliver, D. O., *Appl. Microbiol.*, 28, 295, 1974.

identified as *Vibrio marinus.* This property of the bacterium was maintained only if the bacterium was subcultured at a temperature of 4 to 12°C; after a number of subcultures at 25°C, the antiviral property disappeared. Kelly et al.[204] isolated from activated sludge pure cultures of *Flavobacterium* and *Klebsiella* which rapidly inactivated polioviruses. Shuval et al.[150] showed that all seawater samples with MAVA contained viable marine bacteria at an average density of 10^3/mℓ. These bacteria grew when samples were stored at 22°C; the time at which rapid inactivation of the virus started appeared to be closely associated with the end of the logarithmic growth phase and maximal bacterial concentration in the seawater samples. The studies of Mitchell[205] on bacteriophage ΦX174 led to the conclusion that the native marine microflora are involved in viral inactivation. Katzenelson[104] recently isolated a marine *Flavobacterium* that showed an activity identical to MAVA and retained inactivating capacity after eight passages. Fujioka et al.[176] suggested that high-salt-requiring marine microorganisms were the source of antiviral activity. In their experiments, poliovirus 1 remained stable for 3 days at 24°C in water samples treated with penicillin (200 U/mℓ) and streptomycin (0.2 mg/ℓ).

Table 8
TIME IN DAYS FOR 99% INFECTIVITY LOSS OF POLIOVIRUS 1 IN DIFFERENT TYPES OF WATER

Input dose/mℓ	Temperature (°C)	Time in days	Ref.
Surface water			
10^6 PFU	19—25	15	108
10^4 PFU	23—27	3	110
10^4 PFU	16—23	6	92
10^4 PFU	20	6—8	190
10^6 PFU	18—20	25	189
Synthetic or autoclaved seawater			
2.10^4 PFU	21—26	8	142
2.10^5 DCP_{50}	18	>75	193
10^3—10^4 PFU	22	12—24	144
Seawater			
$10^{5.8}$ $TCID_{50}$	18—20	2—3	174
10^5 PFU	25	4—5	175
10^5—10^6 PFU	15	7	150
5.2×10^3 PFU	24	10	63
6.10^2 PFU	24	3—4	138
10^5 DCP_{50}	18	23	193
10^3—10^4 PFU	22	3—12	144
10^5 PFU	18—20	3—4	90
10^4—10^6 PFU	24	1—4	176
2.10^4 PFU	20	4—6	190
10^5—10^6 PFU	30—33	1—2	89

The antiviral capacity of receiving waters can be related to the multiplication of certain microorganisms, bacteria, or algae.[206-210] Proteolytic enzymes may be involved in enterovirus inactivation processes. Some viruses, including polioviruses, were very resistant to most proteolytic enzymes while other enteroviruses, particularly coxsackievirus A9, were susceptible. Enzymes from vertebrates and those of microbial origin (*Pseudomonas aeruginosa*) inactivated coxsackievirus A9. These inactivations involved the eventual release of viral RNA. The use of membrane diffusion chambers or dialysis bags allows *in situ* studies which are closer to reality.[89,108,110,142,149,189,211,212] By using membranes of different porosity, Plissier and Hugues[188] showed that the factors that inactivated poliovirus 1 and adenovirus 5 had a molecular weight of up to 1500.

The results contained in the literature on activation of viruses in water are often difficult to compare. Most investigations have been carried out at different temperatures, with different quantities of viruses and quantities of destruction have often been expressed as 90 or 99 or 100% of inactivation. Selected results summarized in Table 8 give some idea of the time in days for 99% infectivity loss of poliovirus 1 (T_{99}) in several types of waters. In surface water, irrespective of its origin (artificial stream water,[92] Rio Grande,[110] Lake Wingra,[108] Lake Pavin[189] etc.) the T_{99} of poliovirus 1 was between 3 and 25 days with an average of 11 days. In synthetic or autoclaved seawater, the T_{99} seemed very much higher (from 8 to 75 days) than in raw seawater. In these observations, whatever the source of the samples (Mediterranean Sea, Red Sea, Baltic Sea, North Sea, Atlantic Ocean, Gulf of Mexico,

Pacific Ocean) the T_{99} for poliovirus 1 was between 1 and 12 days (average of 5 days) except for the assays conducted by Pietri and Breittmayer[193] where it was 23 days.

IV. SUMMARY AND CONCLUSIONS

Via sewage, all receiving waters are contaminated by viral particles. Improved viral analysis and increasing interest in solids-associated virions explain the high frequencies of virus isolation. Moreover, quantitative investigations allow a better appreciation of the numbers of flowing viruses and a more objective perception of the risks taken by direct and indirect consumers. However, our results always underestimate the concentration of human viruses in water, on the one hand because of the large diversity of serotypes among the enteroviruses, on the other hand because of the numerous new viruses of interest (rotaviruses, coronaviruses, caliciviruses, astroviruses, etc.).

From several sources of discharge, enteric viruses are discharged to receiving waters. Dilution, aggregation, and sedimentation of viral particles occur spontaneously but are not in themselves inactivating factors. To the contrary, numerous reports demonstrate that adsorption, aggregation, and storage in sediments increase the survival time of viruses. It is really difficult to discern among all factors influencing virus survival those that are most important. However, two parameters, temperature and microbial activity, seem preponderant and favor virus inactivation. Comparisons between the studies carried out in the laboratory and in the field (directly or indirectly by using dialysis bags) are dangerous. First, viral die off curves are often nonlinear and unreproducible; for poliovirus 1, variations in T_{99} between 1 and 23 days have been reported. Second, the absence of standardization both in virus stock preparations and in receiving water characterization limits these comparisons. Even if viral inactivation appears more pronounced in seawater than in freshwater, division between the two water types is arbitrary. In any event, only the amplitude of the inactivating mechanisms may differ.

The biological inactivating factors (MAVA) are in most cases thermolabile and particulate and can be related to bacteria or their products (enzymes, etc.). Bacteria isolated from all types of water have demonstrated inactivating capacity. Other mechanisms by which viruses are inactivated in environmental waters are capsid and/or nucleic acid damage from physical, chemical, or biochemical origin.

Of course, our knowledge about virus survival is still very limited; for example, almost nothing is known of the evolution of viral pathogenicity itself in water; on the other hand we must develop a "virosensitivity test", to attempt to predict virus fate in receiving waters. The actual situation in some countries is characterized by a constant discharge and a slow dieoff of flowing particles. In the absence of adequate epidemiological information, more effective virus removal from sewage effluents is important because environmental water self-purification is limited.

REFERENCES

1. **Aymard, M. and Brigaud, M.,** Organisation de la surveillance de la circulation des entérovirus dans les eaux d'égouts, *Rev. Epidemiol. Sante Publiqué*, 25, 469, 1977.
2. **Sattar, S. A. and Westwood, J. C. N.,** Isolation of apparently wild strains of poliovirus type 1 from sewage in the Ottawa area, *Can. Med. Assoc. J.*, 116, 25, 1977.
3. **Derbyshire, J. B.,** Fate of animal viruses in effluent from liquid farm wastes, *J. Milk Food Technol.*, 39, 214, 1976.
4. **Hinshaw, V. S., Webster, R. G., and Turner, B.,** Water-borne transmission of influenza A virus?, *Intervirology*, 11, 66, 1979.

5. **Srivastava, R. N. and Lund, E.,** The stability of bovine parvovirus and its possible use as an indicator for the persistence of enteric viruses, *Water Res.*, 14, 1017, 1980.
6. **Charrier, F., Beytout, D., and Laveran, H.,** Surveillance des virus dans les eaux usées urbaines. Considérations techniques et épidémiologiques, *Med. Mal. Infect.*, 7, 55, 1977.
7. **Schmidt, N. J., Ho, H. H., Riggs, J. L., and Lennette, E. M.,** Comparative sensitivity of various cell culture systems for isolation of viruses, from wastewater and faecal samples, *Appl. Environ. Microbiol.*, 36, 480, 1978.
8. **Denis, F.,** Incidences épidémiologiques de la contamination virale des eaux, *Rev. Epidemiol. Med. Soc. Santé Publique*, 21, 273, 1973.
9. **Wellings, F. M., Mountain, C. A., and Lewis, A. L.,** Viruses in ground water, in *Proc. 2nd Natl. Conf. Individual Onsite Wastewater System,* National Sanitation Foundation, Ann Arbor, Mich., 1975, 61.
10. **Walter, R. and Rudiger, S.,** Studies on the occurrence of viruses in ground water, *Z. Res. Hyg.*, 23, 461, 1977.
11. **Vaughn, J. M., Landry, E. F., Baranosky, L. J., Beckwith, C. A., Dahl, M. C., and Dellhas, N. C.,** A survey of human virus occurrence in wastewater recharged groundwater on Long Island, *Appl. Environ. Microbiol.*, 36, 47, 1978.
12. **Marzouk, Y., Goyal, S. M., and Gerba, C. P.,** Prevalence of enteroviruses in groundwater of Israel, *Ground Water,* 17, 487, 1979.
13. **Slade, J. and Edworthy, K.,** Virology of wastewater recharge to a chalk aquifer, presented at Symp. Viruses and Wastewater Treatment, Guildford, U.K., September 1980.
14. **Leibscher, S.,** Enteroviruses in swimming pool water, *Z. Ges. Hyg.*, 16, 198, 1970.
15. **Marzouk, Y., Goyal, S. M., and Gerba, C. P.,** Relationship of viruses and indicator bacteria in water and wastewater of Israel, *Water Res.*, 14, 1585, 1980.
16. **Hugues, B., Plissier, M., Mothon, F., and Bocquet, J. P.,** Analyse virologique des eaux récréatives, *Rev. Fr. Santé Publique*, 7, 53, 1979.
17. **Foliguet, J. M., Schwartzbrod, L., and Gaudin, O. G.,** La pollution virale des eaux usées, de surface et d'alimentation: étude effectuée dans le département français de Meurthe et Moselle, *Bull. W.H.O.*, 35, 737, 1966.
18. **Hoehn, R. C., Randall, C. W., Bell, H. F., and Shaffer, T. B.,** Trihalomethanes and viruses in a water supply, *J. Environ. Eng. Div.*, EE5, 803, 1977.
19. **Akin, E. W. and Jakubowski, W.,** Viruses in finished water: the Occoquan experience, in *Proc. AWWA Water Quality Technology Conference,* American Water Works Association, Denver, Colo., 1977, 1.
20. **Drozdov, S. G. and Kasanteva, V. A.,** Detection of viruses in sewage, surface and tap waters: effectiveness of treatment in removal of viruses, in *Abstr. IVth Int. Congr. Virol.*, The Hague, The Netherlands, August 30 to September 6, 1978, 152.
21. **Nestor, I.,** Correlation between the presence of enteroviruses in drinking water and some water quality indicators, *Rev. Roum. Med. Virol.*, 30, 155, 1979.
22. **Rao, V. C., Waghmare, S. W., Lakhe, S. B.,** Detection of viruses in drinking water using magnetic iron oxide, Abstracts 80th Annual Meeting, *Ann. Soc. Microbiol.*, Miami, 1980, 216.
23. **Sekla, L., Stackiw, W., Kay, C., and Vanbuckenhart, L.,** Enteric viruses in renovated water in Manitoba, *Can. J. Microbiol.*, 26, 518, 1980.
24. **Payment, P. and Trudel, M.,** Isolation of viruses from drinking water, *Can. J. Microbiol.*, 27, 417, 1981.
25. **Fujioka, R. S. and Loh, P. C.,** Recycling of water for irrigation: persistence of enteroviruses in sewage effluent and natural waters receiving the effluent, *Water, Air, Soil Pollut.*, 9, 213, 1978.
26. **Sattar, S. A. and Westwood, J. C. N.,** Viral pollution of surface waters due to chlorinated primary effluents, *Appl. Environ. Microbiol.*, 36, 427, 1978.
27. **Buras, N.,** Concentration of enteric viruses in wastewater and effluent: a two year survey, *Water Res.*, 10, 295, 1976.
28. **Rao, V. C., Lakhe, S. B., Waghmare, S. V., and Dube, P.,** Virus removal in activated-sludge sewage treatment, *Prog. Water Technol.*, 9, 113, 1977.
29. **Selna, M. W. and Miele, R. P.,** Virus sampling in wastewater-field experiences, *J. Environ. Eng. Div.*, EE4, 693, 1977.
30. **Kott, Y.,** Ozonized secondary effluents for agricultural use, Report 013-714, National Council for Research and Development, Haifa, Israel, 1978.
31. **Steinmann, J.,** Virus reduction after chemical flocculation in a treatment plant, in *Abstr. IVth Int. Congr. Virol.*, The Netherlands, August 30 to September 6, 1978, 155.
32. **Dahling, D. R. and Safferman, R. S.,** Survival of enteric viruses under natural conditions in a subarctic river, *Appl. Environ. Microbiol.*, 38, 1103, 1979.
33. **Dryden, F. D., Chen, C. L., and Selna, N. W.,** Virus removal in advanced wastewater treatment systems, *J. Water Pollut. Control Fed.*, 51, 2098, 1979.

34. **Fluegge, R. A., Metcalf, T. G., and Wallis, C.**, Virus inactivation in wastewater effluents by chlorine, ozone, and ultraviolet light, USEPA Report, 600/9-79-018, *Proc. Natl. Symp. Prog. Wastewater Disinfect. Technol.*, 1979, 223.
35. **Hugues, B., Bougis, M. A., Plissier, M., André, M., Laurent, D., and Pagliardini, A.**, Evaluation de la contamination virale marine après rejet d'un effluent par un émissaire en mer, *Zentralbl. Bakteriol., Parasitenkd., Infektionskr., Hyg., Abt. 1: Orig., Reihe B*, 169, 253, 1979.
36. **Hugues, B., Plissier, M., André, M., Pagliardini, A., and Laurent, D.**, Evaluation de la charge virale dans les eaux d'une station d'épuration biologique au moyen de deux méthodes de concentration de virus par adsorption — élution sur poudre de verre ou sur microfibre de verre, *Water Res.*, 13, 1117, 1979.
37. **Hugues, B., Plissier, M., Pagliardini, A., Laurent, D., and André, M.**, Application de la méthode d'adsorption — élution sur filtre en fibre de verre pour déterminer la charge virale à l'entrée et à la sortie d'une station d'épuration biologique, *J. Fr. Hydrol.*, 10, 21, 1979.
38. **Payment, P., Larose, Y., and Trudel, M.**, Poliovirus and other enteroviruses in urban sewage from Laval (Canada): presence of non-vaccinal strains of poliovirus, *Can. J. Microbiol.*, 25, 1305, 1979.
39. **Schwartzbrod, L., Lucena, F., and Finance, C.**, Etude quantitative de la population virale dans l'affluent et l'effluent d'une station d'épuration d'eaux résiduaires, *J. Fr. Hydrol.*, 10, 7, 1979.
40. **Gerba, C. P.**, Virus survival in wastewater treatment, presented at Symp. Viruses and Wastewater Treatment, Guildford, U.K., September 1980.
41. **Grabow, W. O. K.**, Comparison of primary vervet kidney cells with the BGM cell line for the enumeration of enteric viruses in water by means of a tube dilution technique, presented at Symp. Viruses and Wastewater Treatment, Guildford, U.K., September 1980.
42. **Joret, J. C., Block, J. C., Lucena-Gutierrez, F., Schwartzbrod, L., Hugues, B., and Plissier, M.**, Virus concentration from secondary wastewater: comparative study between epoxy fiberglass and glass powder adsorbents, *Eur. J. Appl. Microbiol. Biotechnol.*, 10, 245, 1980.
43. **Ward, R. L. and Ashley, C. S.**, Mode of initiation of cell infection with sludge-associated poliovirus, *Appl. Environ. Microbiol.*, 38, 329, 1979.
44. **Berg, G.**, Removal of viruses from sewage, effluents, and waters. 2. Present and future trends, *Bull. W.H.O.*, 49, 461, 1973.
45. **Lund, E., Hedström, C. E., and Jantzen, N.**, Occurrence of enteric viruses in wastewater after activated sludge treatment, *J. Water Pollut. Control Fed.*, 41, 169, 1969.
46. **Malakhova, T. S.**, Presence of enteroviruses in sewage after purification, *Microbiol. Zh.*, 39, 228, 1977.
47. **Joret, J. C., Block, J. C., Hartemann, P., and Foliguet, J. M.**, Virus recovery from wastewater using microfiberglass filters, presented at Symp. Viruses and Wastewater Treatment, Guildford, U.K., September 1980.
48. **Berg, G. and Metcalf, T. G.**, Indicators of viruses in waters, in *Indicators of Viruses in Water and Food*, Berg, G., Ed., Ann Arbor Science, Ann Arbor, Mich., 1978, 267.
49. **Rolland, D.**, Appréciation des flux de virus à l'entrée et à la sortie d'une station d'epuration, M. thesis, Université de Metz, France, 1981.
50. **Berg, G.**, Reassessment of the virus problem in sewage and in surface and renovated waters, in *Progress in Water Technology*, Vol. 3, Jenkins, S. H., Ed., Pergamon Press, Elmsford, N.Y., 1973, 87.
51. **Sarrette, B., Danglot, C., and Vilagines, R.**, Nouvelle méthode permettant la détermination quantitative des virus présents dans les eaux de surface, *C.R. Acad. Sci.*, 285, 1359, 1977.
52. **Slade, J. S.**, Enteroviruses in partially purified water, *J. Water Eng. Sci.*, 31, 219, 1977.
53. **Block, J. C., Joret, J. C., Chauvin, A., and Stoerkel, D.**, Etude virologique des eaux de surface dans le bassin de la Haute Moselle, Final Report, Agence Financière de Bassin Rhin-Meuse, Metz, 1979.
54. **Block, J. C., Joret, J. C., Morlot, M., and Foliguet, J. M.**, Recherche des entérovirus dans les eaux superficielles par adsorption — élution sur microfibre de verre, *Tech. Sci. Munic. — L'eau*, 73, 181, 1978.
55. **Schwartzbrod, L.**, Contamination virale des eaux de rivière et cotières de la région de Barcelone, presented at Symp. Marine Microbiol., Marseille, France, October 1979.
56. **Vaughn, J. M., Landry, E. F., Thomas, M. Z., Vicale, T. J., and Penello, W. F.**, Survey of human enterovirus occurrence in fresh and marine surface waters on long island, *Appl. Environ. Microbiol.*, 38, 290, 1979.
57. **Vilagines, P., Sarrette, B., Danglot, C., and Vilagines, R.**, Détection d'entérovirus par concentration sur poudre de verre en lit fluidisé à partir d'échantillons de 500 litres d'eaux superficielles, *Bull. Acad. Natl. Med., Paris*, 163, 668, 1979.
58. **Dobberkau, H. J.**, Ten years exploitation of the aluminium sulfate flocculation method for quantitative virus determination in the surface waters of the G.D.R., presented at Symp. Viruses and Wastewater Treatment, Guildford, U.K., September 1980.
59. **Edwards E. and Wyn-Jones, P.**, Incidence of enteroviruses associated with wastewater treatment plants on the river Wear, presented at Symp. Viruses and Wastewater Treatment, Guildford, U.K., September 1980.

60. **Morris, R. and Waite, W. M.,** Evaluation of procedures for recovery of viruses from water, II. Detection systems, *Water Res.*, 14, 795, 1980.
61. **Crovari, P., De Flora, S., and Badolati, S.,** The virological monitoring of water, II. Sea water, *Bull. Inst. Sieroter.*, 53, 525, 1974.
62. **Metcalf, T. G., Wallis, C., and Melnick, J. L.,** Virus enumeration and public health assessments in polluted surface water contributing to transmission of virus in nature, in *Virus Survival in Water and Wastewater,* Malina, J. F. and Sagik, B. P., Eds., University of Austin, Texas, 1974, 57.
63. **De Flora, S., De Renzi, G. P., and Badolati, B.,** Detection of animal viruses in coastal sea water and sediments, *Appl. Microbiol.*, 30, 472, 1975.
64. **Farrah, S. R., Goyal, S. M., Gerba, C. P., Wallis, C., and Melnick, J. L.,** Concentration of enteroviruses from estuarine water, *Appl. Environ. Microbiol.*, 33, 1192, 1977.
65. **Gerba, C. P., Smith, E. M., and Melnick, J. L.,** Development of a quantitative method for detecting enteroviruses in estuarine sediment, *Appl. Environ. Microbiol.*, 34, 158, 1977.
66. **Katzenelson, E.,** Concentration and identification of viruses from seawater, *Rev. Int. Oceanogr. Med.*, 48, 9, 1977.
67. **Steinmann, J.,** Detection of viruses in water on the Baltic sea, *Zentralbl. Bakteriol., Parasitenkd., Infektionskr., Hyg.*, 164, 492, 1977.
68. **Edmond, T. D., Schaiberger, G. E., and Gerba, C. P.,** Detection of enteroviruses near deep marine sewage outfalls, *Mar. Pollut. Bull.*, 9, 246, 1978.
69. **Goyal, S. M., Gerba, C. P., and Melnick, J. L.,** Human enteroviruses in oysters and their overlying waters, *Appl. Environ. Microbiol.*, 37, 572, 1979.
70. **Shuval, H. I., Vasl, R., Katzenelson, E., and Fattal, B.,** Survival of enteroviruses and bacterial indicator organisms in the sea, presented at Symp. Viruses and Wastewater Treatment, Guildford, U.K., September 1980.
71. **Tyler, J. M.,** Occurrence and survival of viruses in seawater, in *Industrialised Embayments and their Environmental Problems: A Case Study of Swansea Bay,* Collins, M. B., et al., Eds., Pergamon Press, Oxford, 1980, 343.
72. **Senault, R., Foliguet, J. M., Laurent, R., and Martin, J. M.,** Etude des effluents de fosses septiques comme facteurs de pollution du milieu extérieur par les virus fécaux, *Rev. Hyg. Med. Soc.*, 13, 283, 1965.
73. **Shuval, H. I.,** Detection and control of enteroviruses in the water environment, in *Developments in Water Quality Research,* Shuval, H. I., Ed., Ann Arbor-Humphrey Science, Ann Arbor, Mich., 1970, 47.
74. **Simkova, A. and Wallnerova, Z.,** Isolations of Coxsackieviruses from Danube river water, *Acta Virol.*, 17, 363, 1973.
75. **Bitton, G.,** Adsorption of viruses to surfaces: technological and ecological implications, in *Adsorption of Microorganisms to Surfaces,* Bitton, G. and Marshall, K. C., Eds., John Wiley & Sons, New York, 1980, 332.
76. **Block, J. C., Collin, J. F., Joret, J. C., Rolland, D., and Foliguet, J. M.,** Elimination des entérovirus au cours du traitement par boues activées des eaux usées urbaines, *Tech. Sci. Munic., — l'Eau,* 74, 201, 1979.
77. **Cliver, D. O.,** Virus association with wastewater solids, *Environ. Lett.*, 10, 215, 1975.
78. **Lund, E.,** The effect of pretreatments on the virus contents of sewage samples, *Water Res.*, 7, 873, 1973.
79. **Gerba, C. P., Stagg, C. M., and Abadie, M. G.,** Characterization of sewage solid-associated viruses and behavior in natural waters, *Water Res.*, 12, 805, 1978.
80. **Wellings, F. M., Lewis, A. L., and Mountain, C. W.,** Demonstration of solids-associated virus in wastewater and sludge, *Appl. Environ. Microbiol.*, 31, 354, 1976.
81. **Bartell, P., Pierzchala, W., and Tint, H.,** The adsorption of enteroviruses by activated attapulgite, *J. Ann. Pharm. Assoc.*, 49, 1, 1960.
82. **Bitton, G. and Mitchell, R.,** Effect of colloids on the survival of bacteriophages in seawater, *Water Res.*, 8, 227, 1974.
83. **Gerba, C. P., Goyal, S. M., Hurst, C., and Labelle, R. L.,** Type and strain dependence of enterovirus adsorption to activated sludge, soils and estuarine sediments, *Water Res.*, 14, 1197, 1980.
84. **Goyal, S. M., Gerba, C. P., and Melnick, J. L.,** Prevalence of human enteric viruses in coastal canal communities, *J. Water Pollut. Control Fed.*, 50, 2247, 1978.
85. **Labelle, R. L., Gerba, C. P., Goyal, S. M., Melnick, J. L., Cech, I., and Bogdan, G.,** Relationship between environmental factors, bacterial indicators, and the occurrence of enteric viruses in estuarine sediments, *Appl. Environ. Microbiol.*, 39, 588, 1980.
86. **Carlson, G. F., Woodward, F. E., Wentworth, D. F., and Sproul, O. J.,** Virus inactivation on clay particles in natural waters, *J. Water Pollut. Control Fed.*, 40, R 89, 1968.
87. **Moore, B. E., Sagik, B. P., and Malina, J. F., Jr.,** Viral association with suspended solids, *Water Res.*, 9, 197, 1975.
88. **Gerba, C. P. and Schaiberger, G. E.,** The effect of particulates on virus survival in seawater, *J. Water Pollut. Control Fed.*, 9, 567, 1975.

89. **La Belle, R. L. and Gerba, C. P.**, Influence of estuarine sediment on virus survival under field conditions, *Appl. Environ. Microbiol.*, 39, 749, 1980.
90. **Smith, E. M., Gerba, C. P., and Melnick, J. L.**, Role of sediment in the persistence of enteroviruses in the estuarine environment, *Appl. Environ. Microbiol.*, 35, 685, 1978.
91. **Matson, E. A., Hornor, S. G., and Buck, J. D.**, Pollution indicators and other microorganisms in river sediment, *J. Water Pollut. Control Fed.*, 50, 13, 1978.
92. **Cubbage, C. P., Gannon, J. J., Cochran, K. W., and Williams, G. W.**, Loss of infectivity of poliovirus 1 in river water under simulated field conditions, *Water Res.*, 13, 1091, 1979.
93. **Floyd, R. and Sharp, D. G.**, Aggregation of poliovirus and reovirus by dilution in water, *Appl. Environ. Microbiol.*, 33, 159, 1977.
94. **Gerba, C. P. and Schaiberger, G. E.**, Aggregation as a factor in loss of viral titer in seawater, *Water Res.*, 9, 567, 1975.
95. **Galasso, G. L. and Sharp, D. G.**, Virus particle aggregation and the plaque forming units, *J. Immunol.*, 88, 339, 1962.
96. **Floyd, R., Johnson, J. D., and Sharp, D. G.**, Inactivation by bromine of single poliovirus particles in water, *Appl. Environ. Microbiol.*, 31, 298, 1976.
97. **Wei, J. H. and Chang, S. L.**, A multipoisson distribution model for treating disinfection data, in *Disinfection of Water and Wastewater*, Johnson, J. D., Ann Arbor Science, Ann Arbor, Mich., 1975, 11.
98. **Shuval, H. I. and Katzenelson, E.**, Detection and inactivation of enteric viruses in wastewater, Report 600/2-77-095, Environmental Protection Series, U.S. Environmental Protection Agency, Washington, D.C., 1977.
99. **Floyd, R. and Sharp, D. G.**, Viral aggregation: effects of salts on the aggregation of poliovirus and reovirus at low pH, *Appl. Environ. Microbiol.*, 35, 1984, 1978.
100. **Young, D. C. and Sharp, D. G.**, Poliovirus aggregates and their survival in water, *Appl. Environ. Microbiol.*, 33, 168, 1977.
101. **Foliguet, J. M., Block, J. C., Hartemann, P., and Joret, J. C.**, Recent developments of virological water research in France, in *Abstr. IVth Int. Congr. Virol.*, The Hague, The Netherlands, August 30 to September 6, 1978, 151.
102. **Akin, E. W., Benton, W. H., and Hill, W. F., Jr.**, Enteric viruses in ground and surface waters: a review of their occurrence and survival, Proc. 13th Water Quality Conf., Snoeyink, V. L., Ed., *Univ. Ill. Bull.*, 69, 59, 1971.
103. **Bitton, G.**, Survival of enteric viruses, in *Water Pollution Microbiology*, Vol. 2, Mitchell, R., Ed., John Wiley & Sons, New York, 1978, 273.
104. **Katzenelson, E.**, Survival of viruses, in *Indicators of Viruses in Water and Food*, Berg, G., Ed., Ann Arbor Science, Ann Arbor, Mich., 1978, 39.
105. **Kapuscinski, R. B. and Mitchell, R.**, Processes controlling virus inactivation in coastal waters, *Water Res.*, 14, 363, 1980.
106. **Sattar, S. A.**, Virus survival in receiving waters, presented at Symp. Viruses and Wastewater Treatment, Guilford, U.K., September 1980.
107. **Chang, P. N., Liu, O. C., Miller, L. T., and Li, S. M.**, Multiplication of human enteroviruses in Northern Quahogs, *Proc. Soc. Exp. Biol. Med.*, 136, 1380, 1971.
108. **Herrmann, J. E., Kostenbader, K. D., Jr., and Cliver, D. O.**, Persistance of enteroviruses in lake water, *Appl. Microbiol.*, 28, 895, 1974.
109. **Cords, C. E., James, C. G., and Mc Laren, L. C.**, Alteration of capsid proteins of coxsackievirus A 13 by low ionic concentrations, *J. Virol.*, 15, 244, 1975.
110. **O'Brien, R. T. and Newman, J. S.**, Inactivation of polioviruses and coxsackieviruses in surface water, *Appl. Environ. Microbiol.*, 33, 334, 1977.
111. **O'Brien, R. T. and Dacus, J. M.**, Viral changes associated with inactivation in freshwater and marine environments, *Abstr. Annu. Meet. 78th Am. Soc. Microbiol.*, Las Vegas, Nev., 1978, 167.
112. **Dawe, L. L. and Penrose, W. R.**, "Bactericidal" property of seawater: Death or debilitation?, *Appl. Environ. Microbiol.*, 35, 829, 1978.
113. **Young, D. C. and Sharp, D. G.**, Partial reactivation of chlorine-treated echovirus, *Appl. Environ. Microbiol.*, 37, 766, 1979.
114. **Attree-Pietri, C. and Breittmayer, J. P.**, Etude comparée de germes-tests de contamination fécale et de bactériophages en eau de mer: action du rayonnement solaire, *J. Fr. Hydrol.*, 10, 103, 1970.
115. **Hill, W. F., Hamblet, F. E., Benton, W. H., and Akin, E. W.**, Ultraviolet devitalization of eight selected enteric viruses in estuarine water, *Appl. Microbiol.*, 19, 805, 1970.
116. **Hill, W. F., Akin, E. W., Benton, W. H., and Hamblet, F. E.**, Viral disinfection of estuarine water by U.V., in *Proc. Nat. Specialty Conf. Disinfect.*, American Society of Council Engineers, July 1970, 223.
117. **Bitton, G., Fraxedas, R., and Gifford, G. E.**, Effect of solar radiation on poliovirus: preliminary experiments, *Water Res.*, 13, 225, 1979.

118. **Agres, P. A.,** Coliphages in sewage and the marine environment, in *Aquatic Microbiology,* Skinner, F. A. and Shewan, J. M., Eds., The Society for Applied Bacteriology Symposium Series No. 6, Academic Press, London, 1977, 275.
119. **Zafiriou, O. C.,** Marine organic photochemistry previewed, *Mar. Chem.,* 5, 497, 1977.
120. **Wallis, C. Gerba, C. P., and Melnick, J. L.,** Photodynamic inactivation of viruses and bacteria in sewage effluents, in *Viruses in Water,* Berg, G., et al., Eds., Interdisciplinary Book, Pamphlets and Periodicals for the Professionnal and the Layman, Washington, D.C., 1979, 180.
121. **Berry, S. M. and Noton, B. G.,** Survival of bacteriophages in seawater, *Water Res.,* 10, 323, 1976.
122. **Malherbe, H. H. and Strickland-Cholmley, M.,** Survival of virus in the presence of algae, in *Transmission of Viruses by the Water Route,* Berg, G., Ed., Wiley-Interscience, New York, 1967, 449.
123. **Fukada, T., Hoshido, H., Endo, M., and Shirota, M.,** Photodynamic antiviral substance extracted from *Chlorella* cells, *Appl. Microbiol.,* 16, 1809, 1968.
124. **Clarke, N. A., Stevenson, R. E., and Kabler, P. W.,** Survival of Coxsackie virus in water and sewage, *J. Am. Water Works Assoc.,* 48, 677, 1956.
125. **Gard, S. and Maale, O.,** Inactivation of viruses, in *The Viruses,* Vol. 1, General Virology, Burnet, F. M. and Stanley, W. H., Eds., Academic Press, New York, 1959.
126. **Norman, A. and Veomett, R. C.,** Heat inactivation of poliovirus ribonucleic acid, *Virology,* 12, 136, 1960.
127. **Cioglia, L. and Loddo, B.,** The process of self-purification in marine environment, III. Resistance of some enteroviruses, *Nuovi Ann. Ig. Microbiol.,* 13, 11, 1962.
128. **Darbon, A., Girier, L., and Hervieu, P.,** Contribution à l'étude de la résistance du virus poliomyélitique type 1 dans les eaux, *Rev. Corps Santé Armées,* 3, 506, 1962.
129. **Pohjanpelto, P.,** Oxidation in thermoinactivation of poliovirus, *Virology,* 16, 92, 1962.
130. **Gevaudan, P. and Charrel, S.,** Survie du virus Coxsackie dans différentes eaux, *Rev. Pathol. Gen. Physiol. Clin.,* 63, 747, 1963.
131. **Lund, E.,** Oxidative inactivation of poliovirus at temperature, *Arch. Ges. Virusforsch.,* 13, 375, 1963.
132. **Clarke, N. A., Berg, G., Kabler, P. W., and Chang, S. L.,** Human enteric viruses in water: source, survival and removability, *Proc. Int. Conf. Water Pollut. Res.,* London, 1962, Pergamon Press, Elmsford, N.Y., 1964, 523.
133. **Albano, A., De Donato, S., and Salvaggio, L.,** Sulk soprovvivenza del virus poliomelitico in liquame di fogna e in acqua potabile, *Ann. Sclavo,* 7, 857, 1965.
134. **Dimmock, N. J.,** Differences between the thermal inactivation of picornaviruses at "high" and "low" temperatures, *Virology,* 31, 338, 1967.
135. **Poynter, S. F. B.,** The problem of viruses in water, *Water Treat. Exam.,* 17, 187, 1968.
136. **Bagdasaryan, G. A. and Abieva, R. M.,** Survival of enteroviruses and adenoviruses in water, *Gig. Sanit.,* 36, 10, 1971.
137. **Simkova, A. and Wallneronce, Z.,** Survival of small amounts of coxsackie A 4 virus in Danube river water under laboratory conditions, *Acta Virol.,* 17, 505, 1973.
138. **Akin, E. W., Hill, W. F., Jr., and Clarke, N. A.,** Mortality of enteric viruses in marine and other waters, presented at Int. Symp. Discharge of Sewage from Sea Outfalls, London, August 1974.
139. **Denis, F., Brisou, J. F., and Dupuis, T.,** Recherche sur l'inactivation des entérovirus dans les eaux de mer, *C.R. Acad. Sci.,* 281, 471, 1975.
140. **Schwartzbrod, J., Dixneuf, P., Schwartzbrod, L., Brochet, J. C., and Foliguet, J. M.,** Etude du comportement du virus poliomyélitique de type 1 dans différents milieux hydriques, *Rev. Epidemiol., Med. Soc. Santé Publ.,* 23, 235, 1975.
141. **Niemi, M.,** Survival of *Escherichia coli* phage T 7 in different water types, *Water Res.,* 10, 751, 1976.
142. **Lo, S., Gilbert, J., and Hetrick, F.,** Stability of human enteroviruses in estuarine and marine waters, *Appl. Environ. Microbiol.,* 32, 245, 1976.
143. **Salo, R. J. and Cliver, D. O.,** Effects of acid pH, salts, and temperature on the infectivity and physical integrity of enteroviruses, *Arch. Virol.,* 52, 269, 1976.
144. **Denis, F. A., Dupuis, T., Denis, N. A., and Brisou, J. F.,** Survie dans l'eau de mer de 20 souches de virus à ADN et ARN, *J. Fr. Hydrol.,* 8, 25, 1977.
145. **Kokina, A. G., Lukashewich, N. A., and Nouitskoya, N. V.,** Effect of chemical and microbiological quality of groundwater on the survival of enteroviruses and enterobacteria, *Gig. Sanit.,* 0(5), 111, 1977.
146. **Block, J. C., Joret, J. C., Morlot, M., and Foliguet, J. M.,** Persistance of poliovirus 1 during the freezing-thawing of hydroxide sludges from a water treatment plant under laboratory conditions, *Eur. J. Appl. Microbiol. Biotechnol.,* 5, 331, 1978.
147. **Lund, E.,** Inactivation of viruses, in *Progress in Water Technology,* Vol. 3, Water Quality: Management and Pollution Control Problems, Jenkins, S. M., Ed., Pergamon Press, Oxford, 1973, 95.
148. **Norman, A. and Veomett, R. C.,** Heat inactivation of poliovirus ribonucleic acid, *Virology,* 12, 136, 1960.

149. **Metcalf, T. G. and Stiles, W. C.**, Survival of enteric viruses in estuary waters and shellfish, in *Transmission of Viruses by the Water Route*, Berg, G., Ed., Interscience, London, 1967.
150. **Shuval, H. I., Tompson, A., Fattal, B., Cymbalista, S., and Wiener, Y.**, Natural virus inactivation processes in seawater, in *Proc. Natl. Specialty Conf. Disinfect.*, American Society of Chemical Engineers, New York, 1971, 429.
151. **Liew, P. F. and Gerba, C. P.**, Thermostabilization of enteroviruses by estuarine sediment, *Appl. Environ. Microbiol.*, 40, 305, 1980.
152. **Won, W. D. and Ross, M.**, Persistance of bacteria and virus in seawater, *J. Environ. Eng. Div. Am. Soc. Civ. Eng.*, 99, 205, 1973.
153. **Sörensen, L. H.**, Fixation of enzyme protein in soil by clay mineral montmorillonite, *Experientia*, 25, 20, 1969.
154. **Sörensen, L. H.**, The influence of clay on the rate of decay of amino acid metabolites synthesized in soil during decomposition of cellulose, *Soil Biol. Biochem.*, 7, 171, 1975.
155. **Schaub, S. A. and Sagik, B. P.**, Association of enteroviruses with natural and artificially introduced colloidal solids in water and infectivity of solids-associated virions, *Appl. Environ. Microbiol.*, 30, 212, 1975.
156. **Duboise, S. M., Moore, B. E., Sorber, C. A., and Sagik, B. P.**, Viruses in soil systems, *C.R.C. Crit. Rev. Microbiol.*, 11, 245, 1979.
157. **Mitchell, R. and Jannasch, H. W.**, Processes controlling virus inactivation in seawater, *Environ. Sci. Technol.*, 3, 941, 1969.
158. **Babich, H. and Stotsky, G.**, Reductions in inactivation rates of bacteriophages by clay minerals in lake water, *Water Res.*, 14, 185, 1980.
159. **Sproul, O. J.**, Critical review of virus removal by coagulation processes and pH modifications, EPA-600/2-80-004, U.S. Environmental Protection Agency, Cincinnati, Ohio, 1980.
160. **Mandel, B.**, Characterization of type 1 poliovirus by electrophoretic analysis, *Virology*, 44, 554, 1971.
161. **Fujioka, R. S. and Ackermann, W. W.**, The inhibitory effects of $MgCl_2$ on the inactivation kinetics of poliovirus by urea, *Proc. Soc. Exp. Biol. Med.*, 148, 1063, 1975.
162. **Crowell, R. L. and Philipson, L.**, Specific alterations of Coxsackievirus B3 eluted from Hela cells, *J. Virol.*, 8, 509, 1971.
163. **Bachrach, H. L. and Schwerdt, C. E.**, Purication studies on Lansing poliomyelitis virus: pH stability, CNS extraction and butanol purification experiments, *J. Immunol.*, 69, 551, 1952.
164. **Wallis, C. and Melnick, J. L.**, Effect of organic acids on poliovirus at 50°C, *Proc. Soc. Exp. Biol. Med.*, 111, 305, 1962.
165. **Pozhar, V. N.**, Viability of enteroviruses in alkaline and acid effluents of the Gaysk Ore dressing combine, *Gig. Sanit.*, 38, 101, 1973.
166. **Ward, R. L. and Ashley, C. S.**, pH modification of the effect of detergents on the stability of enteric viruses, *Appl. Environ. Microbiol.*, 38, 314, 1979.
167. **Wallis, C. and Melnick, J. L.**, Cationic stabilization — a new property of enteroviruses, *Virology*, 16, 504, 1962.
168. **Yana Moto, S. and Shingu, M.**, The effect of cations (Mg^{++}, Ca^{++}) on enteroviruses, *Kurumo Med. J.*, 9, 126, 1962.
169. **Skujins, J., Pukite, A., and Mc Laren, A. D.**, Adsorption and activity of chitinase on kaolinite, *Soil Biol. Biochem.*, 6, 179, 1974.
170. **Jones, G. E.**, Effect of chelating agents on the growth of *E. coli* in seawater, *J. Bacteriol.*, 87, 483, 1964.
171. **Babich, H. and Stotzky, G.**, Differential toxicities of mercury to bacteria and bacteriophages in seawater and in lake water, *Can J. Microbiol.*, 25, 1252, 1979.
172. **Jones, G. E.**, Precipitates from autoclaved seawater, *Limnol. Oceanogr.*, 13, 165, 1967.
173. **Wegrzyn, R. J.**, Synergism of cupric ion and phenol in inactivating RNA from poliovirions, Abstr. Annu. Meet. 78th Am. Soc. Microbiol., Las Vegas, Nev., May 1978, 218.
174. **Magnusson, S., Gundersen, K., Brandberg, A., and Lycke, E.**, Marine bacteria and their possible relation to the virus inactivation capacity of seawater, *Acta Pathol. Microbiol. Scand.*, 71, 274, 1967.
175. **Matossian, A. M. and Garabedian, G. A.**, Virucidal action in seawater, *Am. J. Epidemiol.*, 85, 1, 1967.
176. **Fujioka, R. S., Loh, P. C., and Lau, L. S.**, Survival of human enteroviruses in the Hawaiian ocean environment: evidence for virus-inactivating microorganisms, *Appl. Environ. Microbiol.*, 39, 1105, 1980.
177. **Lycke, E., Magnusson, S., and Lund, E.**, Studies on the nature of the virus inactivating capacity of seawater, *Arch. Ges. Virusforsch.*, 17, 409, 1965.
178. **Ward, R. L.**, Mechanism of poliovirus inactivation by ammonia, *J. Virol.*, 26, 299, 1978.
179. **Ward, R. L. and Ashley, C. S.**, Identification of detergents as components of wastewater sludge that modify the thermal stability of reovirus and enteroviruses, *Appl. Environ. Microbiol.*, 36, 889, 1978.
180. **Ward, R. L. and Ashley, C. S.**, Effects of wastewater sludge and its detergents on the stability of rotavirus, *Appl. Environ. Microbiol.*, 39, 1154, 1980.

181. **Ward, R. L. and Ashley, C. S.**, Comparative study on the mechanisms of rotavirus inactivation by sodium dodecyl sulfate and ethylendiamine-tetraacetate, *Appl. Environ. Microbiol.*, 39, 1148, 1980.
182. **Bagdasaryan, G. A., Nedachin, A. Y., and Doskina, T. V.**, The effect of chemicals on some microbial self purification processes in bodies of water, *Gig. Sanit.*, 2, 104, 1977.
183. **Zepp, R. G., Wolfe, N. L., Baughman, G. L., and Mollis, R. C.**, Singlet oxygen in natural waters, *Nature (London)*, 267, 421, 1977.
184. **Squeri, L., Ioli, A., and Garrani, A.**, Osservazioni sul periodo di sopravivenza dei virus Coxsackie typi B1, B3 e B5 nell' a acqua, *Atti Soc. Peloritana Sci. Eis. Mat. Nat.*, 7, 433, 1962.
185. **Duff, M.**, The uptake of enteroviruses by the New-Zealand marine blue mussel *Mytilus edulis aoteanus*, *Ann. J. Epidemiol.*, 85, 486, 1967.
186. **Welke, G., Friedrich, U., and Mai, K.**, Inactivation time determination for echoviruses 6, 11, 30, and 33 in water, *Z. Ges. Hyg.*, 5, 353, 1973.
187. **Blawat, F., Potajallo, U., Dabrouwski, J., Towianska, A., and Jarnuszkiewicz, I.**, Survival of some viruses in the seawater samples collected from the Gulf of Gdansk. Preliminary studies, *Bull. Inst. Mar. Trop. Med. Gdynia*, 27, 331, 1976.
188. **Plissier, M. and Hugues, B.**, Survie comparée *in situ* de poliovirus 1, adénovirus 5, *Salmonella paratyphi B, Escherichia coli, Candida albicans*, en eau de mer et en eau de rivière, Report 237, Ministère de la Culture et de l'Environnement, Paris, 1978.
189. **Laveran, H., Laluque, J. B., Charrier, F., Monghal, M., and Beytout, D.**, Etude sur la perte du pouvoir infectieux d'entérovirus sauvages et vaccinaux dans les eaux naturelles et au laboratoire, *Rev. Fr. Sante Publ.*, 11, 41, 1980.
190. **Hurst, C. J. and Gerba, C. P.**, Stability of simian rotavirus in fresh and estuarine water, *Appl. Environ. Microbiol.*, 39, 1, 1980.
191. **Mahnel, H., Ottis, K., and Herlyn, M.**, Stabilität von neun Viruarten unterschiedlicher Genera in Trunk- and Oberflächenwasser, *Zentralbl. Bakteriol. Parasitenkd., Infektionskr., Hyg., Abt. 1: Orig., Reihe B.*, 164, 64, 1977.
192. **Leogrande, G. and Iandolo, E.**, Effect of seawater on survival of virulent and attenuated poliovirus, *Ig. Mod.*, 62, 705, 1969.
193. **Pietri, C. and Breittmayer, J. P.**, Etude de la survie d'un entérovirus en eau de mer, *Rev. Int. Oceanogr. Mar.*, 4, 77, 1976.
194. **Carlucci, A. and Pramer, D.**, An evaluation of the factors affecting the survival of *Escherichia coli* in seawater — IV Bacteriophages, *Appl. Microbiol.*, 8, 254, 1960.
195. **Berg, G., Scarpino, P. V., and Berman, D.**, Survival of bacteria and viruses in natural waters, *Proc. Nat. Symp. Quality Standards for Natural Waters*, Ann Arbor Science, Ann Arbor, Mich., 1967, 231.
196. **Baross, J. A., Hanus, F. J., and Morita, R. Y.**, Survival of human enteric and other sewage microorganisms under simulated deep-sea conditions, *Appl. Microbiol.*, 30, 309, 1975.
197. **Girier, L., Discamps, P. G., and Richer, G.**, Problèmes hygièniques posés par la résistance des entérovirus dans les eaux de boissons, *Ann. Hyg. Lang. Fr.*, 1, 37, 1965.
198. **Hetrick, F. M.**, Survival of human pathogenic viruses in estuarine and marine waters, *Environ. Lett.*, 44, 300, 1978.
199. **Kabler, P. W.**, Purification and sanitary control of water (potable and waste), *Ann. Rev. Microbiol.*, 16, 127, 1962.
200. **Plissier, M.**, L'inactivation dans l'eau de mer et l'eau d'alimentation de certains entérovirus, *Arch. Ges. Virusforsch.*, 13, 76, 1963.
201. **Plissier, M., Therre, P., Stone, M., Broquet, M., and Broquet, B.**, Recherches sur l'inactivation *in vitro* du poliovirus dans l'eau de mer, II. Expérimentation sur dix mois, *Ann. Inst. Pasteur, Paris*, 103, 665, 1962.
202. **Rhodes, A. J. and Van Rooyen, C. E.**, Prolonged survival of human poliomyelitis virus in experimentally infected river water, *Can. J. Publ. Health*, 54, 146, 1962.
203. **Gunderson, K., Branberg, A., Magnusson, S., and Lyke, E.**, Characterization of a marine bacterium associated with virus inactivating capacity, *Acta Pathol. Microbiol. Scand.*, 71, 281, 1967.
204. **Kelly, S., Sanderson, W. W., and Neid, C.**, Removal of enteroviruses from sewage by activated sludge, *J. Water Pollut. Control Fed.*, 33, 1056, 1961.
205. **Mitchell, R.**, Destruction of bacteria and viruses in seawater, *J. Sanit. Eng. Div., Am. Soc. Civ. Eng.*, 97, 425, 1971.
206. **Sobsey, M. D. and Cooper, R. C.**, Enteric virus survival in algal-bacterial wastewater treatment systems, I. Laboratory studies, *Water Res.*, 7, 669, 1973.
207. **Ehresmann, D. W., Deig, E. F., Hatch, M. T., Di Salvo, L. H., and Vedros, W. A.**, Antiviral substances from California marine algae, *J. Phycol.*, 13, 37, 1977.
208. **Cliver, D. O. and Hermann, J. E.**, Proteolytic and microbial inactivation of enteroviruses, *Water Res.*, 6, 797, 1972.

209. **Herrmann, J. E. and Cliver, D. O.,** Degradation of Coxsackievirus type A9 by proteolytic enzymes, *Infect. Immun.*, 7, 513, 1973.
210. **Grabow, W. O. K., Prozesky, O. W., Applebaum, P. C., and Lecatsus, G.,** Absence of hepatitis B antigen from feces and sewage as a result of enzymatic destruction, *J. Infect. Dis.*, 131, 658, 1975.
211. **Vasconcelas, G. J. and Schwartz, R. G.,** Survival of bacteria in seawater using a diffusion chamber apparatus *in situ*, *Appl. Environ. Microbiol.*, 31, 913, 1976.
212. **Fliermans, C. B. and Gorden, R. W.,** Modification of membrane diffusion chambers for deep water studies, *Appl. Environ. Microbiol.*, 33, 207, 1977.

Chapter 8

VIRUSES IN WASTEWATER SLUDGES

Samuel R. Farrah and Stephen A. Schaub

TABLE OF CONTENTS

I. INTRODUCTION

Enteric viruses are usually associated with infections of the gastrointestinal tract but may also infect other parts of the body. As intestinal pathogens, these viruses may be shed with feces and enter the wastewater treatment system. The number of viruses associated with feces may be high. As many as 10^6 viruses per gram of feces may be excreted by infected individuals.[1] The large number of viruses excreted by infected individuals may produce a relatively high virus load in raw sewage. By inoculating antibiotic and chloroform-treated samples of raw sewage directly onto cell cultures, Burras[2] found between 35,000 and 490,000 plaque-forming units (PFU) of viruses per liter of raw sewage in Israel. Other workers have found variable numbers of viruses in raw sewage from other countries. England and coworkers[3] found 4000 PFU/ℓ of viruses in raw sewage from Santee, Calif. after widespread immunization with Sabin attenuated poliovirus was practiced. After a community-wide program of immunization with the live poliovirus vaccine was started, the level of viruses in the raw sewage increased to approximately 200,000 PFU/ℓ. Cliver[4] found 13 to 170 PFU of enteroviruses per liter and 60 to 170 PFU of reoviruses per liter of raw sewage. Safferman and Morris[5] observed an hourly variation in the numbers of enteroviruses in primary effluent from a Washington, D.C. treatment plant. The numbers varied from 24 to 169 PFU/ℓ over a 20-hr period. A wide variety of enteric viruses have been identified in raw sewage. These include different serotypes of polioviruses, echoviruses, coxsackieviruses, reoviruses, and adenoviruses (Table 1). The viruses detected represent a portion of the 100 or more different serotypes of viruses that could be present in untreated sewage.[1] Certain viruses that are important human pathogens may also be present in untreated wastewater but may escape detection unless special procedures are employed for their recovery and quantitation. Human rotaviruses, the Norwalk agent, and hepatitis A virus cannot be detected with the cell cultures and procedures that are adequate for detection of other enteric viruses.[8,9]

When one considers the potential contamination of sewage by enteric viruses from infected individuals and when one considers the results of field studies, raw sewage entering a wastewater treatment plant seems likely to contain a large number of enteric viruses.

II. PRODUCTION, TREATMENT, AND DISPOSAL OF WASTEWATER SLUDGE

Both laboratory and field studies indicate that viruses rapidly become associated with wastewater sludge (Section IV).

Therefore, the different types of sludge produced, the sludge treatment procedures, and the means of sludge disposal will now be considered (Figure 1). The information in Figure 1 is not inclusive but indicates several possible sources of sludge and different sludge treatment practices that may be employed. Individual wastewater treatment plants vary in the procedures actually employed.

Raw sewage contains particulate matter that can be removed by screening or sedimentation. In general, the larger particles that are removed by screening are called grit while the smaller particles recovered during sedimentation are called primary sludge. The supernatant remaining after sedimentation is usually treated in one or two processes involving biological decomposition of the soluble organic material in the sewage. These biological processes are treatment by activated sludge or trickling filter units. Activated sludge is flocculent mass of diverse microorganisms that is produced during aeration of raw wastewater.[10] The bacteria associated with the flocs are thought to be the main agents responsible for reducing the level of soluble organic carbon in raw sewage. After the aeration step, the treated wastewater from aeration tanks and sludge flocs (mixed liquor) flows to settling tanks. After settling, the supernatant is removed for discharge or additional treatment while the floc is divided

Table 1
VIRUSES ISOLATED FROM RAW SEWAGE

Country	Viruses identified	Ref.
Sweden	Poliovirus 3	6
	Coxsackievirus B1, B3, B4, and B5	
	Echovirus 4, 9, and 11	
	Adenovirus 1 and 3	
Israel	Poliovirus 1, 2, and 3	2
	Echovirus 4, 13, 16, 24, and 25	
	Coxsackievirus A1	
U.S.	Poliovirus 1, 2, and 3	7
	Coxsackievirus B2, B3, and B4	
	Adenovirus, Reovirus	3
	Poliovirus 1, 2, and 3	
	Coxsackievirus A13 and A18	
	Coxsackievirus B1, B2, B3, B4, and B5	
	Echovirus 3, 7, 8, 9, 11, 12, 14, 19, 20, 21	

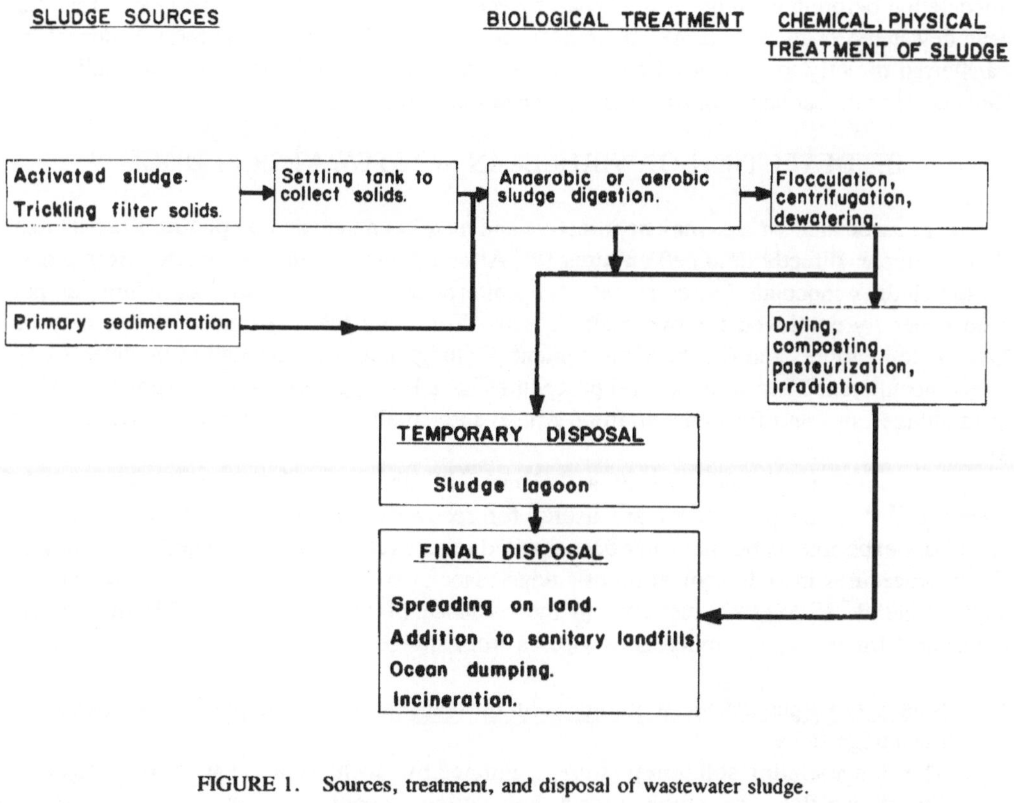

FIGURE 1. Sources, treatment, and disposal of wastewater sludge.

into two portions. One portion (return sludge) is used to inoculate raw wastewater entering the aeration tank while the other portion (wasted sludge) is removed for treatment and disposal. Trickling filters consist of rocks or other inert materials in large cyclindrical containers. As wastewater is passed through the filters by gravity, a slime layer develops on the surfaces of the inert material. This slime layer contains a mixture of microorganisms that perform a function similar to that performed by the microorganisms in activated sludge in oxidizing the soluble organic compounds in the raw sewage. Although sludge is not produced by trickling filters, a portion of the slime layer may be sloughed off during normal operation. These solids may be collected by sedimentation and treated along with solids from the activated sludge unit.

The solids collected from primary settling, wasted activated sludge, and trickling filter solids may be further treated by anaerobic or aerobic digestion. Aerobic digestion of sludge is accomplished by rapidly stirring sludge to promote aeration or by forcing air into the sludge. The process is therefore similar to activated sludge treatment in that aerobic conditions are maintained. It differs from activated sludge treatment in that fresh, soluble organic compounds from raw sewage are not continually introduced into the digesters. The conditions existing during anaerobic digestion of sludges are fundamentally different from those existing during activated sludge treatment or aerobic digestion. Anaerobic digesters are designed to exclude oxygen and a fermentative digestion of sludge occurs in contrast to the oxidative digestion that occurs in aerobic sludge digestion.

After digestion, the sludge may be treated in any of several ways. One of the first steps may be to reduce the amount of water associated with the sludge. This reduces the total sludge volume and makes transport and handling easier. The sludge may be dewatered by vacuum filtration. In other cases, sludge may be mixed with charged polymers to promote flocculation before the sludge is thickened by centrifugation or settling. The sludge that has been dewatered or thickened may be transferred to sludge lagoons for temporary storage or transferred directly to a disposal site. Sludge is disposed of by incineration, spreading on land, addition to sanitary landfills, or by ocean dumping.

III. DETECTION OF VIRUSES IN WASTEWATER SLUDGE

Sludge-associated viruses may be infective and have been detected by plating ether-treated sludge samples directly onto cell cultures.[11,12] Although viruses may be isolated from ether-treated sludges inoculated directly onto cell cultures in some cases, the procedure has not been generally employed for two main reasons. These are that (1) treated sludge may be toxic to cell cultures and (2) the virus content of sludge may be too small to be detected by direct inoculation. Therefore, several procedures have been developed for separating viruses from sludge flocs and for concentrating viruses into small volumes that are convenient for assay.

Methods used to recover viruses associated with sludge particles have been recently reviewed.[13,14] Some procedures are useful for recovering viruses seeded into sludge in laboratory experiments but have not been applied to recovering viruses indigenous to sludge. These procedures include sonication of sludge-associated viruses in the presence of sodium dodecyl sulfate (SDS) and blending in the presence of fetal calf serum.[15,16] In general, procedures for recovering indigenous viruses from sludge include four steps:

1. Sludge is suspended in an eluting solution that promotes desorption of viruses from the sludge flocs
2. The sludge-eluting solution mixture is agitated by mechanical means or by sonication
3. The sludge flocs are removed from the solution by centrifugation
4. Viruses in the supernatant phase remaining after centrifugation are concentrated into a smaller final sample volume that is used to inoculate cell cultures

The eluents, agitation methods, and concentration procedures used in different procedures are summarized in Table 2.

Mixtures of proteins and peptides such as fetal calf serum or beef extract elute sludge-associated viruses. Chemically defined eluents have been used in only a few studies. Hurst

Table 2
PROCEDURES FOR RECOVERY OF SLUDGE-ASSOCIATED VIRUSES

Eluting solutions	Agitation procedures	Concentration procedures
Beef extract[15,17,18]	Sonication[15-19]	Hydroextraction[18]
Fetal calf serum[4,15,19,20]	Stirring[18]	Organic flocculation[17]
Tryptose-phosphate broth[21]	Blending[4,20,24]	Membrane filter adsorption-elution[22]
Gelatin[15]	Shaking[22,24]	Inorganic flocculation[23]
Distilled water[21]		
Sodium dodecyl sulfate[16]		
Glycine[22]		
Urea[23]		

et al.[22] used basic solutions of 0.05 *M* glycine to elute viruses from activated sludge flocs and from dried sludge collected from land application sites. In this study, virus elution was increased by raising the pH of the glycine buffer to pH 11. Ward and Ashly[16] used 0.1% SDS to recover polioviruses adsorbed to sludge floc. Ethylenediaminetetraacetic acid (versene) was a poor eluent while 4 *M* urea buffered at pH 9 was a good eluent for sludge-associated viruses.[23,25]

Suspensions of sludge in eluents have been mixed by magnetic stirring, blending, and mechanical shaking, and they have been subjected to ultrasonic treatment. Following these treatments, sludge particles are usually removed from the eluting solution by centrifugation. In some cases, supernatants remaining after centrifugation have been used to inoculate cell cultures directly.[4,15,16,19-21] Other procedures provide for concentrating the eluted viruses in a smaller final volume. Procedures employed for the concentration step include membrane adsorption-elution, organic flocculation, hydroextraction, and adsorption of viruses to inorganic flocs.[17,18,22,23]

Several of these procedures have been used to detect indigenous viruses associated with wastewater sludges.[4,17,18,22,23] Therefore, the fate of viruses associated with the sludge particles at different stages of treatment may be determined with existing procedures.

IV. ADSORPTION OF VIRUSES TO SLUDGE PARTICLES

A. Laboratory Studies

Laboratory studies have shown that most of the viruses added to a sludge suspension rapidly become associated with sludge flocs and relatively few remain in the liquid phase. Clark et al.[25] found that the association of viruses with activated sludge flocs conformed to the Freundlich adsorption isotherm. These authors considered this evidence that removal of viruses by sludge flocs was an adsorption phenomenon. In this study, greater than 99% of coxsackievirus A9 was removed from the supernatant phase of an activated sludge suspension in 45 min (Table 3).

Malina[21] reported that the adsorption of viruses by sludge flocs occurred rapidly. By using a tritium labeled poliovirus, these workers demonstrated that 60% of the radioactivity became associated with the sludge flocs 1 min after the labeled virions were mixed with sludge flocs. After 10 min of mixing, only 5% of the radioactivity remained in the supernatant phase. Balluz et al.[19] studied the fate of poliovirions in a laboratory model activated sludge unit. The virions were unequally divided between the solids and liquid phases. Only 15% of the infectious virions were in the liquid phase while the remaining 85% were recovered from the sludge solids. While sludge flocs generally adsorb viruses, the extent of virus adsorption may vary with different types of sludge. Pancorbo et al.[26] found that 95% of the virions added to a suspension of aerobically digested sludge became associated with the

Table 3
EFFECT OF ACTIVATED SLUDGE TREATMENT ON VIRUSES IN LABORATORY-SCALE UNITS

	Adsorption of viruses to sludge flocs		Inactivation of viruses		
Viruses added	% viruses remaining in the supernatant fraction	Time after inoculation	% viruses remaining	Time after inoculation	Ref.
Poliovirus 1	7—15	2—4 hr	<1	48 hr	18
Coxsackievirus A9	<1	45 min	NA[a]	NA	25
Poliovirus 1					
Poliovirus 1	34	2 hr	NA	NA	27
Coliphage T2	36	2 hr	10	5 hr	
Poliovirus 1	<5	1 hr	<1	24 hr	21

[a] Data not available.

Table 4
EFFECT OF ACTIVATED SLUDGE TREATMENT OF THE VIRUS CONTENT OF WASTEWATER

Viruses in effluent from activated sludge units		Reduction in virus numbers (compared with the number of viruses in the influent to the unit)	
Virus types	Quantity		Ref.
Reoviruses, polioviruses, coxsackieviruses, echoviruses	6—24 PFU[a]/mℓ	76—90%	3
None detected	<1 PFU/100 mℓ of swab eluate	From 2—5 PFU/100 mℓ of swab eluate	29
Adenoviruses, polioviruses, coxsackieviruses	NA[b]	1.5 to 2.0 $\log_{10}$ (estimated)	28
Enteric viruses	<1 PFU/ℓ	>95%	5

[a] Plaque-forming unit.
[b] Data not available.

sludge flocs after 10 to 60 min of mixing. In contrast, the percent of solids-associated virions on activated sludge solids and anaerobically digested sludge was 57 and 70%, respectively. The rapid adsorption of virions to sludge flocs is followed by inactivation of the adsorbed virions. The rate of inactivation has been determined by measuring the decrease in infectious virions and by measuring solubilization of labeled viral components.[15,21]

B. Field Studies

Removal and inactivation of viruses by activated sludge flocs in pilot and full-scale wastewater treatment plants has been observed in field studies (Table 4). England and co-workers[3] found that activated sludge treatment reduced the levels of polioviruses 1, 2, and 3 in raw sewage by 75 to 79%. Lund and co-workers[28] estimated that the virus content of effluent from an activated sludge unit was 1.5 to 2.0 $\log_{10}$ less than the virus content of the influent to the unit. By seeding primary effluent with a coliphage, Safferman and Morris[5] determined the fate of viruses in several stages of a multi-stage activated sludge unit. Virus removal by the activated sludge module ranged from 90 to 98%. Greater than 95% of the indigenous enteric viruses were removed by the same activated sludge unit.

The virus content of undigested sludge has been determined in several studies (Table 5). Although the numbers and the virus types vary, different authors agree that viruses can be detected in most samples of undigested sludge.

Table 5
DETECTION OF VIRUSES IN RAW, PRIMARY, AND WASTED ACTIVATED SLUDGE

	Viruses detected		
Sludge type	**Types**	**Quantity**	**Ref.**
2/3 primary 1/3 wasted	NI[a]	38—120 PFU[b]/ℓ	30
Wasted	Poliovirus 1 Echoviruses 7 and 17 Coxsackievirus B1	3—49 PFU/g	22
Wasted	NI[a]	1.8 MPN/ℓ	31
Wasted	Polioviruses 1 and 2 Coxsackieviruses B3 and B5 Adenoviruses 1 and 2	0.1—9.0 $TCID_{50}$/mg	32

[a] Not identified.
[b] Plaque-forming unit.

Both laboratory and field studies suggest that most of the viruses entering a wastewater treatment plant become associated with the sludge particles although some of the viruses are found in the effluent. Therefore, a consideration of the fate of viruses during wastewater treatment requires that the level of viruses associated with sludge particles be determined at different stages of sludge production, treatment, and disposal.

V. THE FATE OF SLUDGE-ASSOCIATED VIRUSES DURING SLUDGE TREATMENT PROCESSES

A. Biological Treatment of Sludge

1. Anaerobic Digestion

Laboratory-scale anaerobic digesters or samples from full scale anaerobic digesters maintained under anaerobic conditions in the laboratory have been used to study the survival of enteroviruses in anaerobically digested sludge (Table 6). Bertucci and co-workers[15] studied the survival of four human enteric viruses (poliovirus 1, coxsackievirus A9, coxsackievirus B4, and echovirus 11) and one bacterial virus (MS2) in a laboratory model anaerobic digester. The viruses were inactivated rapidly at rates between 75 and 93%/day. At least a portion of the antiviral factors present in anaerobically digested sludge are soluble and are present in the supernatant fraction remaining after digested sludge is centrifuged. Fenters and co-workers[33] observed inactivation rates of 1.3 to 2.5 log_{10}/day for viruses in supernatant from anaerobically digested sludge incubated at 35°C. These workers found that increasing the temperature or increasing the concentration of ammonia in the supernatant increased the rate of inactivation while addition of antibiotics did not affect the inactivation rate. A relatively rapid rate of inactivation (2 log_{10}/day) was also observed for coxsackievirus B3 added to a pilot scale anaerobic digester.[34] Changes in temperature, detention time, and solids loading all influenced the survival of poliovirus 1 in a laboratory model anaerobic sludge digester studied by Sanders and co-workers.[35] Increasing the detention time from 5 to 15 days and increasing the solids from 6300 mg/ℓ to 21,800 mg/ℓ increased the inactivation rate slightly. In contrast, increasing the temperature from 34 to 50°C resulted in a large increase in the inactivation rate. It was an important finding of this study that the rate of inactivation of viruses changed with time. At 34 or 37°C, the rate of inactivation was 84 to 99%/day for the first 24 hr; then, the rate dropped to between 30 and 60%/day.

The importance of temperature in determining the rate of virus inactivation in sludge was

Table 6
INACTIVATION OF VIRUSES IN ANAEROBICALLY DIGESTED SLUDGE UNDER LABORATORY CONDITIONS

Viruses added	Sludge temperature (°C)	Inactivation rate	Ref.
Poliovirus 1	35	75—97%/day	15
Coxsackieviruses A9, B4	35		
Echovirus 11	35		
Poliovirus 1[a]	35	2.5 $\log_{10}$/day	33
Echovirus 6	35	2.3 $\log_{10}$/day	
Coxsackievirus B4	35	1.3 $\log_{10}$/day	
Coxsackievirus B3	35	2.0 $\log_{10}$/day	34
Poliovirus 1	34, 37	A. 84—99%/day for the first 24 hr	35
		B. 30—60%/day after 24 hr	
Poliovirus 1	Room temperature	3.0 $\log_{10}$/12 weeks	20
	4	<50% in 14 weeks	
Poliovirus 1	28	>1 $\log_{10}$/day	16
	4	1 $\log_{10}$/5 days	

[a] Virus was added to sludge supernatant.

confirmed by Subrahmanyan.[20] Greater than 50% of poliovirus 1 added to digested sludge samples could be recovered after 14 weeks of incubation at 4°C. Approximately 20% of the virus in sludge at room temperature could be recovered after the same length of time.

Ward and Ashley[16] also found that the temperature of incubation had a marked effect on the survival rate of a poliovirus in digested sludge. The inactivation rate was greater than 1 $\log_{10}$/day at 25°C but only 0.2 $\log_{10}$/day at 4°C. These workers identified ammonia as an antiviral agent present in anaerobically digested sludge.[16] Ammonia and related compounds were more viricidal at pH 9.5 than pH 7. The inactivation of viruses by ammonia was associated with changes in the poliovirus RNA.

Indigenous enteric viruses have been detected in sludge from anaerobic digesters in several studies (Table 7). The viruses detected were similar to those detected in raw sewage and in undigested sludge (Table 1 and 5). These included different serotypes of reoviruses, polioviruses, coxsackieviruses, and echoviruses. The results obtained in studies with sludge from full-scale digesters were similar to those obtained in laboratory studies in that increasing temperature or detention time increased the rate of virus inactivation. However, the rate of inactivation observed in field studies was less than that for laboratory produced viruses in laboratory digesters. For example, inactivation rates of 90 to 95% in 20 days have been reported for viruses in full-scale digesters while rates of 75 to 97%/day have been observed in laboratory studies.[15,16,30,31] Two reasons for this discrepancy in the inactivation rates between laboratory and full-scale digesters have been offered. First, fresh undigested sludge is added to full-scale digesters on a regular basis. Therefore, the digested sludge is frequently mixed with undigested sludge that is likely to contain viruses. The failure of anaerobic digesters to produce virus-free sludge may therefore be a function of the recontamination of digested sludge before it is discharged. And second, survival of viruses under laboratory conditions has been determined for relatively short times of 24 to 28 hr. Sanders[35] and co-workers reported that the long-term inactivation rate for viruses in digested sludge may differ from the short-term rate.

2. Aerobic Digestion

A limited amount of information is available on inactivation of viruses during aerobic digestion of sludge. Preliminary data have been obtained by Scheuerman and co-workers[35a] on survival of viruses during aerobic digestion of wasted activated sludge in laboratory

Table 7
INDIGENOUS VIRUSES IN ANAEROBICALLY DIGESTED SLUDGE

Detention time in digesters	Operating temperature of digesters (°C)	Virus types detected	Number of viruses detected	Reduction in numbers of viruses initially present in raw sludge during digestion	Ref.
NG[a]	30—32	Enteroviruses Reoviruses	0.8—4.5 PFU[b]/mℓ	NG	4
20 days	Approximately 35	NG	30—40 PFU/100 mℓ	90%/20 days	30
20 days	Approximately 49	NG	0—17 PFU/100 mℓ	>99%/20 days	30
40 days	33	Reovirus Poliovirus 3 Echovirus 11	0.85 MPN[c]/100 mℓ	95%/40 days	31
20 days	35	Reovirus Coxsackievirus B3, B4	ND	ND	24

[a] Not given.
[b] Plaque-forming unit.
[c] Most probable number.

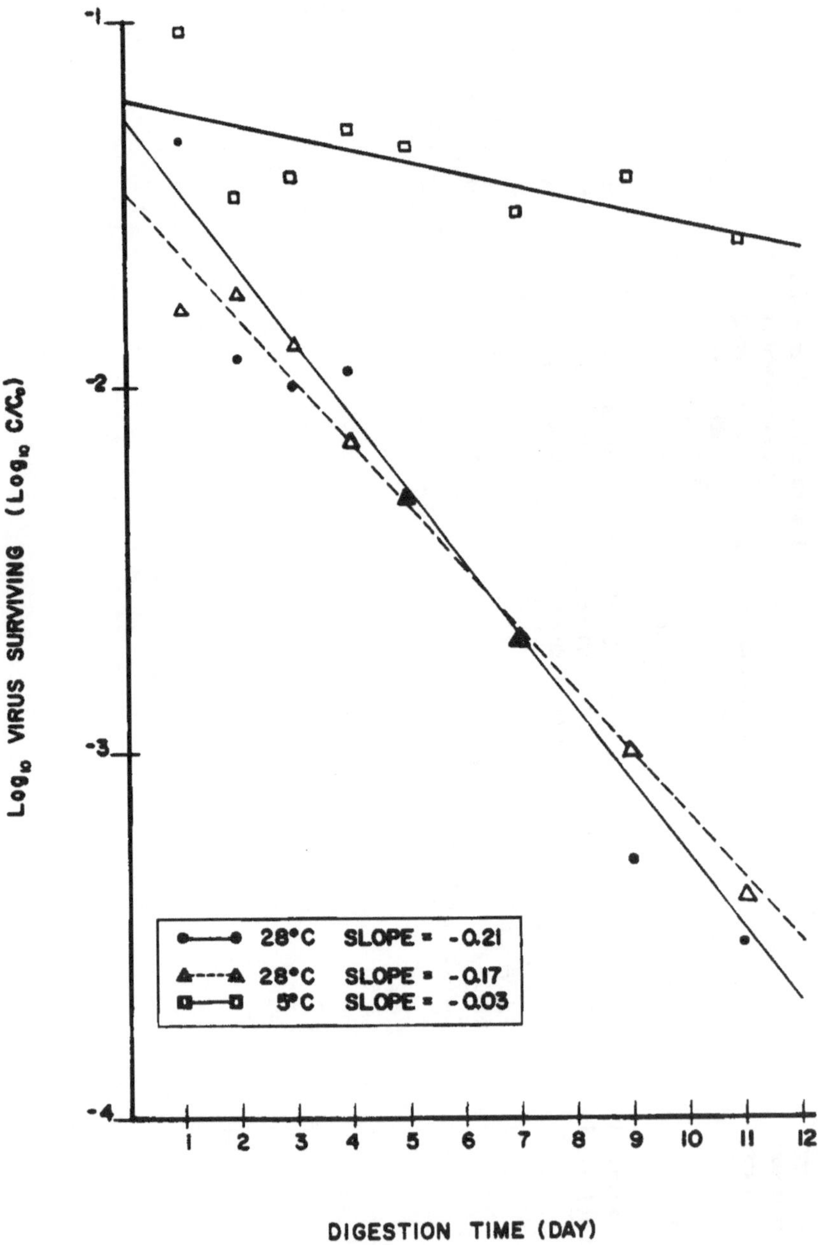

FIGURE 2. Survival of poliovirus 1 in laboratory scale aerobic sludge digesters.

digesters. Temperature was a major factor influencing the inactivation rate of a poliovirus in these studies. Increasing the temperature from 5 to 28° increased the long-term inactivation rate from less than 0.1 $\log_{10}$/day to approximately 0.3 $\log_{10}$/day (Figure 2). A more rapid inactivation rate occurred during the first 24 hr. In laboratory studies on survival of viruses during aerobic and anaerobic digestion of sludge, two distinct inactivation rates occurred with time. In a comparative study on inactivation of viruses under anaerobic conditions and during aerobic digestion with 5 mg/ℓ dissolved O_2, a higher rate of inactivation of a poliovirus was observed in the sludge under aerobic conditions even though the anaerobic digester was

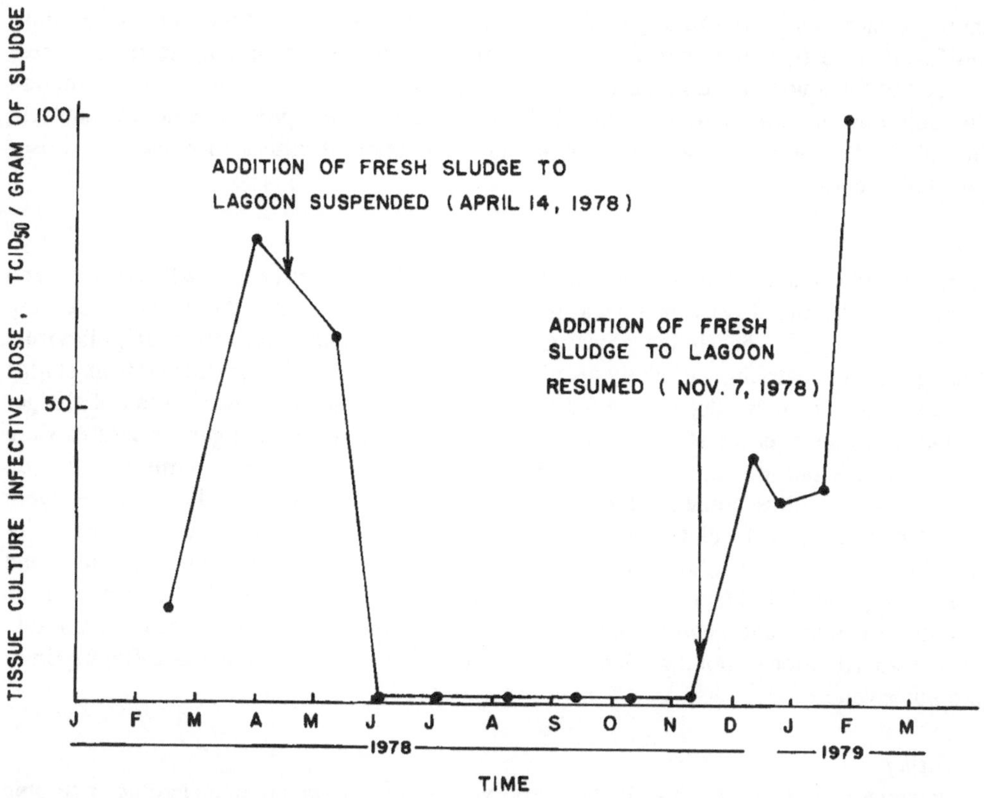

FIGURE 3. Survival of enteroviruses in lagooned sludge. (From Farrah, S. R., Bitton, G., Hoffmann, E. M., Lanni, O., Pancorbo, O. C., Lutrick, K. C., and Bertrand, J. E., *Appl. Environ. Microbiol.*, 41, 459, 1981. With permission.)

operated at 37°C compared to 28°C for the aerobic digester. Varying the detention time from 15 to 40 days and obtaining sludge from three different sources did not significantly affect the inactivation rate for the poliovirus. Although quantitative studies on survival of viruses during aerobic digestion of sludge in full-scale aerobic digesters are not complete, it appears that the virus level is reduced during aerobic digestion of wasted sludge.

3. Sludge Lagooning

Survival of viruses during temporary storage in sludge lagoons has been followed in two studies.[24,37] In one study, viruses were detected in samples of lagooned sludge for up to 8 months.[24] This work was done in Canada and the lagoon was frozen over for 4 months of the study. Therefore, the low temperatures may have been a factor in permitting virus survival. In another study, viruses were found not to survive for long periods in lagooned sludge in Florida.[37] As shown in Figure 3, once addition of fresh sludge to a lagoon was suspended, the level of viruses associated with the sludge dropped to low or undetectable levels. Viruses were again isolated from the sludge shortly after addition of fresh digested sludge to the lagoon was resumed.

B. Chemical and Physical Treatment of Sludge

1. Dewatering

Following biological treatment, sludge is frequently thickened by removal of a portion of the water associated with the sludge. This reduces the sludge volume and makes sludge handling easier. Sludge thickening or dewatering can be accomplished by vacuum filtration, centrifugation, or by simply permitting the sludge to settle. Synthetic polyelectrolytes may be added to the sludge to promote flocculation during settling or centrifugation. Although information on survival of solids-associated viruses during these process is not available, it is not likely that these processes result in either inactivation of viruses or release of viruses from sludge particles.

2. Drying

Drying sludge under laboratory conditions inactivated polioviruses, coxsackieviruses, and reoviruses.[38] The rate of inactivation was related to the solids content of the sludge. Increasing the solids content of the sludge above 65% resulted in rapid inactivation of poliovirus. Although sufficient drying will likely inactivate viruses, echovirus 7 was isolated from sludge that had been on a sludge drying bed for 13 days.[18] In two studies on inactivation of sludge-associated viruses after the sludge was applied to land for disposal, indigenous viruses were inactivated at a rate of approximately 2 $\log_{10}$/week.[22,37] In both cases, virus inactivation occurred as the solids content of the sludge increased with drying. These studies were conducted in areas with relatively warm climates (Texas and Florida). A much lower inactivation rate of 0.2 $\log_{10}$/week was observed for coxsackievirus B3 added to sludge and placed in lysimeters in Denmark.[39] This study was conducted during the winter months so virus survival may have been favored by the lower temperatures that reached a maximum of less than 10°C for 4 months. Data on the solids content of the sludge at different times were not given.

3. Heating

The increased rate of virus inactivation observed as the temperature of anaerobic or aerobic sludge digestion is increased was not discussed in section V.A. In general, viruses are not stable at higher temperatures unless they are protected by agents such as divalent cations or cystine.[40,41] Heat may therefore be used to inactivate viruses in sludge. Foliguet and Doncoeur[42] found that poliovirus 1 and coxsackievirus B3 in sludge were rapidly inactivated as the sludge temperature was raised to the 60 to 80°C range. Maintaining the sludge at the elevated temperatures for 10 min was sufficient to inactivate viruses. Sludge components may reduce or increase the inactivation rate[43] of viruses in sludge.

4. Composting

Composting of sludge can be accomplished by mixing dewatered sludge with a source of cellulose such as sawdust or wood chips and forming the mixture into windrows or aerated piles to promote aeration. The aerobic transformations that result may increase the temperature to greater than 70°C. At these temperatures viruses are rapidly inactivated.[44] Composting may be carried out under cold winter conditions with production of temperatures in excess of 55°C.[45] One advantage of composting is the production of a marketable product that can be used to offset the cost of sludge treatment.[46]

5. Irradiation

Ionizing radiation alone or with heat treatment inactivates viruses along with other infectious agents. Inactivation of viruses in sludge subjected to heat and/or ionizing radiation has been considered by Ward.[47]

C. Sludge Disposal

Viruses in raw sewage may survive both wastewater treatment and subsequent sludge digestion processes. Therefore, it is necessary to consider the fate of viruses during sludge disposal to determine the possible health hazards associated with different disposal methods. Possible problems with pathogens in sludge have been considered in other reviews.[48-50]

1. Ocean Dumping

Dumping of sludge at ocean sites has been practiced by coastal cities such as New York, Boston, and Los Angeles. This dumping was scheduled to end by December 31, 1981, according to the Marine Protection Resources Sanctuaries Act (MPRSA) passed by Congress. The fact that this method of disposal was being phased out was the probable reason for the lack of studies on the survival of sludge associated viruses in the ocean.

2. Incineration

Incineration of sludge produces an ash as a final product that is free of viral as well as other pathogens. The advantages of a pathogen-free final product may be offset by the increasing cost of energy and the problem with air pollution. Approximately 35% of the sludge produced in the U.S. is disposed of by incineration.[51]

3. Sanitary Landfills

Approximately 50% of the sludge produced in the U.S. is disposed of in sanitary landfills (25%) or by application to land (25%).[51] The chemical environment in sanitary landfills and in leachate from the landfills is apparently not conducive to virus survival. Viruses added to solid wastes were not detected in leachate from a model sanitary landfill[51] and were inactivated rapidly when added to leachate.[52,53] Finding suitable sites for sanitary landfills and avoiding chemical contamination of groundwater may be more of a problem than viral contamination of surface or groundwater by leachates.

4. Spreading on Land

With the increasing popularity of disposing of sludge on land, studies have been undertaken to determine the fate of sludge-associated viruses after application of sludge to land. Lysimeter studies with sludge seeded with laboratory produced viruses have been conducted under different climatic conditions. Viruses in sludge applied to lysimeters survived for up to 23 weeks during a Danish winter[39] but only for 3 weeks during a dry autumn in Florida.[54] Although temperatures were higher during the summer than in the autumn in the Florida study, viruses survived longer in the wetter summer than in the drier autumn. Viruses were not detected in leachates from the lysimeters in either study. Since the numbers of viruses added to the sludges greatly exceeded the numbers that would likely be indigenous to sludge, movement of sludge-associated viruses into groundwater does not appear likely. As described in Section IV.B, drying of sludge on land reduces the numbers of indigenous viruses by about 2 $\log_{10}$/week.[22,37]

VI. SUMMARY AND CONCLUSIONS

A wide variety of viruses may be found in raw wastewater. Although a large number of these viruses are inactivated during aerobic wastewater treatment, the surviving viruses are likely to be associated with the solids produced during the wastewater treatment. Digestion of sludge reduces the number of viruses but will usually not totally eliminate viruses. This is especially true in continuous-feed digesters where undigested sludge is regularly or continuously added to the digesters. Physical treatment of sludge such as drying, heating, and irradiation may reduce the numbers of viruses in sludge to undetectable levels. These treatments require special treatment and storage facilities at wastewater treatment plants and increase the cost of sludge processing. With composting, a portion of the increased cost can be recovered by sale of the final product. As ocean dumping is eliminated and the cost of incineration increases, sludge disposal on land will likely increase. It may be that proper management of sludge addition will permit increased application of sludge to land without greatly increasing the problem of viral contamination of groundwater.

REFERENCES

1. **Gerba, C. P., Wallis, C., and Melnick, J. L.,** Viruses in water: the problem, some solutions, *Environ. Sci. Technol.*, 9, 1122, 1975.
2. **Buras, N.,** Concentration of enteric viruses in wastewater and effluent: a two year survey, *Water Res.*, 10, 295, 1976.
3. **England, B., Leach, R. E., Adame, B., and Shiosaki, R.,** Virological assessment of sewage treatment at Santee, California, in *Transmission of Viruses by the Water Route*, Berg, G., Ed., Wiley-Interscience, New York, 1967, 401.
4. **Cliver, D. O.,** Virus association with wastewater solids, *Environ. Lett.*, 10, 215, 1975.
5. **Safferman, R. S. and Morris, M. E.,** Assessment of virus removal by a multi-stage activated sludge process, *Water Res.*, 10, 413, 1976.
6. **Lund, E., Hedström, C. E., and Strannegård, O.,** A comparison between virus isolations from sewage and from fecal specimens from patients, *Am. J. Epidemiol.*, 84, 282, 1966.
7. **Kelly, S., Winsser, J., and Winkelstein, W., Jr.,** Poliomyelitis and other enteric viruses in sewage, *Am. J. Publ. Health*, 47, 72, 1957.
8. **DuPont, H. L., Portnoy, B. L., and Conklin, R. H.,** Viral agents and diarrheal illness, *Ann. Rev. Med.*, 28, 167, 1977.
9. **Kapikian, A. Z., Feinstone, S. M., Purcell, R. H., Wyatt, G., Thornhill, T. S., Kalica, A. R., and Chanock, R. M.,** Detection and identification by immune electron microscopy of fastidious agents associated with respiratory illness, acute nonbacterial gastroenteritis, and hepatitis A, in *Perspectives in Virology*, Vol. IX, Pollard, M., Ed., Academic Press, New York, 1975, 9.
10. **Taber, W. A.,** Wastewater microbiology, *Ann. Rev. Microbiol.*, 30, 263, 1976.
11. **Schaub, S. A. and Sagik, B. P.,** Association of enteroviruses with natural and artificially introduced colloidal solids in water and infectivity of solids-associated virions, *Appl. Microbiol.*, 30, 212, 1975.
12. **Lund, E. and Ronne, V.,** On the isolation of virus from sewage treatment plant sludges, *Water Res.*, 7, 863, 1973.
13. **Hurst, C. J.,** Recovering viruses from wastewater sludges and from solids in water, in Methods for Recovering Viruses from the Environment, Berg, G., Ed., CRC Press, Boca Raton, Fla., in press.
14. **Farrah, S. R.,** Detection of viruses associated with wastewater sludge in, *Methods in Environmental Virology*, Gerba, C. P. and Goyal, S. M., Eds., Marcel Dekker, New York, 1982, 161.
15. **Bertucci, J. J., Lue-Hing, C., Zenz, D., and Sedita, S. J.,** Inactivation of viruses during anaerobic sludge digestion, *J. Water Pollut. Control Fed.*, 49, 1642, 1977.
16. **Ward, R. L. and Ashley, C. S.,** Inactivation of poliovirus in digested sludge, *Appl. Environ. Microbiol.*, 31, 921, 1976.
17. **Glass, J. S., Van Sluis, R. J., and Yanko, W. A.,** Practical method for detecting poliovirus in anaerobic digester sludge, *Appl. Environ. Microbiol.*, 35, 983, 1978.
18. **Wellings, F. M., Lewis, A. L., and Mountain, C. W.,** Demonstration of solids-associated virus in wastewater and sludge, *Appl. Environ. Microbiol.*, 31, 354, 1976.
19. **Balluz, S. A., Jones, H. H., and Butler, M.,** The persistence of poliovirus in activated sludge treatment, *J. Hyg. Cambridge*, 78, 165, 1977.
20. **Subramanyan, T. P.,** Persistence of enteroviruses in sewage sludge, *Bull. W.H.O.*, 55, 431, 1977.
21. **Malina, J. F., Jr., Ranganathan, K. R., Sagik, B. P., and Moore, B. E.,** Poliovirus inactivation by activated sludge, *J. Water Pollut. Control Fed.*, 47, 2178, 1975.
22. **Hurst, C. J., Farrah, S. R., Gerba, C. P., and Melnick, J. L.,** Development of quantitative methods for the detection of enteroviruses in sewage sludges during activation and following land disposal, *Appl. Environ. Microbiol.*, 36, 81, 1978.
23. **Farrah, S. R., Scheuerman, P. R., and Bitton, G.,** Urea-lysine method for recovery of enteroviruses from sludge, *Appl. Environ. Microbiol.*, 41, 455, 1981.
24. **Sattar, S. A. and Westwood, J. C. N.,** Recovery of viruses from field samples of raw digested, and lagoon-dried sludges, *Bull. W.H.O.*, 57, 105, 1979.
25. **Clarke, N. A., Stevenson, R. E., Chang, S. L., and Kabler, P. W.,** Removal of enteric viruses from sewage by activated sludge treatment, *Am. J. Public Health*, 51, 1118, 1961.
26. **Pancorbo, O. C., Scheuerman, P. R., Farrah, S. R., and Bitton, G.,** Effect of sludge type on poliovirus association with and recovery from sludge solids, *Can. J. Microbiol.*, 27, 279, 1981.
27. **Kelly, S., Sanderson, W. W., and Neidl, C.,** Removal of enteroviruses from sewage by activated sludge, *J. Water Pollut. Control Fed.*, 33, 1056, 1961.
28. **Lund, E., Hedstrom, C. E., and Jantzen, N.,** Occurrence of enteric viruses in wastewater after activated sludge treatment, *J. Water Pollut. Control Fed.*, 41, 169, 1969.
29. **Kelly, S. and Sanderson, W. W.,** The effect of sewage treatment on viruses, *Sewage Ind. Wastes*, 31, 683, 1959.

30. **Berg, G. and Berman, D.,** Destruction by anaerobic mesophilic and thermophilic digestion of viruses and indicator bacteria indigenous to domestic sludges, *Appl. Environ. Microbiol.*, 39, 361, 1980.
31. **Palfi, A.,** Survival of enteroviruses during anaerobic sludge digestion, in *Advances in Water Pollution Research*, Proc. 6th Int. Conf., Jerusalem, Jenkins, S. H., Ed., Permagon Press, Elmsford, N.Y., 1973, 99.
32. **Nielsen, A. L. and Lydholm, B.,** Methods for the isolation of virus from raw and digested wastewater sludge, *Water Res.*, 14, 175, 1980.
33. **Fenters, J., Reed, J., Lue-Hing, C., and Bertucci, J.,** Inactivation of viruses by digested sludge components, *J. Water Pollut. Control Fed.*, 51, 689, 1979.
34. **Eisenhardt, A., Lund, E., and Nissen, B.,** The effect of sludge digestion on virus infectivity, *Water Res.*, 11, 579, 1977.
35. **Sanders, D. A., Malina, J. F., Jr., Moore, B. E., Sagik, B. P., and Sorber, C. A.,** Fate of poliovirus during anaerobic digestion, *J. Water Pollut. Control Fed.*, 51, 333, 1979.
35a. **Scheuerman, P., et al.,** unpublished data.
36. **Ward, R. L. and Ashley, C. S.,** Comparative effects of ammonia and related compounds on poliovirus, *Appl. Environ. Microbiol.*, 36, 198, 1978.
37. **Farrah, S. R., Bitton, G., Hoffmann, E. M., Lanni, O., Pancorbo, O. C., Lutrick, K. C., and Bertrand, J. E.,** Survival on enteroviruses and coliform bacteria in a sludge lagoon, *Appl. Environ. Microbiol.*, 41, 459, 1981.
38. **Ward, R. L. and Ashley, C. S.,** Inactivation of enteric viruses in wastewater sludge through dewatering by evaporation, *Appl. Environ. Microbiol.*, 34, 564, 1977.
39. **Damgaard-Larsen, S., Jensen, K. O., Lund, E., and Nissen, B.,** Survival and movement of enteroviruses in connection with land disposal of sludges, *Water Res.*, 11, 503, 1977.
40. **Wallis, C., Melnick, J. L., and Rapp, F.,** Different effects of $MgCl_2$ and $MgSO_4$ on the thermostability of viruses, *Virology*, 26, 694, 1965.
41. **Pohjanpelto, P.,** Stabilization of poliovirus by cystine, *Virology*, 6, 472, 1958.
42. **Foliquet, J. M. and Doncoeur, F.,** Inactivation assays of enteroviruses and *Salmonella* in fresh and digested wastewater sludges by pasteurization, *Water Res.*, 6, 1399, 1972.
43. **Ward, R. L. and Ashley, C. S.,** Identification of detergents and components of wastewater sludge that modify the thermal stability of reovirus and enteroviruses, *Appl. Environ. Microbiol.*, 36, 889, 1978.
44. **Kawata, K., Cramer, W. N., and Burge, W. D.,** Composting destroys pathogens in sewage solids, *Water Sewage Works*, 124, 76, 1977.
45. **Olver, W. M., Jr.,** Cold weather sludge composting works in Maine, *Compost Sci.*, 21, 20, 1980.
46. **Heaman, J. D.,** Composting — an approach to using sewage waste, *Compost Sci.*, 18, 28, 1977.
47. **Ward, R. L.,** Inactivation of poliovirus in wastewater sludge with radiation and thermoradiation, *Appl. Environ. Microbiol.*, 33, 1218, 1977.
48. **Cliver, D. O.,** Surface application of municipal sludges, in *Virus Aspects of Applying Municipal Waste to Land*, University of Florida, Gainesville, 1976, 77.
49. **Lund, E.,** Disposal of Sludges, in *Viruses in Water*, Berg, G., Bodily, H. L., Lennette, E. H., Melnick, J. L., and Metcalf, T. G., Eds., American Public Health Association, Washington, D.C., 1976, 196.
50. **Sagik, B. P., Duboise, S. M., and Sorber, C. A.,** Health risks associated with microbial agents in municipal sludge, in *Sludge — Health Risks of Land Application*, Bitton, G. and Damron, B. L., Eds. Ann Arbor Science, Ann Arbor, Mich., 1980, 15.
51. **Johnson, M. J.,** Does U.S. need national sludge management policy?, *Water Wastes Eng.*, 16, 44, 1979.
52. **Sobsey, M. D., Wallis, C., and Melnick, J. L.,** Studies on the survival and fate of enteroviruses in an experimental model of a municipal solid waste landfill and leachate, *Appl. Microbiol.*, 30, 565, 1975.
53. **Engelbrecht, R. S. and Amirhor, T.,** Disposal of municipal solid wastes by sanitary landfill, in *Virus Aspects of Applying Municipal Waste to Land*, University of Florida, Gainesville, 1976, 59.
54. **Bitton, G., Farrah, S. R., Pancorbo, O. C., and Davidson, J. M.,** Fate of viruses following land disposal of sewage sludge, I. Survival and transport patterns in core studies under natural conditions, in Int. Symp. Viruses and Wastewater Treatment, University of Surrey, Guilford, U.K., September 1980.

Chapter 9

VIRUSES IN SOILS AND GROUNDWATERS

James M. Vaughn and Edward F. Landry

TABLE OF CONTENTS

I. INTRODUCTION

Within the confines of normal access routes, soil and groundwater systems should play no significant role in the transmission of disease to humans. It is only through exogenous contamination resulting from human activity that viral pathogens may find a temporary reservoir in these systems and actively threaten human health. During the millennia that preceded the development of high population areas, waste disposal practices were dictated by individual convenience. Societal structures based upon small, widely scattered, family or tribal groups (often nomadic in nature) had little need to concern themselves with the long-term ramifications of biological waste accumulation. However, the eventual evolution of these units into fixed village, town, and urban groupings necessitated the creation of measures for the removal and efficient disposal of waste material from living quarters. The two most commonly used regimes involved disposal into surface waters and disposal on land. Historically, land application may be traced to a number of ancient civilizations.[1] While rarely the principal mode of disposal, these methods have survived the rigors of time and presently find themselves in a most unaccustomed demand. The purpose of this chapter is to identify methods of land application which may serve as sources of human viral pollution to soil and groundwater. These will be discussed in terms of their rationale, likely benefits, and potential hazards to human health. In addition, factors which may enhance or inhibit viral accumulation in soils will be discussed. It is such processes (adsorption, desorption, mobilization, and survival) that ultimately determine the extent of hazards posed by land application.

II. WASTEWATER DISPOSAL METHODS WHICH MAY CONTAMINATE SOIL AND GROUNDWATER SYSTEMS

The revival of interest in land application and land disposal schemes may be linked to a series of population-induced developments which have been accelerated over the past few centuries. Included among these are (1) an increased awareness of the hazards associated with sewage pollution of surface waters, (2) the increasing demand placed on present sources of potable water, (3) the realization that valuable nutrients must be recycled, and (4) the need to recondition depleted soils in order to meet ever increasing demands for human and animal crop production. To some, land treatment represents an attainable panacea, capable of satisfying most, if not all, of the above needs. Unquestionably, land application represents a significant disposal resource which, if properly managed, would abate pollution of surface fresh- and saltwater systems.[2,3] In addition, the reuse of even a fraction of the billions of gallons of wastewater produced annually in the U.S. for the replenishment of groundwater reserves (which currently satisfy the drinking water needs of approximately half the population),[4] would contribute significantly to the alleviation of demands for potable water.[2,5] Recharge is presently considered imperative in many regions served exclusively by groundwater, where future demands for potable water could not likely be met without severely jeopardizing the local ecological and hydrological balances.[6] The irrigation potential of wastewater and its promise of a nearly inexhaustible nutrient supply for soil enhancement and crop production have also been recognized.[2,5,7] Treatment by land application is also considered by many to be less costly than any advanced wastewater treatment alternative.[8,9] As a result of the 1972 amendments to the Water Pollution Control Act of 1965, Congress emphasized the use of land treatment systems (reviewed by Larkin).[7] By these amendments, the practicality of land treatment must be considered for all wastewater treatment systems constructed since 1974. Furthermore, by 1983, the best practical treatment must be used which will not cause surface water pollution. Land application has subsequently been encouraged in the Federal Clean Water Act (PL 95-217) passed in 1977. The enactment of laws favorable to land application has significantly increased the participation of federal regulatory agencies in the development of appropriate management practices.[10] At the present time, more than 700 cities are disposing of a portion, or all, of their wastewater effluents and sludge in some land-based system such as forests, golf courses, pastures, and agricultural soils used for animal crop production.[7]

A complete listing of land treatment methods should include both "intentional" and "unintentional" modes of disposal on or in soil. This latter group is of no small consequence as some 800 billion gal of raw sewage from septic tanks and cesspools leach into the ground on an annual basis, with an additional 250 billion gal of raw and treated sewage reaching the soil each year from leaking municipal sewage systems.[11] As will be seen in a later section, such sources of "unintentional recharge" have often been identified as the contamination source in a number of human virus disease outbreaks.

The selection of an appropriate "intentional" land treatment method from among the three basic systems available is usually dictated by the prevailing soil conditions, with soil permeability being the most notable determinant.[1] Irrigation/spraying of sewage liquids and settled solids (sludges and sludge slurries) to agricultural lands are usually carried out on soils which only permit low-rate infiltration of associated liquids. Chemical and biological treatment in these systems is accomplished by physical filtering, adsorption, metabolic activity of soil microbes, and natural uptake mechanisms in plants.[12] Land treatment by overland flow also uses soils with low permeability that often have an underlying confining clay or artificial barrier to prevent percolation through the soil.[13] Here, wastewater applied to a vegetated slope travels along the soil-vegetation interface.[12] Treatment is accomplished by both physical (adsorption, sieving) and biological processes (metabolic activity of ambient microbes and plants). Effluents produced by this system may receive additional treatment through recycling or may be disposed of via discharge to surface waters, or groundwater recharge basins.[13] It has been recommended that such discharges be preceded by disinfection in order to reduce the numbers of pathogens surviving the original treatment process.[14,15] The third type of land treatment, low rate infiltration, is recommended for areas having moderately permeable soils. It is the most cost and land effective of the three, requiring only the construction of a series of shallow basins.[12] The mode of treatment is through physical filtration of large particulates and suspended solids at the soil surface, and adsorption of free particles to specific components at sub-surface depths. Modification of this method in highly permeable soils has been useful in wastewater treatment; however, this rapid infiltration mode presents the greatest potential for abuse when the primary intent of the process shifts from treatment to high volume disposal. In this situation, the maintenance of excessively high soil infiltration rates compromises the normal treatment potential of the soil percolation process (treatment mechanisms include adsorption to soil, sieving, microbial metabolic activity in upper soil layers, etc.). Thus, with effective treatment by-passed, the danger of significant groundwater contamination may be markedly increased.

A listing of the above methods is presented in Table 1 in terms of the overall benefits of each method, potential mechanisms for virus removal, and possible viral risks associated with their long-term use. Although not a land disposal method in the strictest sense, the increasing use of deep well injection systems for the disposal of chemicals and sewage (some 650,000 currently in use in the U.S.) poses a similar potential threat for widespread contamination of groundwater resources,[4] warranting its inclusion in the table.

Foremost among questions demanded of any land treatment scheme are those pertaining to potential hazards to human health. The "answers" to these questions, as provided over the past century, have supported all possible viewpoints ranging from "no conceivable hazard" to pronouncements of the irrevocable poisoning of all land and groundwater systems. During the 19th century, the First, Second, and Third Royal Commission on Sewage Disposal advised Victorian England that sewage purification could only be accomplished by its application to land, a practice which would simultaneously abate the pollution of surface waters.[16] A review of land application usage from the late 19th to early 20th century indicates that a number of experimental projects were initiated in several western European and American cities.[17] In many cases projects were abandoned, often for aesthetic reasons or as a result of apprehensions caused by the fear of potential health hazards. While the health aspects of land treatment have not as yet been adequately studied, many authors have expressed confidence in the impunity of its use. In 1968, Dunlop[18] indicated that no known disease outbreak associated with land treatment of properly treated sewage had ever been reported. In the same year, Krone[19] described land disposal as both feasible and reasonable for widespread use. A U.N. report published in 1975[20] suggested that groundwater recharge with adequately treated wastewater would provide a considerable degree of safety (in terms of human health) as a result of natural purification, the likely long detention time of pathogens in soil, and the dilution of potentially hazardous material in the aquifer. The report further indicated the technological feasibility of the removal of all or most human pathogenic bacteria and viruses by a "carefully managed" reuse system, but cautioned that additional toxicological information was needed. More recently, Uiga[21] suggested that the land application of raw sludge provided equal or better protection against pathogens than activated sludge treatment. A 1980 review, prepared by Reynolds and Cissell,[9] maintained that land treatment of raw wastewater was as effective as conventional secondary treatment in the removal of many wastewater constituents.

Wolman,[3] in a 1977 review of the public health aspects of land treatment, summarized the public's basic prejudice against land treatment, noting that "the intuitive distrust of man's human wastes has a long heritage, not always epidemiologically validated." The concerns, however, of those who would question the large-scale use of land treatment do not arise from mere prejudice. Wellings et al.[22] and Roy[23] warned that land disposal of wastes posed an increased threat to human health through chemical and biological contamination of potable water supplies. Duboise et al.[1] also noted the distinct possibility of groundwater contamination by pathogens as a result of the use of land disposal methods. Sorber and Sagik,[2] while acknowledging the potential of land treatment in surface water pollution abatement, cautioned that land treatment would bring humans into closer contact

Table 1
LAND DISPOSAL METHODS: POTENTIAL BENEFITS AND RISKS WITH RESPECT TO HUMAN VIRUSES

Application method	Overall benefit(s) of disposal method	Potential virological benefits	Potential virus risks
Overland flow	Additional wastewater treatment Soil renovation Crop production	Virus adsorption/removal at soil-vegetation interface Nonspecific viral inactivation by products of bacterial/plant metabolism and other phys/chem effects	Virus contamination of food crops Virus penetration of soil-groundwater contamination Surface water contamination via runoff
Rapid infiltration	Speedy disposal of wastewater Recharge of groundwater Moderate level of additional wastewater treatment Minimum land requirements	Virus adsorption/removal during soil percolation Nonspecific virus inactivation due to microbial processes in upper soil layers	Viral contamination of groundwater
Spray irrigation	Rapid wastewater disposal Recharge of groundwater aquifer Crop production Additional wastewater treatment	Virus inactivation/adsorption at soil-vegetation interface and during soil percolation	Aerosolization of virus particles Contamination of crops Contamination of groundwater Contamination of surface water via runoff
Sludge application	Soil renovation Crop production	Localization/inactivation of virus in sludge or upper soil layers	Crop contamination Groundwater contamination Surface water contamination via runoff
Deep well injection[a]	Rapid wastewater disposal Groundwater recharge	Little likelihood contamination of soil, surface water, crops, or production of infectious aerosols	Groundwater contamination

[a] While not usually considered a "land disposal" method, injection is included here on the basis of its possible risk for groundwater contamination.

with a variety of pathogens. In addition to groundwater pollution, concerns have been expressed over the possible contamination of food crops grown in sewage-irrigated soils.[7] Addressing the likelihood of soil contamination through land disposal methods, Akin[24] observed that "the literature clearly indicates that members of each group of sewage pathogens can survive sewage treatment, albeit in reduced numbers, and can, at times, be recovered from the receiving soils." Duboise et al.[1] calculated that with a modest wastewater application rate of 2 in./week, the daily virus load would be 1.6×10^4 PFU/acre.

While these contentions do not conclusively prove or disprove the existence of a qualified health hazard associated with the use of land treatment, they underscore the inadequacy of present knowledge to provide a basis for a reliable assessment of risks. A number of recommendations have been proposed which would mitigate much of the potential microbial health threats associated with land treatment schemes. One of the major health-oriented considerations in planning any sewage reuse operation is the level of contamination of the wastewater itself.[25] It follows therefore that the simplest means of assuring safety from pathogenic organisms is the total removal of pathogens from wastewater effluents prior to their introduction into land treatment system.[26] Unfortunately, the attainment of complete pathogen removal (especially with respect to human viruses), while highly desirable, is not yet readily within the reach of most conventional sewage treatment methods. As a result, the recommendations for the operation of land treatment systems that have been developed are based upon the assurance that wastewater inputs will contain varying numbers of pathogens. Shuval[27] suggested some form of sewage pretreatment as a means of reducing the pathogen concentration before land application. Schaub et al.[28] pointed out that prolonged application of wastewater (overland flow systems) may modify the physical/chemical components of soil, with respect to available sites for virus adsorption, and suggested that continuously saturated soils may be less available for the adsorption of pathogens. These findings supported an earlier recommendation by Bell and Bole,[29] who proposed a drying period between wastewater applications. Krone[19] proposed the use of clay soils in overland flow systems to prevent the movement of pathogens to groundwater. Vaughn and Landry[15] proposed guidelines for rapid infiltration-aquifer recharge systems that used secondarily-treated wastewater. These guidelines included:

1. Maintenance of coliform levels to conform to standards prescribed for secondary effuents including a fecal coliform MPN of no greater than 200/100 mℓ; (this recommendation was in approximate agreement with that of Krishnaswami[30] who suggested a more stringent wastewater coliform MPN of no greater than 100/100 mℓ prior to land treatment).
2. The placement of wastewater recharge basins should be dictated by ambient soil characteristics and local depth to groundwater in order to assure maximum pathogen removal.
3. Basins should not be constructed in areas which abut surface water systems where saturated soil conditions might facilitate the subsurface movement of pathogens.
4. Recharge basins should not be constructed within close proximity to public water supply wells.

(Note: similar guidelines for recharge basin siting have recently been recommended by the Environmental Protection Agency [EPA]).[5] The application of settled sewage solids (sludge) may represent an additional risk since they contain the greater proportion of sewage pathogens.[23] Addressing this problem, Moore et al.[31] advocated sludge treatment (digestion) to reduce the overall pathogen load to the land. Hunt[14] and Vaughn and Landry[32] indicated the need to disinfect effluents resulting from overland flow treatment prior to disposal of the effluents into recharge basins. Jelnick[33] identified the Food and Drug Administration guidelines on crop production in sludge-fertilized soil. These stipulated that (1) crops normally eaten raw should not be planted within 3 years after the last application of sludge and (2) crops which may contaminate other foods when brought into the home should not be grown in recently treated soil unless the sludge has been shown to be free of pathogens. Larkin[7] and Wolman,[3] supporting a more conservative approach, suggested that no crops consumed raw should be grown in sludge-treated soils.

Chief among concerns associated with land treatment of pathogen-containing wastewaters and sludge is the integrity of the aquifer. Groundwater protection was stipulated an essential by-product of all land application methods in a recent state-of-the-art review by Wright and Rovey.[12] Keeley[11] underscored the need to protect existing pristine groundwater systems before they become contaminated, rather than to attempt to restore them after they are contaminated. Given the numerous information gaps which presently obstruct the adequate evaluation of pathogen fate in the various land treatment modes, the prudence of establishing microbial standards for wastewater entering land systems might be argued. Before attempting to provide such standards, one would do well to heed Stokinger's advice[34] concerning the considerations important in standard setting: "1) standards should be based on scientific facts, not on political feasibility, expedience, emotion of the moment, or unsupported information; 2) avoid establishing an unnecessarily severe standard; 3) determine realistic levels, haste to "put a number on it" has yielded criteria developed by a "look what I found" attitude and acceptance of questionable information by harried standard setters grasping at straws to comply with an unreasonable legal deadline." The application of these criteria to the unique problems posed by human viruses suggests the wisdom of avoiding such standards at the present time. Existing information on such varied aspects as virus-soil interactions, virus movement in soils, virus movement and survival in groundwater, epidemiology, and dose responses do not as yet provide the appropriate data base from which the necessary decisions indicated by Stokinger could be made. Attempts to impose an artificially derived viral standard would likely result in the chaotic situation described in Stokinger's third tenet. For the present, then, it would appear that the only recourse to the inability to produce virus-free effluents, and the undesirability of establishing unsupportable standards is to use land treatment management practices with the proven ability to reduce the number of viruses. A discussion of specific measures for reducing pathogen concentration during land treatment appears in a later section.

III. VIRUS OCCURRENCE IN SOILS, PLANTS, AND GROUNDWATER

Despite the historical use of land treatment methods, concerted efforts to identify the impact of human viruses on these systems have only recently been initiated. The delay has largely been the result of technical problems associated with the still embryonic field of Environmental Virology. Techniques for the extraction, concentration, and enumeration of naturally occurring human viruses from soils, foods, and water, while undergoing tremendous advances during the past decade, remain only marginally efficient for a few "marker" virus types (e.g., some enteroviruses), and totally inadequate for an alarming number of other environmentally important groups (e.g., hepatitis A, Norwalk agent, human rotaviruses). This dictates that virus recoveries reported in field studies (such as those discussed below) should be assessed on a semi-qualitative, rather than a strict quantitative basis. In addition, the inability to detect viruses in particular samples does not preclude the possibility of viral presence in low concentrations. Great care should always be exercised before assigning a "zero virus" designation to any sample, or group of samples. Conclusions based upon such results should always identify the limitations of the testing methods in order to prevent misinterpretation and over-generalization of data.

A. Virus Occurrence and Distribution in Soils

Several authors have described viral distribution in soils receiving domestic wastewater. Brown et al.[35] studied the adsorption of septage-associated coliphage f2 to three different soil types containing 80, 41, and 7.6% sand, respectively. Initial distribution patterns indicated that the greatest percentage of viruses were located within the first 15 cm of soil below the septic application lines. Several isolates were found at 85 cm, with single isolations observed at 100 and 120 cm depths. Hurst and Gerba[36] identified the distribution pattern of a seeded poliovirus in Flushing Meadows soils. Of the viruses recovered, 91% were in the top 2.5 cm of soil. The remaining 5.1 and 3.4% were isolated from the 2.5 to 10 and 10 to 25 cm depths, respectively. Similar distribution profiles were reported by Landry et al.[37] in 10.1 × 75 cm *in situ* soil cores challenged with a guanidine-resistant strain of poliovirus[1] (LSc-2ab); 77% of the seed virus was found in the top 5 cm of soil and an additional 11% was recovered from the 5 to 10 cm depth. There was 8% of the seed virus recovered from the next 15 cm, and the remaining 4% was uniformly distributed through the lower 50 cm of soil. All of these studies clearly indicate the tendency of most viruses to be retained initially in the upper soil layers. Mechanisms by which viruses may later desorb and migrate through soils will be the subject of Section III.C in this chapter.

While most research emphasis is placed on movement of viruses to groundwater, several workers have identified viruses in soils following land treatment. Kerfoot et al.[38] observed the penetration of seeded coliphage MS2 to depths of 1.2 m in the saturated soil of a sand filter bed located on Cape Cod. Derbyshire and Brown[39] recovered porcine and bovine enteroviruses from soils which had been treated with animal waste slurries. Sagik et al.[40] isolated bacteriophages and human viruses (67% of which were poliovirus 1) from lysimeters placed in saturated soils at a spray irrigation site and reported virus recoveries to soil depths of 1.4 m. Schaub et al.,[28] investigating virus removal in 11 × 36 m overland flow test plots, recovered seed coliphage f2 from wastewater-saturated topsoil.

B. Virus Association with Crops Grown in Wastewater-Treated Soil

Because of the known persistence of human viruses in soil and on the surface of vegetables,[7,41] the potential for disease transmission via crops grown in soils used for the treatment of wastewater and sludge must be considered. Reports published over the past 20 years indicate the major health risk to be associated with virus occurrence on vegetable surfaces. In 1964, Bagdasaryan[42] investigated viral presence on vegetables collected from agricultural spray irrigation fields and recovered human viruses from 3.3% of the samples tested. Artykov[43] reported the isolation of group B coxsackieviruses from vegetables grown in wastewater-irrigated soils. He confirmed the sewage-origin of the isolates by also recovering them from the irrigation wastewater, and from irrigated soils. Bondarenko[44] reported the isolation of high concentrations of coxsackieviruses and echoviruses from carrots, beets, and potatoes which had been grown in soils extensively contaminated with these virus types. Sadovski et al.[45] proposed a means of alleviating surface contamination of crops by placement of wastewater drip lines 10 cm below the soil surface, which was then covered with a plastic sheet. In extensive field tests, the authors were unable to recover any viruses from vegetable surfaces, even when large numbers of viruses were seeded into the irrigation water.

The likelihood of virus transmission within crops appears to be remote. Murphy and Syverton[46] reported the adsorption and infrequent acropetal transfer of FA mouse encephalomyelitis virus in root systems of hydroponically grown tomato, pea, and lettuce plants. In the same experiment, poliovirus 1 adsorbed but was not transferred through roots. The data led the authors to conclude that plants would not likely serve as reservoirs or carriers of human viruses. On the basis of these results, and the fact that no ensuing studies have identified virus uptake by plants, it would appear that crop-contamination in land treatment systems is solely a function of surface accumulation from spraying, direct contact with contaminated soils, or indirect soil contact through splashing during rainfall.[47]

C. Occurrence of Viruses in Groundwater

Gerba et al.[48] in a 1975 article advanced the position that the mere presence of viral pathogens in groundwater should be viewed as comprising a potential health hazard. Considering the enormous number of people relying solely on groundwater for their drinking water source and the unknown disease potential of certain human viruses, this view should not be dismissed as overly alarmist. During the past decade, an increasing number of reports have identified the presence of human viruses in groundwater as a direct result of intentional, or unintentional soil discharges. Several studies have also identified the potential for lateral transfer of viable virus particles through aquifer entrainment.

1. Groundwater Contamination Resulting from "Unintentional" Land Treatment

Wellings et al.[49] demonstrated echovirus 22/23 in a series of 10.6 to 12.1 m deep groundwater wells which were clustered in an area surrounded by septic tanks. A septic tank source was also proposed by Vaughn and Landry[32] following the isolation of echoviruses 11 and 23, and coxsackievirus A16 from a groundwater observation well located in the center of a cluster of single family dwellings which discharged wastes to individual septic systems. Sobsey and Scandura,[50] in a recent interim report to the North Carolina Department of Human Resources, noted the appearance of small numbers of marker viruses in an aquifer following a seeding of bovine enterovirus into septic system leach lines. More recently, Vaughn and Landry[51] observed groundwater contamination with naturally-occurring viruses emanating from an underground septic disposal source. In addition to indicating virus occurrence in the glacial aquifer immediately beneath and adjacent to the septic source, this latter study has identified a 30.5 m lateral migration of aquifer-entrained viruses.

Further discussions of virus contamination from unintentional recharge sources and their association with disease outbreaks will appear in Section II.E.

2. *Virus Occurrence in Groundwater following Intentional Land Treatment*

Studies assessing virus contamination of groundwater resulting from land treatment practices have produced contrasting results. Reporting on a spray irrigation project conducted on Cape Cod, Kerfoot and Ketchum[52] found no evidence of seeded coliphage MS2 in groundwater monitoring wells located 14.6 to 16.7 m below the experimental plot. In a brief study conducted at the Flushing Meadows recharge facility, Gilbert et al.[53] were unable to recover viruses from groundwater observation wells located beneath the recharge basins. Vaughn et al.,[54] reporting the results of a 1-year monitoring study of groundwater located 30.5 m below a secondary effluent recharge system, also failed to isolate human viruses. Koerner and Haws,[55] studying viral impacts in an area used for land treatment of domestic wastewater over a 33-year period found no evidence of viruses in groundwater. Factors such as slow infiltration rates, appreciable depth to groundwater, and clay-containing soils were credited with efficient removal of percolating viruses.

Several authors have investigated the potential for viral contamination of groundwater resulting from sludge application to soil. Derbyshire and Brown[39] were unable to recover bovine and porcine enteroviruses from an aquifer located directly beneath soil layers receiving animal waste slurries contaminated with these virus types. Similarly, no detectable viruses were noted in the groundwaters beneath an area used for land application of anaerobically digested sludge collected from the city of Chicago.[56] Recently, Farrah and Bitton[57] reported good viral retention in Florida soils receiving sludge applications. Aerobically and anaerobically digested sludges were lagooned for a brief period and applied to respective agricultural fields. Viruses were routinely isolated from lagoons and sporadically recovered from sludge-soil mixtures following application. However, no viruses were detected in the groundwater monitoring wells located on the disposal site, or near the sludge lagoons.

Contrasting (but not necessarily conflicting with) the above studies are a growing number of reports which have identified virus contamination of groundwater as a direct result of some form of land treatment. In addition to confirming the vertical movement of viruses through soils, several of the studies described below have presented evidence for lateral movement of viruses through aquifers. A recent review by Keswick and Gerba[58] contains a comprehensive summary of those reports indicating virus occurrence in groundwater.

Perhaps the earliest evidence for the presence of viruses in groundwater based upon actual recovery, rather than an epidemiological premise, was that provided by Mack et al.[59] in 1972. With a relatively inefficient virus concentration method, Mack and colleagues isolated poliovirus 2 from a 30.5 m deep drinking water well located some 91.5 m from the edge of a wastewater drain field. The data indicated both vertical soil penetration and significant lateral viral movement through the aquifer. Similar findings were reported by Wellings,[60] who described soil penetration and lateral movement of viruses through groundwater. The source of the viruses was a discharge of secondary wastewater through a cypress dome. In this study, coxsackievirus B4 and polioviruses 1 and 2 were isolated. In a later study, Wellings et al.[49] recovered echovirus 22/23 from a well located 30 lateral meters from a wastewater disposal site. Vaughn et al.[54] repeatedly recovered enteroviruses from two rapid wastewater infiltration sites located on Long Island, a region where groundwater serves as the sole source for potable water. At the first site, echovirus 12 and several unidentifiable isolates were recovered from a groundwater observation well adjacent (3.0 m) to a sand recharge basin that received daily applications of secondary wastewater. Virus recoveries at the second site provided additional evidence for lateral movement in the aquifer. Here, the observation well was located 45.7 m down gradient from sand basins receiving tertiary wastewater effluents. Among isolates identified were echoviruses 6, 9, 21 and 25. Vaughn and Landry[61] also reported poliovirus 2 isolations from groundwater at another rapid infiltration site on Long Island which recharged tertiary treated effluents. Further evidence for groundwater contamination resulting from the use of rapid infiltration treatment/disposal of wastewater was provided by Schaub and Sorber,[62] Koerner and Haws,[55] and Keswick and Gerba.[58] Schaub and Sorber reported virus isolations in groundwater beneath rapid infiltration units receiving secondary effluents. Viruses were recovered at depths of 30 m and at lateral distances of 182.8 m. Koerner and Haws observed the results of rapid infiltration of primary effluent through course gravel and sand. Polioviruses, coxsackieviruses, and echoviruses were recovered from groundwater at a depth of 16.8 m. The authors also reported virus isolations at 250 m down gradient distances. Earlier reports that there was no viral contamination of groundwater beneath the Flushing Meadows rapid infiltration site[53] have been controverted by the recent recovery (reported by Keswick and Gerba[58]) of coxsackievirus B3 from an 18.3 m deep well located on a land application site situated near the now defunct Flushing Meadows facility.

With the exception of rapid infiltration studies, few reports to data have presented evidence for viral contamination of groundwater resulting from the use of "intentional" methods. Vaughn and Landry[31] isolated unidentifiable virus types from a groundwater observation well located 10 ft down gradient from a subsurface leaching pool receiving secondary effluents. In the same account, these investigators reported a single isolation of coxsackievirus B3 from an observation well located some 402 m down gradient from a sanitary landfill. Marzouk et al.[63,64] isolated enteroviruses in 20% of groundwater samples tested in Israel. While the precise source of sewage contamination was not identified, the authors identified the likely sources as subsurface seeping from septic tanks and underground sewer lines, and wastewater-land application systems.

From the above review, it would appear that the greatest potential for groundwater contamination with sewage-borne human viruses is associated with the use of rapid infiltration systems. Conversely (and admittedly based on limited data), sludge application practices appear to promote efficient virus retention in the sludge/soil matrix,[57] thereby determining the passage of viruses to groundwater. As of now, no groundwater isolations of human viruses have been reported as the result of soil conditioning with aerobically or anaerobically digested sludge.

D. Virus Survival in the Aquifer

Comprehensive *in situ* studies of virus survival in groundwater have not as yet been conducted. It has therefore been necessary either to extrapolate survival rates from laboratory studies, or to approximate them on the basis of the measured movement of viable viruses within an aquifer. Laboratory experiments reported by Keswick and Gerba[58] indicated a 200-day virus survival period in drinking water. On the basis of data collected in the field, Wellings et al.[65] suggested survival through a 28-day residence time. Recently, Vaughn and Landry[51] observed a 30.4 m lateral movement of viable viruses through a glacial aquifer.

The average groundwater velocity in the study area was 15.3 cm/day; and approximate minimal survival period of 199 days was calculated.

Factors that control viral survival in groundwater are likely to be similar to those that influence viral survival in saturated soils.[58] An extensive discussion of these factors is presented in Section IV.

E. Virus Disease Outbreaks Associated with Wastewater Contamination of Groundwater

At the present time, the precise assessment of waterborne viral disease outbreaks is hampered by several factors:

1. Mild to moderate cases of acute gastroenteritis are often unrecognized, or go unreported.[66] McDermott[67] suggested that the 245,000 cases of waterborne diseases reported between 1961 and 1970 probably represented only 10% of those which had actually occurred. Application of his belief would result in a revised figure of 2.5 million cases. While surface water contamination probably makes up a major fraction of this total, contaminated groundwater sources would also be expected to contribute significantly to the disease rate.
2. Since many of the diseases which may be waterborne can also be transmitted via direct contact, it is often difficult to accurately trace the origin of a particular outbreak (this is particularly true for most human enteroviruses).
3. Many of the virus types responsible for waterborne outbreaks (see below) are difficult or impossible to isolate and identify.

The most recognized virus group associated with waterborne outbreaks is hepatitis A (infectious hepatitis). For many years, it has been the only viral agent conclusively proven to be transmitted by the water route.[68,69] Recently, efforts have been made to determine the etiology of the numerous cases of waterborne, acute, nonbacterial gastroenteritis which occur each year. Craun[70] indicated that as many as 57% of the waterborne outbreaks occurring between 1971 and 1977, categorized as acute gastroenteritis, were of unknown etiology. He further suggested that many of the outbreaks were consistent with a viral etiology, identifying rotaviruses and the three known serotypes of Norwalk agent as the most likely causative agents. Holmes[71] previously identified these groups as the primary viral agents of gastroenteritis in humans. Richmond et al.[72] recently reported an outbreak of gastroenteritis in children, implicating an adenovirus etiology. Evidence for waterborne transmission of rotaviruses and Norwalk infection are presented below. At present, there are no data identifying waterborne outbreaks of adenoviral gastroenteritis.

Human enteroviruses have frequently been isolated from contaminated groundwater and soils, and appear to represent a reasonable indicator of domestic wastewater contamination. While enterovirus etiologies have rarely been documented in waterborne disease outbreaks, their disease-potential cannot be dismissed. It may be that their propensity towards transmission by direct contact may mask their waterborne transmission character. On the basis of their disease potential, their occurrence in sewage, and their relative handling ease in the laboratory, investigators have frequently used members of this group as a monitoring and research tool in assessing the relative hazards posed by land application systems.

Craun[70,73-75] has presented an annual review of all reported waterborne disease outbreaks for the past decade. The following will focus on those disease outbreaks suspected to be associated with viral penetration of soil layers and their subsequent entrance and movement through the subsurface water table.

Microbial contamination of poorly protected groundwater systems may represent a major source mechanism for waterborne disease outbreaks.[67,70,73,76] In most cases, contamination of shallow aquifers has been the direct result of wastewater disposal at or near the soil surface.[77] The single most often identified source of contamination has been subsurface septic systems and cesspools.[78] Several reports published between 1942 and 1959 specifically identified these sources in outbreaks of waterborne hepatitis.[79-81] More recently, two infectious hepatitis outbreaks involving 184 cases in Polk County, Ark. were attributed to the consumption of groundwater contaminated by lateral drainage leachate from septic tanks.[82,83] Vilim et al.[84] reported on an outbreak of waterborne hepatitis in a small Czechoslovakian community which affected 24 of the 180 adult population. Subsurface wastewater penetration of a drinking water well was believed to have been initial source of the viral agent. Septic tank seepage was also indicated in a hepatitis outbreak which occurred in Alabama.[85] Here, springs which provided drinking water were apparently contaminated from nearby septic tanks following a period of heavy rain. While the rain-induced runoff of surface seepage was the most probable mechanism for virus transmission to the springs, the possibility of subsurface, lateral movement of viruses abetted by the heavy rainfall cannot be ignored (the role of rainwater in the desorption and movement of viruses through soil has recently been described by Wellings et al.[22]).

As previously noted, a virus etiology has been proposed for a number of groundwater-associated outbreaks of acute gastroenteritis when no other agent could be detected in contaminated water or patient stool specimens. An outbreak reported in Alaska, involving over 100 cases, was linked to the consumption of sewage-contaminated groundwater.[86] The causative agent was never identified. Gastrointestinal outbreaks involving 759 persons, occurring in southern Missouri and portions of Arkansas, were associated with wastewater leakage from municipal oxidation lagoons into porous limestone, facilitating movement to groundwater.[87] Again, the etiological agent could not be determined. Subsequent cultures of stool specimens from affected individuals yielded no isolations of enteric bacterial pathogens. In a review of this outbreak, Craun[73] supported the possibility of a virus etiology. A December 1980 outbreak of gastroenteritis affecting 50% of the residents of a South Carolina trailer park was traced to a contaminated drinking water well.[88] While coliforms were detected in the water during the outbreak, no etiological agent could be demonstrated.

Several reports of soil/groundwater-associated, gastroenteritis outbreaks have provided more direct evidence for a virus etiology. Morens et al.[89] investigated an outbreak at a Colorado campsite which involved some 418 clinical cases. The local source of drinking water was apparently contaminated by seepage from a leaking septic system. Immune electron microscopy of diarrheal stool filtrates revealed the presence of 27 nm virus-like particles. Oral administration of bacterial-free filtrates containing these particles to volunteers resulted in the onset of gastrointestinal illness. The morphology of the organism and the symptoms of the disease were consistent with previous descriptions of Norwalk virus outbreaks. Members of this virus group were also implicated in a waterborne gastroenteritis outbreak that occurred in a northeastern Pennsylvania summer camp.[90] Bacteriological studies of the camp's drinking water well showed the presence of coliforms, suggesting sewage contamination. Cultures of stool samples collected from ten patients revealed no bacterial pathogens. However, fourfold or greater rises in Norwalk agent serum antibody levels were demonstrated by radioimmunoassay in several patient serum specimens. Wastewater contamination of a water supply was linked with a gastroenteritis outbreak thought to be caused by a rotavirus.[91] The outbreak involved 3172 cases in a small Swedish town. Virus-like particles of typical rotavirus morphology were detected in patient stool specimens in which bacterial and enteroviral pathogens could not be detected.

To date, no outbreaks of waterborne human disease have been directly linked to the use of the large-scale land-treatment methods described in Section I. This may be due in part to the relatively small numbers of systems currently in use, and their location in remote or low population-density areas. Other factors which may have hampered the identification of land treatment-associated outbreaks have been discussed earlier in this section. In spite of the present lack of unequivocal evidence, the above studies clearly indicate the potential for such systems to contribute significantly to groundwater-borne disease outbreaks. This premise is reinforced by the data presented in Section II.C, which demonstrate the proclivity of some land treatment modes to allow the distribution of viruses to regional aquifers. Caution cannot, therefore, be over-stressed in the selection, planning, and operation of wastewater treatment and disposal methods which may ultimately jeopardize groundwater reserves.

IV. MECHANISMS CONTROLLING ADSORPTION AND MOVEMENT OF VIRUSES THROUGH SOILS

Currently practiced sewage treatment methods are inadequate for the removal of all human viruses from domestic wastewaters.[92] It is therefore reasonable to expect that some level of viral contamination will be applied to soils when wastewater is used in land treatment operations. While the concentration of applied viruses varies with the type and efficiency of pretreatment,[93,94] the mere presence of viruses in effluents targeted for land application dictates that all potential health problems associated with this type of sewage reuse be assessed. The degree of hazard posed ultimately depends on the retention and survival of viruses in soils. Major portions of two recent symposia were devoted to discussing and assessing the hazards posed by viruses in soils and groundwater.[95,96] Generally, when viruses are applied to soils they will (1) adsorb to soils and lose infectivity, (2) adsorb to soils but remain viable for long periods, (3) remain unadsorbed and percolate through the entire soil profile to the aquifer, or (4) initially adsorb but later desorb and percolate through the soil column. It is obvious that the viruses which present the greatest threat to groundwater contamination are those which can desorb, percolate through the entire unsaturated zone and reach the aquifer. Within the past few years, an increasing number of reports have indicated that these viruses can indeed travel considerable horizontal and vertical distances to reach groundwater.[58] It therefore becomes imperative to provide an understanding of the factors which govern viral adsorption, desorption, and migration in soils. The primary objective of this section is to review the nature of the adsorption/desorption process and to discuss those factors which promote or antagonize both processes. The topic of virus survival will be the subject of a later section.

A. Mechanisms of Virus Adsorption to Soil Surfaces

1. Adsorption and Filtration

Viruses released in treated sewage effluents can exist as adsorbed particles bound to or embedded in suspended solids, or as discrete unassociated particles. In the former instance, the major viral removal mechanism in soil is via entrapment or filtration of virus-laden solids at the soil surface. In the unassociated state, the key removal factor is adsorption of the virus particles to specific soil components.[97,98] Vilker et al.[99] recently suggested that a majority of viruses released from treatment plants exist in the unassociated state. He observed that under conditions typical of most treatment plant effluents (i.e., low suspended solids concentration, neutral pH, and moderate ionic strength), 80% of viral particles were free and unbound. A similar finding was also reported by Wellings et al.,[100] who noted that while the total number of sewage-borne viruses decreased with an increase in the number of treatment steps, the ratio of free to associated viruses actually increased. It would therefore appear that in land application systems such as spray irrigation, overland flow, rapid and slow infiltration, the main concern should be the specific adsorption of free viruses to soil.

2. Nature and Formation of the Electrochemical Double-Layer (ECDL)

The general nature of virus adsorption to soils has been discussed in a number of excellent reviews.[1,97,98] Nevertheless, it is important to restate some key facets of this process in order to fully understand those factors which govern virus adsorption and desorption. The capsid, or surface of a virus particle, is composed of numerous ionizable proteins which are subject to protonation and deprotonation reactions, depending upon the pH and ionic strength of the surrounding medium. As a result, these colloidal particles possess an overall electrical charge which determines their behavior in solution. The magnitude and sign of this charge is dependent not only on the concentration and type of surface ionizable groups on the virus, but also the number and type of adsorbed and/or loosely-associated ions originating from the surrounding solution.[101] This entire region of bound and associated ions is termed the electrochemical double-layer (ECDL). While a complete understanding of this layer and its interactions with other colloids portends a modest understanding of complex physical-chemical laws, it is sufficient for this discussion to highlight some of its basic features. Excellent descriptions on the molecular nature of the ECDL have been recently published by Taylor,[102] Murray,[103] and Murray and Parks.[104] Generally, the ECDL consists of an inner, or Stern layer, which is formed by an attraction of electrolytes (cations and anions) and water molecules from the bulk aqueous solution to the ionized surface of the virus particle. These ions, which partially neutralize the surface charge of the ionized viruses, are specifically bound via hydrogen or covalent bonds, and form an immobile layer. The outer, or Guoy layer is formed by the attraction of excess counter-ions (opposite charge), and repulsion of coions (same charge) from the surrounding solution. The ions in this layer exist in a diffuse, unbound state and constantly remain in thermal motion between the fixed layer and the bulk solution. The boundary between the fixed, immobile layer (Stern) and the bulk fluid solution is characterized by an electrical potential called "zeta", which generally reflects the overall charge of the virus particle. This potential is related to the electrophoretic mobility of the particle and ultimately appears to govern its behavior in a specific aqueous environment.

a. Effect of pH and Ionic Strength on ECDL

The pH and the ionic strength of the aqueous environment play a key role in determining the overall type and strength of the virus charge. Since pH influences the extent of surface group ionization, it will have a direct effect on both the zeta potential and the electrophoretic mobility of the virus. The pH at which a particle has a net zero mobility is termed the isoelectric point (IEP). At a pH below the IEP, the virus has an overall positive charge, while at values above the IEP the virus has an overall negative charge. Since each virus type has different surface characteristics, its specific IEP would be expected to be unique to that particular virus. For example, reovirus 3 (Dearing) has an IEP of pH 3.8.[102] Poliovirus 1 (LSc), on the other hand, appears to have two IEP, one at approximately pH 7.0 and the other near pH 4.5.[105] Each of these IEP appears to define a specific conformational state of the virus. Between the IEP, the overall charge is probably weak and heterogeneously distributed around the surface. Below pH 4.5 the particle behaves as a positively charged species, while above 7.0 it possesses a net negative charge. A listing of the IEPs of several enterovirus types has recently been published by Murray and Parks.[104] An analysis of the reported range of IEPs (3.8 to 8.5) indicates how variable charges are on different virus types.

In addition to pH, electrolyte concentration is an important determinant for the strength and charge characteristics of the ECDL. Higher ionic concentrations, especially in the form of multivalent ions, result in a partial charge neutralization closer to the virus surface, causing a reduction in the thickness of the double-layer. As a result, the zeta potential is also reduced. Alternately, lower electrolyte concentrations lead to higher zeta potentials and thicker double-layers.

Since soils are colloidal materials and possess surface ionizable groups, they can establish an ECDL in the same manner as viruses. In this case, the extent of ionization of surface hydrous oxides and hydroxide groups on soils determines their overall charge and governs their behavior in solution.[102] As with differing virus types, the IEP of soil varies with the specific type of soil involved. For example, Taylor[102] showed that aluminum hydroxides have an IEP at approximately pH 9.0, while allophane, an amorphous eluminosilicate, has an IEP close to pH 6.5. Murray and Parks[104] tabulated the IEP of a number of substances including quartz (pH 2 to 3.5), kaolinite (pH <2 to 4.6) and montmorillonite (pH ≤ 2.5). As with virus particles, the IEP of these solids determines their relative ability to participate in adsorptive interactions.

b. ECDL Interactions

Since viruses and soil particles both possess an ECDL and an overall charge characteristic, they respond in an attractive or repulsive manner when placed in close proximity. The overall response is proportional to a number of forces including: coulombic (ECDL); London van der Waals; solvation effects; and specific bonding.[102] While the first three forces operate over distances about the size of the double-layer, the specific bonding force required a much closer approach between the virus and soil particles. Murray[103] and Murray and Parks[104] suggested that the total free energy of adsorption is the sum of all the free energies of the double-layer, van der Waals, hydration, and covalent-ionic interactions. Coulombic interactions are electrostatic in nature and are probably one of the most important forces in double-layer interactions. They are a function of the overall sign and magnitude of the zeta potential, with opposite charges attractive and like charges repulsive. Components with the same charge but different magnitudes are also attractive. London van der Waals forces are a major inherent attractive force between colloids, generated from dipole oscillations or charge fluctuations in each material. According to Murray and Parks,[104] these interactions are the major attractive forces in the virus-soil adsorption process. In this study, the authors observed that poliovirus 1 (LSc 2ab) adsorbed strongly to a number of oxide materials at a specific pH and ionic strength. However, when they calculated the contribution of the electrostatic interactions to the overall adsorption process, they found the forces to be repulsive in nature. To account for strong adsorption occurring in the presence of repulsive electrostatic interactions, they postulated that other types of attractive forces (e.g., van der Waals) were involved. Quantifying the role of van der Waals interactions, they observed that these electrodynamic forces were the major contributing forces to the adsorption process. They found their calculations, based on mass-action free energy, to be in close agreement with potentials derived from the DLVO-Lifshitz theory of colloidal stability. From these observations, they concluded that most of the free energy of adsorption could be accounted for by van der Waals forces and, to a lesser extent, by electrostatic double-layer interactions. While other forces, such as specific bonding (covalent or hydrogen), solvation effects, and hydration forces may contribute to the general adsorption processes, Murray and Park[104] found their contribution to the adsorption of poliovirus 1 to various solid oxides to be negligible.

The adsorption of a virus particle to a soil surface is favored if the particle is positioned near a soil surface where the potential energy of adsorption, that is, the sum of all the energy interactions, is lower than that in the bulk solution.[102] Murray[103] indicated that the adsorption of poliovirus 1 was stronger to CuO (ΔG_{ads} [free energy of adsorption] = -65.8 kJ/mol) than to SiO_2 (ΔG_{ads} = -32.4 kJ/mol). Furthermore, he predicted that materials with high dielectric susceptibilities, such as metals and metal sulfides would develop higher van der Waals attractive forces thereby exhibiting virus adsorption properties which were superior to those of silicates and organic materials. He also suggested that other methods for increasing van der Waals interactions, such as increasing the electrolyte concentration, would stimulate the adsorption process.

3. Adsorption Kinetics

The adsorption of viruses to soils is a reversible process which can best be described in

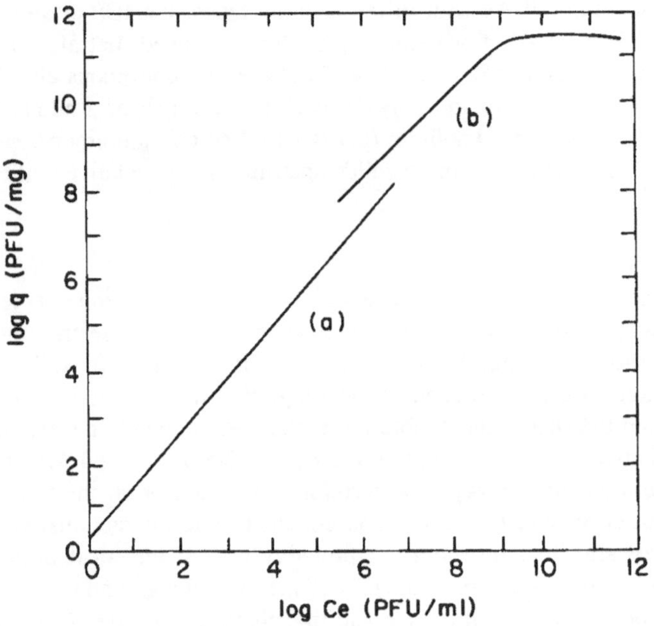

FIGURE 1. Two types of adsorption equilibrium isotherms: a) Freundlich isotherm, b) Langmuir isotherm.

terms of an equilibrium isotherm. Such isotherms can be used to define a number of adsorption characteristics including: the concentration of viruses adsorbed per milligram solid; the relative rate of viral adsorption; and a measure of the surface area involved in the adsorption process.[106] Isotherms are experimentally determined in batch cultures by challenging a constant solids concentration with a range of virus concentrations under a defined set of conditions. If the relationship between the amount of adsorbed and unadsorbed viruses at equilibrium is linear, it is described in terms of a Freundlich isotherm (Figure 1a). This type of relationship has been used to describe the adsorption kinetics of many viruses to a variety of particle types including: bacteriophage R17 to allophane;[107] coliphages f2 and T2 to silty loam;[108] poliovirus 1 to loamy sand,[109] magnetite,[106] and activated carbon.[110] The isotherm is best defined by the equation:[111,112]

$$q = mC_e^{\,n} \tag{1}$$

where: q = the amount of adsorbed viruses per milligram soil; m = the Freundlich constant (y intercept on log q vs. log C_e plot): C_e = the amount of unadsorbed viruses at equilibrium (per milliliter); and n = the slope of the line.

Recently, Vilker[112,113] reported that most adsorption processes were probably saturation limited and followed a Langmuir, rather than a Freundlich isotherm (Figure 1b). This isotherm indicates that there are a large but finite number of adsorption sites on a solid. Once all available sites are saturated by viruses no further adsorption occurs. Vilker suggested that few previous investigations had been conducted with sufficiently high concentrations of viruses to observe this effect. Most of these studies employed relatively low concentrations of viruses in liquid phase and thus the resulting isotherm was well within the linear portion of the saturation curve. In this region, the relationship would be identical to the Freundlich isotherm. A number of studies have shown saturation-limited adsorption including: the adsorption of coliphage T4 to activated carbon;[114] coliphage ΦX174 to silt loam;[115] poliovirus

1 to Ottawa sand[116] and coliphage MS2 to Indian soils.[117] The isotherm itself is defined by the equation:[112]

$$q = QK_LC_e / 1 + K_LC_e \quad (2)$$

where: q and C_e are the same units previously described; Q is the maximum number of available viral adsorption sites; and K_L is the ratio of kinetic coefficients of the forward to reverse reaction k′/k′.

Using these relationships, Vilker[116] determined that poliovirus 1 adsorbed efficiently to an Ottawa sand. Although this sand possessed a total of 3.56×10^8 virus adsorption sites per milligram, he calculated that these adsorption sites represented only 1.5% of the total surface area of the sand. In addition, the measured values he obtained for K_L from the above equation for the poliovirus (3.60×10^{-9} mℓ/virus), for coliphage T4 (4×10^{-7} mℓ/virus) and for coliphage ΦX174 (4.97×10^{-11} mℓ/virus) indicated that the adsorption equilibrium actually favored the liquid rather than the solid phase. However, with the relatively small concentration of viruses used in most experiments and the availability of a large number of adsorption sites, most of the viruses would still be found in the adsorbed state.

In summary, the adsorption of viruses to soils and other surfaces is clearly a complicated and dynamic process. Studies have suggested that many factors are involved in establishing and maintaining the adsorbed state, among which are the IEP of the soil and virus, and the pH and ionic strength of the suspending fluids. In the next section, we will review how these and other factors effect virus adsorption under both field and laboratory conditions.

B. Factors Influencing Virus Adsorption to Soils

1. Introduction

From the foregoing discussion, it is obvious that the adsorption of viruses to soil is a complex reaction which depends upon a number of variables. In the past, it was often presumed (sometimes without basis) that these reactions were invariable and that the application of viruses to soil presented no risk of groundwater contamination. In many cases, these presumptions were correct and planned (or unplanned) application schemes proved adequate for groundwater protection.[118,119] Unfortunately, these presumptions were sometimes unfounded, and the application of virus-containing sewage effluents to soil inadvertently led to groundwater contamination.[54,55,61,65,120] Most investigators now concede that there is an inherent threat of groundwater contamination in any type of land application operation.[40] Since an increasing number of municipalities are considering land treatment, it is clear that there must be a better understanding of virus-soil interactions. In this section, we will attempt to identify and discuss the major factors affecting the adsorption of viruses to soils. While each factor will be considered individually, the reader must be reminded that they do not act independently, but rather are often interrelated. Although the emphasis will be on the most recent studies, the reader is referred to a number of excellent reviews which cover many of the earlier studies.[1,97,98,121]

In a separate section, we will discuss those factors which appear to influence the desorption and mobilization of viruses. While this division is somewhat arbitrary, it is useful in cataloging effects. A complete list of the factors which appear to influence adsorption processes is presented in Tables 2 and 5.

2. Wastewater Factors which Affect Virus Adsorption to Soil

a. pH

The pH of both the percolating fluids and the soil micro-environment strongly influence the adsorption of viruses to soil. As noted previously, pH is one of the major factors responsible for the type and strength of surface charges. With a measure of these charges,

Table 2
WASTEWATER CHARACTERISTICS THAT AFFECT VIRUS ADSORPTION TO SOIL

Characteristics	Ref.
pH	97, 98, 107
Virus type	123—125
Concentration of ions	97, 98, 123, 126, 134
Infiltration rate	40, 61, 129, 135, 144
Presence of organic substances	28, 137, 138, 140

such as the IEP, investigators can generally predict the type of electrostatic interaction that will occur between two charged surfaces under controlled pH conditions. Generally, when the pH is above the IEP of a particle, it behaves as a negatively charged entity, while at pH values below the IEP, the particle is positively charged. By altering the pH and, concomitantly, the overall charge of particles, interactions that are either attractive or repulsive can be created.

Taylor[102] illustrated how pH manipulations influence the adsorption of reovirus 3 (IEP = 3.8) to allophane clays (IEP = 6.3). At pH 5.0, he found adsorption interactions to be favored as viruses were negatively charged and the clay positively charged. Under conditions where the pH of the suspending solution was greater than 6.3, adsorption was considerably reduced due to the mutual repulsive forces of negatively charged particles. The adsorption of other virus types has also been shown to be dependent on pH.[122]

Under the less-controlled conditions of field operations, such precise adsorptive predictions may not be possible. Generally speaking, at the pH of most treatment plant effluents (pH 6 to 8), enteroviruses would be expected to have a net negative charge, presumably due to their low IEP.[104] On the other hand, the charge on the receiving soil particles would vary considerably, depending on the particular soil type (for example, Gerba and Goyal[123] studying the adsorption of viruses to nine different soil types found soil pH values to range from 4.5 to 8.2). Based on pH value alone, electrostatic attractions would appear to be favorable only under conditions where the pH of percolating fluid was between the IEP of the virus and that of the soil. Optimal adsorption would therefore occur best in slightly acidic to alkaline soils. However, since acid soils are strong virus adsorbers,[123] adsorption interactions must depend on other attractive forces (e.g., van der Waals) and associated factors (e.g., ions, virus type, etc.).

b. Virus Type

One of the more significant recent advances in the study of virus/soil interactions was the discovery that the efficiency of virus adsorption to soil particles was directly related to the type and strain of virus adsorbed. Most of the adsorption studies conducted prior to 1978 used reference strains of polioviruses, particularly type 1, strain LSc. The ease of laboratory handling and the avirulent nature of this virus made it a logical choice for such studies. Only recently have investigators begun to study the adsorptive behavior of other human enteroviruses which have varied adsorptive capacities.

One definitive study documenting the strain-dependent behavior of viruses was conducted by Gerba and Goyal.[123] Gerba and Goyal investigated the adsorption of 27 different enteroviruses to 9 different soil types with a 30-min batch-adsorption procedure. The results showed that all three major poliovirus serotypes, including both reference strains and field isolates, adsorbed extremely well to all soil types tested. The adsorption process was generally rapid, with 99% of the viruses adsorbing within 1 min. Other enteroviruses demonstrated a varied adsorption pattern, depending on their origin. Reference strains of coxsackieviruses B1 through B6 adsorbed readily to soil, whereas field isolates of coxsackievirus B4 were poorly adsorbed. Strain-dependency was also observed with echoviruses, where reference strains of echoviruses 1 and 29, as well as natural isolates of echovirus 1 were poor adsorbers, while the remaining types tested demonstrated good soil adsorption. The authors also showed that simian rotavirus (SA-11) was poorly adsorbed to most soils.[124] In a series of field studies, Schaub et al.[28] and Sorber[62] determined that the movement of viruses through a rapid infiltration site, and an overland flow recharge area was dependent upon virus type. These investigators observed that coliphage f2 adsorbed poorly to soil and could move rapidly through both systems, while polioviruses and indigenous enteroviruses were adsorbed to a greater degree. Laboratory experiments have confirmed this variability in virus adsorption. Landry et al.[125] attempting to determine the factors responsible for the migration of enteroviruses through Long Island soil, reported that the extent of virus adsorption and migration depended upon the specific strain of virus studied. Using 4.3 × 12.5 cm soil cores collected from operating recharge basins, the investigators observed that all poliovirus types tested, including both field and reference strains, adsorbed well to soil. However, the adsorption capabilities of nonpolio enteroviruses were quite varied. Echovirus 1 and coxsackievirus B3 adsorbed well to soil, while echovirus 6 demonstrated a reduced adsorptive capacity. The investigators concluded that: (1) virus adsorption to soil was dependent on the particular strain of virus studied, presumably due to some unique viral protein characteristic and (2) since polioviruses failed to show any variability in adsorption, they did not appear to truly represent the behavior of other enterovirus types, particularly those recently isolated from environmental systems.

One theory which might explain this strain-dependent adsorption behavior involves the varied IEP of viruses. Since the protein capsid of a particular virus type is fairly unique, its subsequent ionization should result in a characteristic IEP. The compiled data presented by Murray and Parks[104] appears to corroborate this view. They reported different IEPs for a number of enteroviruses, including some closely related poliovirus types, such as Brunhilde (7.0 and 3.5), CHAT (7.5 and 4.5), Sabin T2 (6.5 and 4.5), and Mahoney (8.5). Mandel[105] had previously reported the IEPs of strain LSc 2ab (4.5 and 7.0). The IEPs of nonpoliovirus enteroviruses have been reported by Taylor[102] and Murray and Parks.[104] Together these data suggested that viruses with varied IEPs might behave differently in the same soil, under the same conditions. Data presented by Sobsey et al.[126] seem to substantiate this claim. They reported that 82% of reovirus 3 adsorbed to a loamy Ponzer soil, but only 34% of poliovirus 1 (LSc 2ab) adsorbed to the same soil. Adjustment of the pH of the suspending media to 5.5 increased reovirus adsorption to 90%, but a pH decrease to 3.5 was required to promote a similar degree of adsorption of the poliovirus. Further analysis of their data indicated that these differences were not evident in soil with a high clay content. Thus, the strain-dependent behavior might be countered by other adsorptive factors such as the presence of cations or the clay content. Recently, Gerba et al.[127] and LaBelle and Gerba[128] presented evidence which indicated that the strain-dependent behavior of enteroviruses applied not only to soil, but was also evident in sewage sludges and sediments. Polioviruses adsorbed well to soil, sediments, and sludges, although three echovirus field isolates exhibited variable adsorption to soil and sludges. All viruses adsorbed well to sediments, presumably facilitated by high salt concentrations. The authors concluded that this behavioral variability was due to differences in the electronegativity (or IEP) of the viruses.

Since any type of capsid difference may result in a varied adsorptive capacity, mutant virus types with capsid alterations might be expected to possess adsorptive capacities different than their parental types. Landry et al.,[125] searching for an avirulent strain of poliovirus with adsorptive features that better represented the majority of enteroviruses, observed that guanidine-resistant (g^r) mutants of poliovirus 1 (LSc 2ab) possessed soil-adsorbing properties which were quite different from those of the parental strain. Elution properties of the mutant also differed from the parent virus. While IEP differences were not pursued, the authors suggested that differences in the electronegativities of the parent and mutant strains might have affected their adsorption behavior.

In addition to the type-dependent adsorption characteristics, there appear to be adsorption differences within a single purified population of viruses. Lance and Gerba[129] observed that the vertical distribution of poliovirus[1] (LSc 2ab) in soil columns (10 by 250 cm) was essentially unchanged over wide concentrations of applied viruses (9×10^1 to 2.6×10^4 PFU/mℓ). While the majority of viruses adsorbed within the top 5 cm of soil, a certain fraction always adsorbed at greater soil depths. As the concentration of applied viruses increased, the number of viruses at lower levels also increased. However, the percentage of viruses observed at each depth remained uniform. The investigators suggested that each population of viruses was composed of several sub-groups which possessed different surface charge characteristics. Those with the optimal adsorbing charge were removed immediately near the soil surface, while those with a somewhat less optimal charge moved to greater soil depths. Thus, adsorption profile appeared to be proportional to charge strength. Burge and Enkiri[130] also demonstrated charge heterogeneity in a virus population. Analyzing the adsorption kinetics of highly purified coliphage ΦX174 to Kranzburg soils, the investigators observed that at least two populations of viruses existed and could be differentiated on the basis of differing soil-adsorption rates. They found that if unadsorbed viruses were collected and mixed with fresh soil, they adsorbed at a progressively decreasing rate. These differences were attributed to varied charge strengths.

In summary, the variability observed in virus adsorption to soil is apparently dependent on the virus type and its unique charge characteristics. This conclusion is substantiated by the following evidence:

1. Different strains of viruses, including mutants of the same strain, adsorb variably to the same soil.
2. Within a purified population of virus particles, sub-groups have been found which adsorb to soil at different rates.
3. The adsorption profile of a virus population is constant over a wide range of concentrations.
4. The presence of other adsorptive factors (e.g., ions) which alter the ECDL of a virus particle reduces strain adsorption variability.

An analysis of the above leads to the conclusion that no single virus type can be used as a model for adequately predicting the general pattern of enterovirus adsorption.[54,124,125,129] Many virus types must be analyzed in order to define completely the virus-retention characteristics of a particle. This appears to be important, since variability in adsorption could lead to deep soil penetration of at least some virus types.[129] The maximum distance traveled might therefore be determined by the slowest adsorbing fraction.[130]

c. Concentration of Ions

The importance of ions in the adsorption process has been well-documented and many of the key studies have been thoroughly reviewed by Duboise et al.,[1] Gerba et al.,[97] and Bitton.[98] Some of the more crucial findings are highlighted below.

The valency of an ion is important in viral adsorption. Carlson et al.,[131] observed that divalent cations were more effective than monovalent ions in promoting the adsorption of coliphage T2 and poliovirus 1 to clay soils. A tenfold higher concentration of monovalent ions was required to achieve the same adsorption efficiency noted with divalent ions. Later, Bitton et al.[106] determined that trivalent aluminum ions were more effective in viral adsorption that divalent ions, while monovalent ions were the least effective. Higher ion concentrations promote improved viral adsorption. Lefler and Kott,[132] studying the movement of poliovirus 1 and coliphage f2 through soil columns, reported that greater than 99% of the viruses were retained when the concentration of Ca^{++} in the percolating solution was 10 m*M*. Reducing the concentration to 1 m*M* resulted in retention of only 37% of the polioviruses. The authors also noted that the type of ion was critical to adsorption, with 0.5 *M* Na^{+} not being as effective as 10 m*M* Ca^{++}. High concentrations of ions were also important in the adsorption of poliovirus 1, coliphage T7[133], and encephalomyocarditis viruses[101] to clay minerals. A number of more recent studies confirmed and extended some of the earlier findings. Gerba and Goyal[123] reported that of 10 m*M* Ca^{++} increased the adsorption efficiency of a number of viruses to Flushing Meadow soil. They noted that the adsorption of a field strain of echovirus 1 increased from 0 to 80% in the presence of this ion. Similar increases were also noted for field strains of coxsackievirus B4. Enhancement was not observed at Ca^{++} concentrations of >1 m*M*. Burge and Enkiri[130] also reported an increased adsorption efficiency with increasing ionic strength. Studying the adsorption of coliphage ΦX174, they noted that phage adsorption increased as the Na^{+} concentration was increased from 0.1 to 20 m*M*. However, as the concentration approached 1 *M* Na^{+}, adsorption dramatically decreased, indicating that very high salt concentrations interfered with the adsorption process. Sobsey et al.[126] reported that 10 m*M* Mg^{++} facilitates the adsorption of poliovirus 1 and reovirus 3 to sandy soils. Although the addition of salts enhanced the adsorption of poliovirus at both pH 4.5 and 7.5, such addition was only effective at pH 4.5 for reoviruses. Working with the same reoviruses, Stotsky et al.[134] found that both Mg^{++} and Ca^{++} promoted the adsorption of the virus to clay minerals, with concentrations of 10 m*M* being more effective than 1 m*M*. They also reported enhanced viral adsorption in estuarine waters, presumably due to the higher ionic strength. Similarly, Gerba et al.[127] reported that the higher concentrations of ions in estuarine sediments were responsible for promoting the adsorption of various echovirus field isolates.

As noted in a preceding section, adsorptive interactions between viruses and soil are chiefly governed by electrostatic and electrodynamic forces, and these forces depend on the overall charge of the respective ECDL. It is at the double-layers that the ions appear to exercise their greatest influence. Taylor[102] suggested that the addition of ions to the suspending medium altered the charge of the ECDL, causing a partial neutralization to occur close to the particle surface. Higher electrolyte concentrations resulted in a thinner ECDL and a lower zeta potential. The presence of ions also appeared to reduce electrostatic repulsive forces and enhance adsorption chiefly, by promoting attraction via London-van der Waals forces.[106,130,135] Other investigators suggested that cations promote adsorption by creating an attractive clay-ion-virus bridge.[101] Kessick and Wagner[136] suggested that such cross-complexes were important in the adsorption of viruses to negatively charged filter surfaces.

d. Presence of Organic Substances

The presence of organic substances in wastewaters may also influence the extent of virus adsorption to soil. As noted in earlier studies,[101,131] materials such as serum proteins and egg albumin inhibited the adsorption of viruses to organic and inorganic surfaces. Bitton et al.[137] reported that organic components in secondary effluents inhibited the adsorption of poliovirus to magnetite. The inhibitory effect was removed by filtration through activated carbon, or by the addition of Ca^{++} ions. Later, Scheuerman et al.[138] reported that the adsorptive capacities of soil could be enhanced by leaching with 10 volumes of groundwater. The leached fluids had a brownish-colored appearance and presumably contained the interfering substances. When this leachate was applied to sandy soil columns, which had previously retained >99% of applied viruses, the adsorption efficiency was dramatically decreased to 1.5%. As had been noted in previous studies, activated carbon pretreatment of the leachate removed the inhibitory substance and restored the adsorption capacity of the soil. Analysis of the colored leachate indicated that the inhibitory activity resided in the <50,000 molecular weight fraction, the approximate size of both humic and fulvic acids. This led the authors to conclude that water and soil containing these compounds would probably not be effective in virus removal. They suggested that these organic compounds interacted with either virus or soil particles, blocking adsorption sites. Several additional studies[139,140] presented evidence that soluble organic substances interfered with viral adsorption processes. Bixby and O'Brien[140] found the adsorption of coliphage MS2 to loamy soil inhibited by fulvic acid in the suspending wastewater. They suggested that the presence of these substances in wastewater effluents would lead to increased mobility of viruses in soil. Other investigators noted the same effects. Lo and Sproul[141] reported that organic matter in secondarily-treated effluents competed with polioviruses for adsorption sites on silicate minerals. They found virus adsorption favored when tapwater was used as the suspending medium, rather than sewage effluent. Burge and Enkiri[115] noted that coliphage ΦX174 adsorbed poorly to Aastad soils, which had an organic carbon content of 3%. In the same study, soil with a lower organic carbon content was a more effective virus-adsorber. Schaub et al.,[28] studying the removal of polioviruses in an overland flow system, observed that viruses adsorbed better to underlying, rather than surface soil. They suggested that during continual recharge, the surface soil had adsorbed a variety of organic substances which blocked available virus adsorption sites.

Although the above studies indicate that organic substances in wastewaters interfere with viral adsorption, Gerba and Lance[142] and Lance et al.[143] presented evidence demonstrating that this need not always be the case. They reported that polioviruses suspended in either primary or secondary effluents efficiently adsorbed to loamy-sand soil. Apparently, the higher concentrations of organic substances in primary effluents did not adversely affect virus adsorption. The virus distributions in 10 × 250 cm columns and their adsorption rates were similar in both effluent types. The investigators suggested that the nature of the soil was perhaps a more important factor, as it apparently had sufficient adsorbing sites for both organic matter and viruses.

e. Soil Infiltration Rate

Regardless of the affinity that a virus particle may have for a particular soil, if the two fail to come into close contact, little or no adsorption occurs. One determinant of virus-soil contact is the rate of fluid percolation through the soil. In coarse sandy soil with a high permeability (e.g., >10 cm/hr), percolating fluids travel a path of least resistance and seek out large pore spaces. This results in little close contact between virus and soil particles. Reducing the infiltration rate not only allows percolating viruses to move through smaller soil pores, but also increases their residence time in the soil. In this instance, the probability of adsorptive interactions increase. In a series of field experiments conducted on an operating rapid infiltration basin, Vaughn and Landry[61] and Vaughn et al.[144] demonstrated the importance of infiltration rate in the adsorption of polioviruses to soil. In the initial experiments, 4000 ℓ of unchlorinated sewage effluent, seeded with poliovirus 1 (LSc) at a concentration of 7.8×10^5 PFU/ℓ, were allowed to percolate through the basin at an average infiltration rate of 1 cm/hr. Composite samples of the percolating fluids were collected from gravity samplers located 0.75, 2.25, and 5.34 m below the basin surface. At the same time, large volume samples (400 to 800 ℓ) were collected from the aquifer located some 7.5 m below the surface. The investigators reported that the peak concentrations of viruses recovered at each level represented a 3-log reduction in the input virus titer (Table 3). In spite of this encouraging removal, the investigators observed that viral particles penetrated the entire

Table 3
THE MOVEMENT OF POLIOVIRUS[1] (LSc 2ab) THROUGH A RECHARGE BASIN DURING VERY LOW RATE RECHARGE (1 cm/hr)

Depth below surface of basin floor (m)	Sample number	Time after seeding (hr)	Virus recovered (PFU/ℓ)
0.75 (Level 1)	1	8—24	0
	2	24—48	4.07×10^2
	3	48—72	2.29×10^2
	4	72—98	2.09×10^1
	5	98—118	2.09×10^1
2.25 (Level 2)	1	34—48	0
	2	48—72	4.13×10^2
	3	72—97	5.71×10^2
	4	97—120	0
	5	120—168	2.47×10^1
5.34 (Level 3)	1	74—99	0
	2	99—122	0
	3	122—144	8.15×10^1
	4	144—170	1.94×10^2
7.62 (Level 4)	1	73	0
	2	98	0
	3	121	0
	4	143	4.21
	5	170	1.37×10^{-2}
	6	194	6.63×10^{-2}
	7	218	1.24×10^{-1}

From Vaughn, J. M., Landry, E. F., Beckwith, C. A., and Thomas, M. Z., *Appl. Environ. Microbiol.*, 41, 1981. With permission.

length of the unsaturated zone and entered the aquifer. While the highest concentration of viruses observed in the aquifer was minimal (0.124 PFU/ℓ), their presence clearly established the potential for extensive viral movement in soils. The second series of experiments was designed to study virus movement in soil under conditions of high rate recharge. Percolation rates (75 to 100 cm/hr) chosen by the investigators accurately reflected the rates commonly encountered at a number of operating recharge sites in the study region. In these experiments, 4000 ℓ of unchlorinated tertiary effluent seeded with the poliovirus 1 to a final concentration of 2.3×10^6 PFU/ℓ were applied to the test basin. The results, summarized in Table 4, indicated that poor virus retention occurred at high infiltration rates, particularly at the shallow soil depths. The peak concentration of viruses occurring in the upper levels (9.7×10^5 PFU/ℓ) represented a 58% removal efficiency when compared to input levels. Peak viral concentrations at the 5.34 m level (1×10^5 PFU/ℓ) demonstrated that large concentrations of viruses could move appreciable vertical distances. Viruses were also capable of moving into the groundwater in large numbers (1.1×10^3 PFU/ℓ). The authors concluded that high infiltration rates failed to allow sufficient virus-soil contact to occur. They cited these high rates of recharge as one of the likely causes of the groundwater contamination previously reported in the experimental region.[32,54] Lance and Gerba[129] also observed the effects of soil application rate on virus retention in soil. In experiments which utilized 10 × 250 cm soil columns, they reported significant virus breakthrough when infiltration rates were raised from 0.6 to 1.2 m/day (2.5 to 5 cm/hr). At the lower application rate, viruses were only observed at the 160 cm depth. Increasing the flow rates to 1.2 m/day resulted in virus breakthrough, and the appearance of small concentrations of viruses (<1%) in the 250 cm

Table 4
THE MOVEMENT OF POLIOVIRUS[1] (LSc 2ab) IN A RECHARGE BASIN DURING HIGH RATE RECHARGE (75 TO 100 CM/HR)

Depth below surface of basin floor (m)	Sample number	Time after seeding (hr)	Virus recovered (PFU/ℓ × 10^4)
0.75 (Level 1)	1	0.60—0.81	78.00
	2	0.86—0.91	97.50
	3	1.03—1.06	26.50
	4	1.20—1.23	1.46
	5	1.41—1.45	1.94
	6	1.58—1.61	1.22
	7	1.75—1.78	0.79
	8	2.00—2.03	0.90
	9	2.41—2.45	0.03
2.25 (Level 2)	1	1.20—1.30	2.44
	2	1.41—1.50	8.58
	3	1.58—1.63	6.70
	4	1.75—1.81	5.40
	5	2.00—2.08	1.81
	6	2.25—2.31	1.38
	7	2.50—2.56	0.31
	8	2.58—3.86	0.05
5.34 (Level 3)	1	1.78—2.15	8.82
	2	2.16—2.28	0.54
	3	2.33—2.41	0.27
	4	2.50—2.55	1.81
	5	2.81—2.85	10.10
	6	3.00—3.03	9.80
	7	3.28—3.31	3.32
	8	3.50—3.53	1.96
	9	3.53—4.03	0.12
7.62 (Level 4)	1	2.50—2.66	0.11
	2	4.66—4.83	0.001

From Vaughn, J. M., Landry, E. F., Beckwith, C. A., and Thomas, M. Z., *Appl. Environ. Microbiol.*, 41, 1981. With permission.

column effluents. They reported that once breakthrough was achieved, further increases in the flow rate, even to a maximum of 12 m/day (50 cm/hr), resulted in no further increases in the number of viruses in the column effluents. The investigators suggested that at high flow rates (>1.2 m/day), all percolating fluids traveled through the same large soil pores with the same decreased liklihood of significant virus-soil contact. Although higher flow rates did not increase virus breakthrough, it seems likely that extended use of such high rates would contribute to soil saturation, ultimately effecting overall viral adsorption and mobilization. Sagik et al.[40] observed a similar relationship between hydraulic loading rates and the efficient removal of microorganisms during land treatment. In their studies, the application of 62 cm of effluent over a 6-day period led to poor removal of bacteria, coliphages, and other viruses. When the fluid loading rate was decreased to 45 cm applied over a 10-day period, soils exhibited improved retention of both coliphages and bacteria. However, human viruses were still detected in both 3 ft (0.9 m) and 4.5 ft (1.37 m) lysimeters.

The method, or schedule of wastewater application, also appears to be critical in the removal of viruses. Lance et al.[135] observed that there were no differences in retention when sewage effluent was continuously applied to soil columns at rates of either 15 or 55 cm/

day. Continuous application of seeded-sewage for 27 days at these rates, failed to saturate the virus adsorbing sites in the soils. Subsequent applications of deionized-distilled water however, caused detectable desorption and redistribution of viruses in the columns. When viruses were applied on an intermittent schedule of 9 days loading and 5 days drying, no viral desorption occurred with subsequent distilled water applications. The investigators suggested that the drying conditions permitted the drainage of free water trapped by the soil, allowing better contact between the virus and soil particles. This strong adsorption coupled with the rapid inactivation caused by drying apparently minimized the vertical movement of active virus particles. Recently, Lance and Gerba[129] suggested a number of ways by which the rate and schedule of wastewater application could be altered to improve viral adsorption. These included: (1) reducing the depth of applied waters, (2) compacting the soil surface to increase bulk density and decrease void space, (3) increasing the total suspended solids in the applied sewage in order to clog the surface layer, and (4) the employment of alternating cycles of loading and drying.

Several investigators have shown that some of these methods enhanced virus retention. Sobsey et al.[126] reported that twice-weekly applications of 2.5 cm of poliovirus-seeded effluent to 2.6 × 10 cm columns of Fripps or Lakeland soil resulted in excellent virus removal, even when extended over a 34 day loading period. These low clay soils were earlier found to be poor virus-adsorbers. Duboise et al.[145] observed that sandy forest soil adsorbed poliovirus 1 (CHAT) more efficiently when seeded effluents were applied with intermittent, rather than continuous loading. Green and Cliver[146] reported that dosing 60 cm soil columns with single daily 5 cm applications resulted in better virus retention than did a single 50 cm dose of seeded effluent. Additional information addressing the importance of flow-rate in virus adsorption can be found in the reviews of Duboise et al.[1] and Gerba et al.[97]

The preceding discussions indicate the importance of controlling hydraulic loading during land application. Since some of these systems can be used for wastewater disposal, as well as for additional treatment, the optimal rate of recharge depends upon the type of treatment required. For the recharge of high quality effluents containing no viral contaminants, rate control is not crucial and effluents can be disposed of rapidly. However, for recharge of lower quality effluents, the choice of application rate should be of paramount importance, particularly if the receiving aquifer represents a source of untreated drinking water.

3. *Soil Factors Affecting Virus Adsorption*

A number of soil factors appear to be involved in virus adsorption. While the reader is probably aware that many of these factors overlap with previously discussed factors, it is pertinent to discuss them under this heading. A complete list of contributing soil characteristics is found in Table 5.

a. Type and Amount of Clay

A number of studies have shown virus/soil adsorptive interactions to be dependent on soil type and composition.[123,125] The type and the amount of clay material within soil are key characteristics governing adsorption. Studies have generally shown that the higher the soil clay content, the more efficient the virus adsorption. Most field studies demonstrating extensive virus movement have been conducted in soil with little clay content.[61,62,144] As early as 1968, Drewry and Eliassen[108] recognized the importance of clay in viral adsorption. Other studies demonstrated the adsorption of viruses to specific clay minerals such as kaolinite, illite, and montmorillonite.[101,141] Recently, Koya and Chaudhuri[117] reported the adsorption of coliphage MS2 to Indian soil dependent on clay type. Lateritic soils (32% clay) containing kaolinite, quartz, iron, and aluminum oxides adsorbed the phage better than Black Cotton soils (28% clay), which were composed of montmorillonite and quartz. A

Table 5
SOIL CHARACTERISTICS THAT AFFECT VIRUS ADSORPTION

Characteristics	Ref.
Clay content	123, 124, 126
Ratio of charged sites	134
Cation exchange capacity	115, 134, 147
Specific surface area	40, 115, 141
Soil pH	115, 123, 124, 126
Depth to groundwater	40, 54, 118, 147
Miscellaneous factors	
Soil continuity	65, 154
Soil conditioning	125, 155
Total organic content	147
Others	124, 156

third soil, Kanpur silt, with the lowest clay content (10%) proved to be a better adsorber than Black Cotton, presumably because of its higher illite and kaolinite content. The investigators suggested that the type, as well as the concentration of clay materials was important. Sobsey et al.[126] reporting the adsorption of poliovirus 1 and reovirus 3 to eight different soil types, observed that adsorption was dependent on the amount of clay present in each soil. Gerba and Goyal[123] and Goyal and Gerba[124] also showed that adsorption was dependent upon the clay content of soils. Soil, such as Vernon and Windthorst, with high clay contents (39 and 53%, respectively) promoted the adsorption of most viral types while lower clay soil (FM — 3%; Rubicon — 4%) exhibited variable adsorption. A number of enteroviruses, including field strains of echovirus 1, were poor adsorbers to low clay soil while adsorbing extremely well to high clay soil. The investigators also demonstrated that the clay content might not be the sole determinant of viral adsorption, since Chigley soil, with a clay content of 28%, was a poor adsorber of field strains of echoviruses and coxsackieviruses.

One of the more interesting studies of the relationship between viral adsorption and clay content has recently been reported by Stotsky et al.[134] These investigators studied the adsorption of coliphages T1 and T7, and reovirus 3 to kaolinite (K) and montmorillonite (M) clays in an effort to determine their specific virus-adsorbing sites. They observed that coliphage T7 had a greater affinity for M clay than coliphage T1, but that both phages adsorbed to M better than to K. They also found that both phages could be simultaneously adsorbed to the same clay, suggesting that the clay had different virus adsorbing sites. Pretreatment of kaolinite with sodium metaphosphate, which binds to the positive charges on the clay surface, substantially reduced the adsorption of coliphage T1, but not to coliphage T7. This suggested that coliphage T1 bound preferentially to the more positive sites on the clay surface, while coliphage T7 appeared to be associated with the more negative sites. Since both coliphages T1 and T7 adsorbed to M clay, even after this treatment, the investigators concluded that this clay had a higher percentage of negatively-charged sites. Other treatments which reduced the ratio of positive to negative sites (e.g., nutrient broth and egg albumin) also reduced coliphage T1 adsorption. Treatments which increased the number of positive sites (e.g., lysozyme) stimulated the adsorption of both coliphages T1 and T7. Reovirus 3 adsorption appeared to be more favorable in soil with more negatively-charged sites, since pretreatment of the clay with chymotrypsin, which binds to negative sites, inhibited the adsorption of this virus. The authors concluded that the ratio of positive to negative charges on a clay soil was an important factor in determining virus adsorption efficiency.

Funderburg et al.[147] reported a rather surprising positive correlation between high clay content and the presence of viruses in soil column effluents. Since this analysis contradicted many previous studies, the investigators suggested that the method for determining clay content was important. In their study, clay analyses had been based on a particle-size distribution which yielded the total quantity, rather than the ion adsorption capacity of the clay. They suggested that the latter factor was more important in viral adsorption. However, a careful analysis of the data showed that a general trend of better virus retention with increasing clay content actually existed. The greatest number of unadsorbed polioviruses were observed in Eufaula soil columns, which had the lowest clay content (<1%). Fewer viruses were seen in column effluents of Click, Victoria Clay, and Axtell-Tabor soil which had clay compositions ranging from >1 to 8%. The least amount of virus appeared in effluents from Katemy, Webb, Austin Chalk, and Venus soils which had clay levels ranging from 3 to 28%. These same general trends were also observed for reoviruses.

Two additional soil factors which appear to be closely related to overall clay content are cation exchange capacity (CEC) and the specific surface area (SSA). Generally speaking, both factors increase with increases in percent clay. Burge and Enkiri[115] reported the adsorption of coliphage ΦX174 to be dependent on both soil SSA and CEC, with virus adsorption capacity increasing with increasing CEC and SSA. Lo and Sproul[141] also suggested that surface area was important to viral adsorption. They found that sandy soils with small surface areas were less efficient virus adsorbers than clay minerals with greater surface area. Sagik et al.[40] suggested that as soil particle size decreased, soil surface area would exponentially increase, thereby increasing the available number of virus adsorption sites. Stotsky et al.[134] found viral adsorption to increase with increasing cation exchange capacity. Coliphage T7 adsorbed better to montmorillonite (CEC = 97 meq/100 g) than to kaolinite soils (CEC = 5.8 meq/100 g). They found that reovirus 3 exhibited this same trend, but coliphage T1 did not. Funderburg et al.[147] also reported a direct relationship between the CEC of a soil and its virus-adsorbing potential. These investigators concluded that soil with a CEC greater than 23 meq/100 g would adsorb large numbers of viruses. Contrarily, Goyal and Gerba,[124] studying the factors which promoted enterovirus adsorption to nine different soil types, reported that they could find no correlation between virus adsorption and surface area or cation exchange capacities. They also reported that no clear-cut correlation existed between viral adsorption and other soil factors such as percent silt, conductivity, total phosphorous, total and exchangeable iron, and calcium and magnesium concentrations.

In field studies, a number of investigators have reported that the general type of adsorbing material was important in virus removal. For instance, Moore et al.[31] found that sludges were effective in retaining seeded and indigenous enteroviruses. Other investigators have reported that sludges applied to field or column soils efficiently adsorbed enteroviruses.[57,148,149] Other materials, such as landfill refuse, containing a mixture of solid wastes and soils, adequately retained seeded enteroviruses.[150-153]

b. Soil pH

The pH of soil is also an important factor affecting the adsorption of viruses. Gerba and Goyal[123] recently reported that soils with a low pH (<5.0) were generally better viral adsorbents than soils of higher pH. A later statistical analysis of their data indicated that a strong correlation existed between soil pH and viral adsorption.[124] The lower the pH, the greater the adsorption of viruses. One exception was noted with Rubicon soil, which had a pH of 5.5, but failed to adsorb viruses effectively. The trends observed with many other kinds of soil led the investigators to conclude that soil pH was the single most important factor influencing viral adsorption to soil. Burge and Enkiri[115] observed that soils with higher pH values were less likely to adsorb coliphage ΦX174. When these investigators plotted the virus adsorption rate constants of five soil types against their ambient pH values, a highly significant correlation was observed. Sobsey et al.[126] also observed this same relationship. They demonstrated that as the pH of a low-clay Lakeland soil was reduced from 7.5 to 4.5, the adsorption of both a poliovirus and a reovirus was increased. They also noted that soil with a higher clay content (30 to 99%) adsorbed viruses better over a wider pH range (pH 4.5 to 7.5). Soil pH was also important in the studies of Drewry and Eliassen.[108]

c. Soil Depth

The greater the distance a virus particle must travel through soil, the greater the likelihood that it will be retained. Thus, the distance from soil surface to static groundwater level is an important factor in determining the potential for groundwater contamination. In field studies, Vaughn et al.[54] observed that high rate recharge of virus-containing effluents through 7 to 10 m soil depths was not sufficient to prevent groundwater contamination. However, recharge through 24 m of soil with a similar profile resulted in no aquifer contamination. In laboratory studies employing 33, 66, and 100 cm soil columns, Funderburg et al.[147] showed that virus removal was proportional to soil depth. Except very sandy soils, the retention of poliovirus 1, reovirus 3, and coliphage ΦX174 exceeded 90% in 100 cm cores. Moore et al.[118] and Sagik et al.[40] reported that distance travelled was important in virus removal. Studying an irrigation site in Texas, these investigators found that weekly applications of 7.6 cm of wastewater to soil with a permeability of 1.5 to 5 cm/hr and CEC of 25 to 50 meq/100 g, resulted in a number of virus isolations in 0.4, 0.9, and 1.3 m lysimeters. However, no confirmed isolates were detected in wells located 11 to 20 m below the application site. The authors concluded that travel through these soil depths was sufficient to remove enteroviruses.

d. Miscellaneous Factors

There are a number of other soil factors that appear to be important to the removal of viruses which have not been studied in great detail. Soil continuity is an obvious factor that affects the extent of virus-soil contact, and the rate of percolation through soil. If a soil contains cracks and fissures, extensive channeling of recharged wastewaters occurs and there is little chance of virus retention. Such channeling has been suggested by Wellings et al.[65] as a possible cause of groundwater contamination. They speculated that the placement of construction pilings in a recharge basin might have compromised a confining clay layer, permitting the passage of viruses into the aquifer. Drewry and Eliassen[108] and Hori et al.[154] also suggested that cracked and fissured soil strata could contribute to groundwater contamination.

Landry et al.[125] and Berg[155] reported that soil conditioning was important for virus removal, especially in sandy soil. They observed that clean unused soil adsorbed few viruses, while soil conditioned by application of treated effluents enhanced virus-removing capacities. In both investigations the soil possessed low-clay content and thus probably lacked sufficient virus-adsorbing sites. Conditioning appeared to result in the deposition of organic constituents in the soil, which then promoted virus-adsorption. Funderburg et al.[147] suggested that the presence of some organic compounds in the soil was required for good virus adsorption. They indicated that good agricultural soil containing at least 0.5 to 1% organic matter would adequately remove viruses. However, Burge and Enkiri[115] demonstrated that higher concentrations of this material (3%) interfered with the adsorption of coliphage ΦX174 to soil.

A number of other factors have been suggested as possible effectors of virus adsorption, but little information is available on their action. Goyal and Gerba[124] found that the amount of exchangeable aluminum in the soil correlated with the adsorption of five different virus types. They further observed that percent sand, total aluminum, total phosphorous, total iron, exchangeable aluminum and magnesium, and resin-extractable phosphorous also influenced the adsorption of a few virus types. However, no consistent adsorption correlation

Table 6
FACTORS THAT INFLUENCE THE DESORPTION AND MOVEMENT OF VIRUSES

Characteristics	Ref.
Concentration of ions	37, 125, 135, 145, 159
Virus type	125, 160
Miscellaneous factors	
pH	125,126,160, 163
Hydrogeological conditions (soil depth, soil continuity)	35, 37, 65, 129, 154
Degree of saturation	58, 165, 166

could be established. Bitton et al.[156] recommended that other factors, such as bulk density, iron oxide content, and soil horizon should be considered when measuring the viral adsorption potential of soil.

In summary, it is clear that the adsorption of viruses to soil is affected by a variety of factors which likely act in concert. Analysis of any adsorption process should therefore include consideration of many, if not all, of these factors. Many investigators are convinced that careful application of these criteria to land treatment practices can virtually eliminate viral groundwater problems.

C. Desorption and Movement of Soil-Bound Viruses

1. Introduction

Virus adsorption is clearly not an irreversible process and under appropriate conditions, particles desorb and become entrained with the percolating fluids. The extent of groundwater contamination therefore depends on the distances traveled by mobilized viruses. If particles resorb quickly, the threat is minimal. However, if viruses move significant distances in relation to the depth to groundwater then vertical migration may present a significant problem. A complete list of the factors influencing virus desorption from soils is shown in Table 6. While some of these factors have already been discussed in relation to adsorption processes (Tables 2 and 5), it is important to review their mode of action in regard to viral desorption.

2. Factors Affecting Virus Desorption from Soil

a. Concentration of Ions

The ion concentration or conductivity of percolating fluids appears to play a major role in virus desorption from soils. One of the first studies noting these effects was performed by Duboise et al.,[157] who reported that when soil columns containing adsorbed coliphage T7 were rinsed with deionized water, a number of phage apparently desorbed and were recovered in column effluents. The investigators theorized that the low conductivity fluids desorbed the phage and prevented resorption through the length of the columns. They suggested that natural rainfall with its low ion concentration may act in the same manner, thus triggering virus migration at land application sites. This mechanism was, in fact, proposed by Wellings et al.,[158] to explain the presence of enteric viruses in groundwaters beneath a spray irrigation system. These investigators isolated a number of enteroviruses including poliovirus 1, echovirus 7, and coxsackievirus B4 from 10 and 20 m sampling wells following a period of heavy rainfall (71 cm). The investigators suggested that the low conductivity rainfall triggered desorption and in combination with the high water to soil ratio prevented resorption and allowed the viruses to move vertically into the aquifer. They later suggested that this mechanism might also have been a factor in groundwater contamination at other recharge sites.[65,158] Desorption by low ionic strength fluids has also been reported by Lance and Gerba[129] and Lance et al.[135] These investigators observed that during the initial passage of poliovirus seeded-sewage effluent through a 250 cm column of sandy loam soil, most (90%) of the viruses adsorbed in the top 2 cm of soil. A somewhat lower adsorption rate was detected below this level since percolation through an additional 38 cm of soil was necessary for a further 90% reduction. However, overall column retention was high, with few viruses detected in the composite column effluents, and none observed in the soil below the 160 cm column depth. When deionized water was applied to the columns, a small percentage of viruses desorbed and moved to greater column depths, with significantly greater numbers of viruses found at both the 80 and 160 cm depths. Viruses resorbed at lower levels, and few were detected in column effluents. The investigation suggested that the desorbing conditions were temporary, and that the slug of deionized water rapidly mixed with the salts from the ambient waters and soils and restored adsorbing conditions. They further reported that other measures which restored high ionic levels, such as the addition of exogenous salts (e.g., 1 m*M* Ca^{++}) also reduced viral desorption. In the later study[129] the investigators indicated that viral adsorption could be confined to the top 5 cm by the addition of 5 m*M* Ca^{++}. Lance et al.,[135] also reported that immediate application of sewage effluent to columns flooded with deionized water prevented extensive virus mobilization. Apparently, the high concentration of resident ions in applied sewage rapidly mixed with the deionized water and facilitated the resorption of viruses. The investigators suggested this technique could be employed as a practical and effective method for the management of operating recharge basins. The effects of rainfall would be countered by simply flooding the affected basins with sewage effluent. In each case, the investigators theorized that fluids of high ionic strength enhanced the adsorptive process by decreasing the thickness of the soil-virus particle ECDL, thus increasing the chance of virus-soil contact and promoting stronger binding via London van der Waal forces.

In perhaps the most pivotal study on the influence of low ionic strength fluids on virus desorption, Duboise et al.[145] reported that poliovirus 1 (CHAT) desorbed from soils after a series of distilled water rinses. The investigators demonstrated that peak viral elution from 5.6×10 cm cores of natural sandy forest soils corresponded with the presence of low conductivity levels in the core effluents. After a series of three cyclical applications of sewage effluent, followed by distilled water rinses, a total of 22% of the adsorbed viruses were recovered in core effluents; 16% were removed following the first wash, while 4.2 and 1.7% were eluted after the second and third applications, respectively. This suggested that during initial percolation of sewage effluent through the cores, a decreasing ionic gradient formed in the soil with maximal ion retention occurring near the surface. Once these sites became saturated, the remaining ions moved to the next level. Since fewer ions were available for adsorption at greater soil depths, the lower soil levels possessed a suboptimal adsorbing environment. When viruses were applied to the soils, a majority adsorbed near the surface at the maximum ion adsorbing sites. However, due to the heterogeneous nature of the virus population, some traveled further and were not adsorbed until lower soil levels. Subsequent application of a rainwater rinse simply diluted the ionic gradient, causing the release of viruses, presumably from the less optimal adsorbing areas (e.g., deeper soil portions). Funderburg et al.[159] extended the early studies of Duboise, demonstrating similar trends in larger sandy forest soil cores (33, 66, 100 cm). Here again, peak virus elution directly corresponded with low conductivity level in the core effluents. While $<1\%$ of the viruses were observed in the larger core effluents (66 and 100 cm), the investigators observed that a major redistribution of the population occurred after rainwater rinses. Viruses were reported at all core depths with peak concentrations at the 33 and 35 cm depths.

Recently, Landry et al.[37,125] reported on two studies which addressed the adsorption and migration of several enterovirus types in Long Island sandy soils. They observed that a variety of enteroviruses were desorbed from 4.3×12.5 cm cores by rinsing the cores with artificial rainwater (the rainwater used in the study had a pH of 4.4 and a conductivity of

40 umho and accurately reflected the anion and cation contents of rainfall in the northeastern U.S.). Elution appeared to vary with virus type. High concentrations (>40%) of a poliovirus reference strain (Leon) and a field strain of poliovirus 3 were eluted from soil cores with a 100 mℓ rainwater rinse, while lower levels (15 to 40%) of a second poliovirus 3 field strain, echovirus 6 (D'Amori) and two guanidine-resistant strains (g^r) of poliovirus 1 (LSc) were also eluted. Little or no desorption was noted for reference strains of poliovirus 1 (LSc 2ab) and echovirus 1 (Farouk). The investigators concluded that the percolating rainwater induced major changes in the adsorbing environment of the virus and soil particles and interfered with the subsequent adsorbing interactions. The studies indicated that conductivity rather than pH was the more important factor in viral elution. A second series of *in situ* field experiments was conducted with 10 × 75 cm soil cores set in an operating recharge basin in an effort to determine the extent of rainwater-induced vertical migration.[37] The virus employed in these studies was a guanidine-resistant strain of poliovirus[1] (LSc) which had soil adsorption/elution characteristics similar to a number of nonpoliovirus types.[125] Analysis of the initial virus distribution in experimental cores (Figure 2) indicated that 77% of the viruses were retained in the upper 5 cm of soil. An additional 11% were adsorbed in the next 5 cm, while a total of 96% were retained before the 25 cm soil level. The remaining 4% were uniformly distributed over the next 50 cm of soil with each 5 cm section containing a minimum of 0.23% of the total viruses recovered. Additional experiments showed that few viruses (<0.22%) percolated through the entire core under the experimental loading conditions. The distribution of viruses in the top soil layers was not very different from that reported for indigenous enteroviruses by Hurst and Gerba[36] and Hurst et al.[160] To determine whether applications of sewage effluent or artificial rainwater would alter the initital distribution pattern, virus-seeded cores were challenged with 21 cm rinses of either artificial rainwater or sewage effluent. The investigators reported that the overall viral distribution in all cores (control and rinsed cores) was essentially the same, regardless of treatment. From this they concluded that no large-scale virus migration occurred during either sewage or rainwater percolation. However, on careful analysis of the data, they noticed the presence of statistically significant regions of localized virus movement after each rinse. A comparison of the percentage of viruses found in each level of all three core types (Table 7) indicated that the adsorption in the upper 5 cm of soil was statistically identical in all situations. The investigators noticed, however, that a higher percentage of viruses were located throughout the lower sections of the rainwater-treated cores (25 to 75 cm) than in the control or sewage-rinsed cores. These desorbed viruses appeared to have originated from the core areas above the 20 cm level. Most desorbed viruses had apparently resorbed at greater soil depths since few (0.16%) were recovered in core effluents. A smaller, but still significant redistribution of the virus population was also observed in sewage-rinsed cores. However, in these cores a majority of viruses appeared to resorb at higher soil depths than observed in rainwater rinsed cores.

The above compiled results suggest that any reduction in ionic strength clearly weakens the virus-soil adsorbing forces, ultimately leading to virus desorption and ensuing entrainment in percolating fluids. The extent of that movement appears to depend upon the strength of the adsorbing environment in the lower soil depths. If extensive flooding with low conductivity fluids occurred, it might be sufficient to dilute ionic gradients in the soil profile, thus favoring extensive viral movement. In addition, a poor virus-adsorbing environment could also be produced by employing soils with a poor ion adsorbing capacity, or by insufficient conditioning to set up the proper ion environment. In most environmental situations, the low-conductivity desorption phenomenon seems readily reversible by simple application of fluids of high ionic strength in the form of treated sewage effluent, or by the addition of exogenous cations.

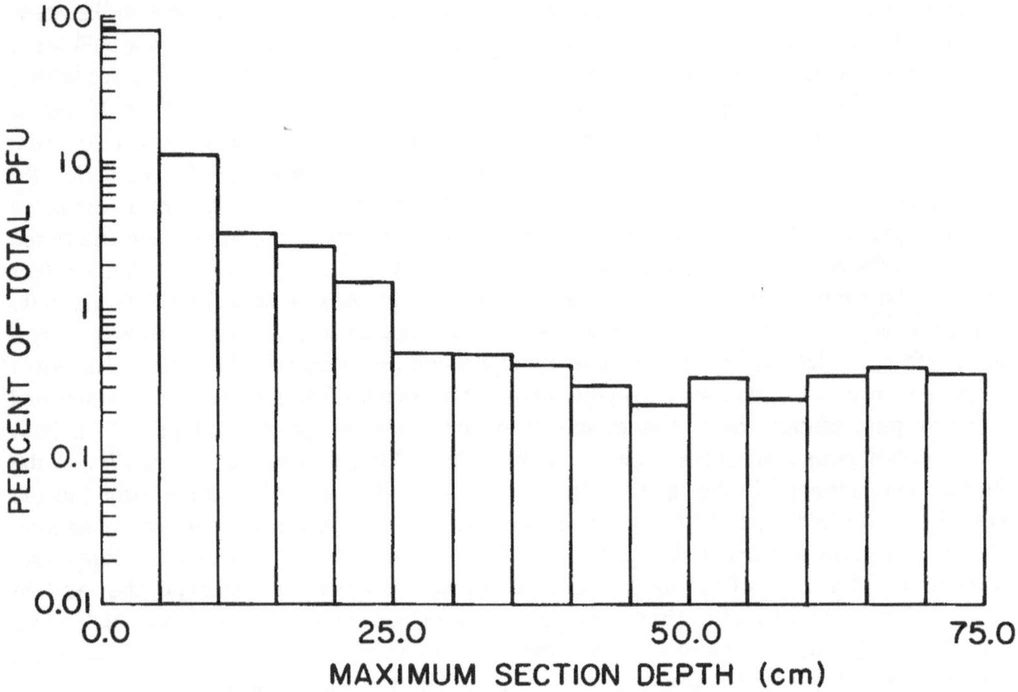

FIGURE 2. Distribution profile of poliovirus 1 (LSc-g^r) in 75 cm sandy soil cores. (From Landry, E. F., Vaughn, J. M., and Penello, W. F., *Appl. Environ. Microbiol.*, 40, 1032, 1980. With permission.)

b. Virus Type

As noted in Section III.B, virus type-dependent adsorption appeared to be governed by the overall electronegativity of the virus particle which appeared to be a unique characteristic of the virus. It follows then, that virus desorption should be similarly type-dependent. Recent evidence appears to indicate that this is the case. Landry et al.[125] observed that the extent of viral elution from soil cores, by either rainwater and sewage effluent rinses, varied with virus type. Relatively low levels (<2%) of vaccine strain poliovirus 1 (LSc 2ab) were eluted from small cores with either type rinse, while large concentrations of guanidine-resistant mutants of this strain were extensively eluted with rainwater. Rainwater elution of field and reference strains of poliovirus 3 was substantial, while its sewage elution by sewage was poor. Based on these data, the investigators concluded that the desorption behavior of enteroviruses in soil was not uniform and was related to the specific virus type. They cautioned against over-extrapolation from data derived from the study of a few virus types. This appeared to be particularly true for poliovirus[1] (LSc) which had poor elution characteristics. Vaughn et al.[54] suggested that all polioviruses may elute poorly, since members of the group were the least encountered during a field study of virus-contaminated aquifers. Hurst et al.[160] also recently demonstrated strain-dependent virus desorption from *in situ* field cores buried in a recharge basin. Following 5 days of basin flooding, they found that about 18% of the poliovirus 1 had moved from the core surface to a depth of 15 cm while <0.1% of echovirus 1 migrated to any depth. Stotsky et al.[134] also reported that the desorption of viruses from clay minerals depended on virus type.

c. Miscellaneous Factors

A number of other factors likely influence the desorption of viruses from soil. Some of these have already been discussed with respect to their relationship to adsorption, and will

Table 7
GROUPINGS[a] OF STATISTICALLY SIMILAR CORE SECTIONS BY PERCENT FOR CONTROL, SEWAGE- AND RAINWATER-RINSED CORES[b]

Section number	Maximum core depth (cm)	Core type R	Core type S	Core type C
1	5	51.11	81.02	63.08
2	10	18.25	3.68	7.08
3	15	3.17		
4	20			1.25
5	25			
6	30		1.32	
7	35			
8	40			
9	45	1.67		
10	50			0.59
11	55			
12	60			
13	65		0.20	
14	70			
15	75			

[a] Designated section groupings are not significantly different based upon the Least Significant Difference technique. Note that other subgroups are possible, but only those that do not violate the continuity of the cores are presented.

[b] R,S,C refer to rainwater-, sewage-rinsed or control cores, respectively. Values represent the mean percent values over the distances defined by the groups or lines.

receive cursory mention here. Alkaline pH effectively removes viruses from filter media,[161,162] and has been used by a number of investigators to elute viruses from soil. High pH fluids appear to create strong electrostatic repulsive forces between negatively charged soil and virus particles causing their desorption. Among eluting fluids moderately successful in extracting viruses from soils are 0.25 *M* glycine — 0.05 *M* EDTA, pH 11.5,[36] 0.5% isoelectric casein pH 9.0,[163] 3% beef extract — 0.05 *M* tris pH 9.5,[37] and 0.3% beef extract — 0.05 *M* glycine pH 9.5.[164] Since wastewater effluents are typically acidic in nature, alkaline-influenced desorption processes should have little impact on virus movement in environmental systems.

Once desorbed, several soil characteristics may contribute to virus movement to groundwater. As previously noted, a number of investigators suggested that virus migration was more extensive in soils which were riddled with cracks or fissures.[35,65,154] These conditions apparently created localized zones of high percolate flow and interfered with the resorption of viruses. The depth of the soil profile also appears to be important in virus migration.[40,118] While movement through small soil cores (10 to 12 cm) was extensive,[125,145] extending the travel distance to 100 to 250 cm significantly reduced virus movement.[37,135,147] The degree of soil saturation appears to be important in determining the extent of virus migration.[58] Noonan and McNabb[165] and Pyle et al.[166] reported extensive migration of coliphage ΦX174 under saturated soil conditions. These investigators indicated that the virus moved laterally over distances of 900 m in a relatively short period of time. The rate of movement was measured at about 350 m/day.

Several of the above factors have been implicated in field studies which have shown viral contamination of the aquifer. The isolates reported by Wellings et al.[158] at a spray irrigation site were recovered from a shallow aquifer (~3 m) following a period of heavy rainfall. Similarly, isolations from groundwaters beneath a recharge site also involved a shallow aquifer (3 m) whose continuity had likely been compromised by a construction project in the basin.[65,120] In addition, the movement of the contaminating viruses may have been influenced by heavy rainfall which occurred during the preceding months. These few examples serve to illustrate that generally more than one factor is involved in aquifer contamination. The threat of groundwater contamination can be substantially reduced by nullifying the effects of one or more of the above factors. For example, rainwater effects can be reversed by flooding application sites with high ionic strength fluids (e.g., sewage effluent), soil saturation can be minimized by decreasing the hydraulic loading rates and employing a scheme of alternate loading and drying, and careful site selection procedures can screen out areas with poor soil continuity and shallow aquifers.

V. VIRUS SURVIVAL IN SOILS

A. Introduction

When assessing the potential health effects of a land application system, it is important to consider not only the ability of a virus to adsorb to receiving soils, but also its ability to survive or retain infectivity in such environments. Ultimately, it is the extent of virus survival which will determine its relative hazard. Thus, for a proper assessment, an investigator must be aware of the survival capacity of a virus in the adsorbed state as well as its survival in percolating fluids and groundwaters. During the past two decades, many reports have addressed the persistence of enteroviruses in soil environments. Many of these were thoroughly reviewed by Duboise et al.[1] and Gerba et al.[97] and only a few will be mentioned here in order to document the problem. Bagdasaryan[42] reported that enteroviruses persisted for long periods of time in seeded-soils and on vegetables. Echovirus 7 survived for up to 170 days in sandy soils during winter months. Larkin et al.[167] observed that viruses survived for periods of up to 96 days in sludge or sewage-irrigated soils under winter conditions. Survival was greatly reduced under spring conditions, but still sufficiently long to overlap the growing and harvesting seasons of some vegetable crops. This caused the investigators to warn of possible contamination of such crops.[168] Wellings et al.[65] reported that viruses could survive for at least 28 days in a soil environment following sewage recharge into a cypress dome. This time period was sufficient to allow the virus to move both vertically and horizontally, and appear in groundwater observation wells. Later, Moore et al.[31] observed that indigenous enteroviruses were able to survive for prolonged periods of time in sub-surface sludge injection sites. These examples serve to illustrate the persistent nature of enteroviruses in soils, sludge, and groundwater.

Although there is limited information on the survival of indigenous enteroviruses in natural soil environments, there are extensive descriptions of the survival of seeded enteroviruses in artificial and natural soil environments under a variety of conditions. Based upon the most recent data, a list of the specific factors which appear to influence the survival of enteroviruses in soils may be assembled. While each factor will be presented individually, the reader should recognize that they do not act as independent variables. Many factors are interrelated and act in concert to enhance or antagonize virus survival in extremely complex and dynamic soil environments. A complete list of survival factors is presented in Table 8. The listing order is arbitrary and does not imply any order of priority.

B. Factors that Influence Virus Survival in Soils

1. State of Adsorption

A number of investigators recently determined that particle-associated viruses survived

Table 8
FACTORS THAT INFLUENCE THE SURVIVAL OF ENTEROVIRUSES IN SOILS

Characteristics	Ref.
State of adsorption	126, 134, 169—171
Temperature	31, 41, 145, 171, 174
Moisture content	45, 57, 160, 171, 175, 176
Virus and soil types	134, 148, 160, 171, 175, 176, 180
Microbial populations	126, 171, 174, 181

longer than those in the unadsorbed state. Bitton et al.[169] demonstrated that adsorption of poliovirus 1 to nontronite clay significantly extended survival time during exposure to solar radiation. In the absence of the clay, the time required for a 1-log reduction in virus titer (T 90) was 75 min while in the adsorbed state the T 90 was 163 min. Foster et al.[170] observed that the association of virus particles with solid surfaces such as feces or cellular material enhanced their survival during ozone disinfection. In the unadsorbed state, viruses were completely inactivated in 10 sec at an ozone concentration of 0.012 mg/ℓ while in the adsorbed state only 45% of the viruses were inactivated during a 30 sec exposure. They reasoned that the key infectivity sites on the virus particles were inaccessible or only partly accessible to the chemical disinfecting agent. The investigators suggested that other materials such as soils might offer the same type of protection. Indeed, soils protect viruses from natural, chemical, and biological degradation. Hurst et al.[171] reported that virus adsorption to soils was one of the four principle factors which enhanced their survival in the environment. These conclusions were made following a study of the survival of a number of enteroviruses and phage including coxsackieviruses A9 and B3, echovirus 1, poliovirus 1, a simian rotavirus, and coliphages T2 and MS2 in nine soil types under the same conditions. After a complex step-wise multiple regression analysis of the various virus survival slopes and 19 independent soil variables, the investigators concluded that the degree of viral adsorption was responsible for almost one half of the total variance in the experiment, indicating that it played a major role in virus survival. They suggested that soils which adsorbed viruses poorly would produce a high rate of viral inactivation, while good virus-adsorbing soils would produce a slow die-off rate. Similar soil protective effects were reported by Sobsey et al.[126] who found extended survival of a poliovirus to be associated with adsorption to certain soils. For example, the time required for a 2-log reduction (T 99) of virus adsorbed to Cecil soils was 167 days compared to a control T99 (unadsorbed) of 86 days. These investigators noted, however, that not all soils offered this protection. Ponzer soils which were poor virus adsorbers under test conditions, apparently accelerated the inactivation time (18 days). Apparently, factors associated with this soil type not only inhibited viral adsorption, but enhanced viral inactivation by some unknown mechanism. Additional protective effects have been associated with adsorption to other surfaces.[172,173] Stotsky et al.[134] found that the adsorption of coliphages T1 and T7, and reovirus 3 to clay minerals markedly enhanced their survival. The survival of coliphage T7 at 4°C was extended from 5 to 31 weeks in the presence of clay material, while survival at 24°C was enhanced from 1 to 9 weeks with montmorillonite and up to 7 weeks with kaolinite. These investigators also noted that clay-bound reoviruses were still capable of infecting a mouse fibroblast cell line. Kaolinite-bound viruses actually had a higher rate of infectivity compared to free viruses, due apparently to a better cellular adsorption rate caused by the sedimentation of the heavier virus particles. A number of other investigators have shown viruses to be infective while adsorbed to surfaces. Moore et al.[133] reported that poliovirus 1 adsorbed to clay could be recovered by direct assay with a 93% efficiency, and Schaub and Sagik[101] reported EMC virus was still infective in both in vitro and in vivo host systems. Lefler and Kott,[132] shunning inefficient elution techniques, used a direct assay of sand particles to detect adsorbed polioviruses. In each case, survival appeared to be enhanced by adsorption.

In summary, the attachment of a virus particle to a solid surface (soils, clay minerals, sediments) protect viruses from hostile environments. Although the exact mechanism of this protection is not known, it has been speculated to involve key infecting or host-adsorbing sites. However, since some viruses are still infective in the adsorbed state, these key infecting sites may not always be bound and rendered inaccessible by the adsorbing surfaces. It is also important that the factors that affected survival in the adsorbed state were complex and variable. Since factors such as pH, ionic strength, and virus and soil types, affect virus adsorption to soils, they also likely have a direct or indirect effect on virus survival. These interrelationships serve to demonstrate the complexity of the survival process.

2. Temperature

The effects of temperature on the survival of viruses in soils have been documented and were recently reviewed by Duboise et al.[1] and Gerba et al.[97] Generally, virus survival is inversely related to soil temperature. Duboise et al.[145] found low temperature (4°C) enhanced the survival of a soil-associated poliovirus, while higher temperature (20°C) promoted inactivation. Damgaard-Larsen et al.,[149] studying the persistence of a seeded coxsackievirus in sludge-soil lysimeters, reported that the virus survived for 23 weeks under winter conditions. They calculated the rate of inactivation to be on the order of 0.2 $\log_{10}$ infective units per week. Tierney et al.[41] observed a similar temperature relationship in the survival of poliovirus 1 in sewage and sludge-irrigated soils. They reported that the rates of inactivation were linear with time for both summer and winter months, with a 1-log decrease occurring about every 20 days in winter, and every 1.5 days during the summer. Recently, Hurst et al.[171,174] reported that the rate of inactivation of poliovirus 1 in Flushing Meadow soils increased with increasing temperature. The rates of inactivation were linear with time, except when temperatures approached 1°C. At this temperature, little inactivation was noted. They found that temperature effects were prominent under aerobic and anaerobic conditions, as well as in sterile and nonsterile environments at a constant moisture content (15%).

The precise mechanism of inactivation at high temperature is as yet undetermined. Several investigators have suggested that such inactivation is related to fluid evaporation. Moore et al.[31] found that the T 90 for a poliovirus in sludge-injected soils was 3 months at 4°C, while at 20 and 30°C it was 1 month and 1 week, respectively. They suggested that the rapid inactivation of viruses at high temperatures was probably tied to a rapid drop in moisture content. Yeager and O'Brien[175,176] after observing an inverse relationship between the survival of a poliovirus and a coxsackievirus and temperature, hypothesized that inactivation might ultimately be related to the amount of moisture in soils.

3. Moisture Content

Recent evidence strongly supports the role of soil moisture in viral inactivation. Sadovski et al.[45] found that during application of seeded-wastewaters to soil plots, viruses persisted for at least 8 to 18 days. However, 5 days after sewage application ceased, they reported a rapid drop in the number of viruses in the soil. Concomitantly, they noted that the soil moisture content had decreased from 15 to 3%. Bitton et al.[148] monitoring the survival of poliovirus 1 in a sludge-soil mix, observed extensive survival (up to 35 days) during the hot, wet summer season. During the dry, autumn season, however, when air temperature was only slightly reduced, survival did not extend past an 8-day period. In a related study, these investigators[57] observed a rapid die off of indigenous enteroviruses in sludge applied to land surfaces. In both studies, they concluded that the inactivation was directly related to loss of soil moisture. Yeager and O'Brien[175,176] also found that moisture content was important in the survival of both poliovirus 1 and coxsackievirus B3 in soils. After failing to recover any infective viruses from dried soils, the investigators attempted to determine whether this inability was the result of irreversible viral adsorption, or if the drying itself had caused inactivation of the adsorbed viruses. In tests with ^{3}H-RNA radiolabeled viruses, they found that both infectious virions and 80% of the ^{3}H-labeled RNA could be recovered from soils saturated with river, ground or septic tank waters. In dried soils, while a major portion of the label was easily eluted, only a few infectious units were recovered, suggesting loss of infectivity by viral inactivation. This observation was consistently observed in eight different soil types including sand, sandy loam, clay, and potting soils. To elucidate the mechanism of inactivation, the investigators attempted to determine whether the inactivation effects resulted from the physical evaporation process itself, or if they were related to exposures to continuous, low-moisture levels. The survival of a poliovirus in naturally evaporating soils was compared with survival in soils held at constant moisture levels. In both cases, the same inactivation trends were observed. Initially, there was a gradual decrease in viral numbers with decreasing moisture until soil moisture content reached 2.9%. Below this level, inactivation effects were greatly amplified as evidenced by greater inactivation rates. Since they were unable to eliminate conclusively the possibility of localized evaporative effects in the continuous moisture tubes, these investigators could not identify the precise mechanism of inactivation. They did note however that viruses could not be recovered from rapidly dried soils. They suggested that under rapid drying conditions, there was little time for moisture to be redistributed. They thus concluded that continuous evaporation was responsible for the higher rates of viral inactivation at low soil moisture levels. Ward and Ashley[177] also suggested that viral inactivation was due to evaporative effects.

In later experimental series, Yeager and O'Brien[176] determined that the mode of virus inactivation in dried soils differed from inactivation occurring in moist soils. They suggested that inactivation was the result of irreversible damage to the virus particle, but the damage in each case was of different origin. In moist soil, they observed that 83% of the ^{3}H-labeled RNA and 72% of ^{14}C-labeled protein could be eluted from the soil with a high pH buffer. However, the RNA could only be recovered in a degraded state, suggesting that inactivation involved dissociation and degradation of both viral genome and capsid. In dried soils, similar amounts of the radiolabeled genome were recovered but in this case the genome appeared to be initially released intact, since under sterile conditions the entire undegraded genome could be recovered. This, coupled with the fact that little ^{14}C-protein was released from the soils, suggested that inactivation under dry conditions involved a slight dissociation of the protein capsid, and subsequent release of an intact genome. Upon release into unsterile moist soils the genome would undergo rapid degradation.

Hurst et al.[160,171,178] also presented evidence relating virus inactivation to moisture content. Analyzing the inactivation of poliovirus 1 in sludge, they noted that a 10% loss of moisture resulted in an 80% reduction in recoverable viruses, while a 0.3% moisture drop led to a 20% viral reduction.[178] A similar conclusion was reached in a later *in situ* basin study.[160] In these studies, the investigators observed that the rate of poliovirus 1 inactivation was more rapid in small, open tubes (0.11 log reduction per day) which were subject to continual moisture loss, than in sealed tubes (0.04 log/day) maintained at constant moisture levels. While the inactivation rates observed in the small tubes (0.11 $\log_{10}$/day) were later found to be less than those reported for larger cores (0.49 $\log_{10}$/day), they still permitted a relative measure of the various moisture effects. The lower rates observed in the smaller tubes were due presumably to the redistribution of soil moisture in the tubes. The investigators also reported that laboratory derived rates of inactivation were actually one half the rates observed under field conditions. Apparently, laboratory experiments failed to incorporate the dynamic nature and multiple-factor involvement associated with field conditions. The investigators proposed that periodic drying of a recharge basin would decrease the moisture content and thereby enhance virus inactivation. This condition would prevent an accumulation of viruses in the soil which might ultimately threaten the groundwater. They conceded, however, that such drying effects would predominate near the soil surface and may not be as drastic in deeper soil areas. This suggests that viruses adsorbed in the lower soil layers may present a hazard due to: (1) their extended survival capacity, (2) the relative ease of desorption, and (3) the relatively short travel distance to the aquifer.

4. Virus and Soil Type

In an earlier section, we noted the enhancement of virus survival via their adsorption to soils. Since adsorption is affected by both virus and soil type, it is reasonable to assume that these same factors should influence virus survival. As early as 1964, Bagdasaryan[42] showed that the survival of poliovirus coxsackievirus, and echoviruses was dependent on soil type. Yeager and O'Brien[175] observed that the survival of poliovirus 1 and coxsackievirus B1 also depended on the soil type. In these studies, they noted that the poliovirus adsorbed to a sandy loam soil (19% clay content) were inactivated at a slower rate than those adsorbed to sandy soils (0.8% clay). Viruses survived longer in soils saturated with septic tank liquor than in river or groundwater. Stotsky et al.[134] observed that coliphage T7 persisted longer at 24°C when adsorbed to montmorillonite rather than kaolinite clay minerals. Hurst et al.[171] recently completed a complex analysis of the survival of seven different viruses in nine different soil types. They concluded that the soil type, as reflected by the pH, the amount of exchangeable aluminum, and resin-extractable phosphorous in the soil, was critical to virus survival. When combined, these three factors accounted for over one half of the total variance in the experiment indicating that they played a major role in survival. Since all three factors had some effect on viral adsorption, the investigators suggested that virus survival actually depended on the extent of adsorption. They noted, for example, that viral adsorption as well as survival in soil increased with decreasing soil pH and resin-extractable phosphorous. Moreover, they observed a positive correlation between the concentration of exchangeable Al^{+++} in the soils and virus survival. Since increasing concentrations of these ions also enhance viral adsorption to a number of surfaces including filters,[179] clays,[131] and magnetite,[106] the investigators suggested that increased survival was the result of better viral adsorption.

The composition of the soil surface also affects the viability of soil-bound virus particles. Murray[103] and Murray and Laband[180] reported no loss of infectivity when viruses were desorbed from SiO_2 or Fe_2O_3 surfaces. However, when viruses were eluted from materials such as CuO, MnO_2, and Al_2O_3, the viruses were physically disrupted, with a resultant loss of infectivity. The investigators suggested that at least two mechanisms were responsible for virus inactivation. In the case of CuO, extensive viral degradation appeared to take place on the surface of the inorganic particles causing the release of large amounts of both the labeled RNA and protein. Alternatively, inactivation on MnO_2 and Al_2O_3 surfaces appeared to be the result of only slight conformational changes in the viral capsid. They concluded that the ability of a virus to desorb from a surface and retain its infectivity depended upon the specific soil surface.

Since the extent of soil adsorption appears to be a key factor in virus survival, it follows that other factors influencing adsorption, such as the virus type, also are important. This has recently been proved true by Hurst et al.,[160] who observed that soil adsorbed echovirus

1 was inactivated at a higher rate than poliovirus 1 under similar conditions. Bitton et al.[148] also found that virus survival was related to the particular type of virus. Presumably, the variability identified in both studies related to the degree of viral adsorption to soils.

5. Microbial Population

Sobsey et al.[126] reported that the survival of poliovirus 1 and reovirus 3 depended upon the microbial activity in the soil. They observed that viruses persisted longer in sterile soils than in nonsterile soils. Hurst et al.[174] also noted that aerobic soil microorganisms had an adverse effect on the survival of poliovirus 1 and coliphage T2 and MS2 in different soils. In a later study,[160] these same investigators found extended survival under aerobic nonsterile conditions, but not under aerobic sterile, anaerobic sterile, or anaerobic nonsterile soil conditions. This difference became less apparent as the temperature increased to a maximum of 37°C. The investigators suggested that at this temperature, thermal and moisture effects were probably more pronounced than microbial effects in all soil environments. Cheo[181] found that the survival of tobacco mosaic virus (TMV) was also related to the presence of aerobic microorganisms in the soil. TMV viruses were relatively stable in sterile soils, but were rapidly inactivated when placed in a nonsterile soil environment. Bacterial activity was the suspected cause of inactivation since the presence of high concentrations of bacteriocidal agents such as streptomycin and cycloheximide reduced the rate of inactivation.

The preceding review has identified a number of factors which influence the survival of viruses in soils. Factors, such as solar radiation,[182] extracellular nucleases and proteases[1] may also exert an influence, but the extent of this influence is unknown. All of the factors which definitely influence survival, i.e., moisture content, virus type, soil type, and state of adsorption, appear to be clearly interrelated. We may conclude that factors which promote virus attachment to soils will also enhance their survival. As Hurst et al.[171] have pointed out, this was somewhat ironic since optimal land treatment practices demand soils which efficiently adsorb viruses. It is these same soil types which would likely abet virus survival.

VI. PREDICTING VIRUS MOVEMENT IN SOILS

Adsorptive interactions between viruses and soils are complex and depend upon a number of interrelated wastewater, soil, and virus characteristics (Table 2 and 5). Due to the many variables involved and the sheer variety of virus types known to be present in sewage, an empirical study of all virus-soil interactions would be difficult, if not impossible. A predictive approach to these interactions is that of mathematical modeling. Such modeling has only recently been undertaken following an increased understanding of adsorptive interactions and the availability of detailed adsorption data. One of the most comprehensive models available was recently described by Vilker.[99,111-113,116] With a series of complex ion exchange/adsorption equations, he has constructed an adsorption-mass transfer model which appears to be useful in describing the breakthrough or concentration of viruses in a soil column. The percolation/adsorption model was based upon three important characteristics:

1. A nonlinear adsorption equilibrium isotherm relationship which describes the distribution of viruses between the solid and liquid phase
2. The virus mass balance, which is the sum of the liquid phase depletion term, the convention term, and the solid phase accumulation term
3. The adsorption mass transfer coefficient which describes the rate of approach to adsorption equilibrium involving the diffusion of the viruses from the liquid to the solid soil surface.[113]

The model incorporates these relationships to calculate a series of concentration or breakthrough curves with which to describe the likely depth of virus penetration in a soil column.

Table 9
SPECIFIC FACTORS REQUIRED FOR SOLUTION OF THE ADSORPTION MASS TRANSFER MODEL

Superficial or percolation velocity of the fluid (cm/min)
Void or empty bed fraction
Interstitial or average fluid velocity between solid particles (cm/min)
Soil particle diameter (cm)
Column bulk density ($8/cm^3$)
Outer surface interfacial area of sorbent particles (cm^{-1})
Virus diffusivity (cm^2/sec)
Viscosity (cp)
Mass transfer coefficient which describes the rate of approach to equilibrium, accounting for the diffusive transport of the virus from a bulk solution to the surface of the solid (cm/min)
Adsorption equilibrium isotherm which describes the distribution of viruses between the liquid and solid phase (mℓ/mg)

It is clear that much information is required for predicting adsorption interactions including: the type and concentration of virus; the type and composition of soil; and the rate of fluid application. In order to solve the adsorption mass transfer model, a number of other measurable characteristics are required, including a detailed description of the soil column bed. A complete list of all the specific factors required can be found in Table 9.[111] Values for each of these characteristics were either assigned, calculated, or experimentally measured. Prior to solution, the model was additionally qualified for a defined set of conditions. Vilker chose to base his model on the movement of single virus particles through a homogeneous column of uniform bulk density at a constant rate of application. Moreover, he assumed that no viral inactivation occurred during percolation. While conceding that these conditions might not be applicable to field systems where soil beds, rates of sewage-application, and virus loads were not uniform, he argued that they were valid considerations for soil columns. He suggested that the moderate ionic strength, neutral pH, and low suspended solids content of most treatment plant effluents, would favor an unassociated or free viral conformation.[99] In spite of these considerations, the model appeared to offer a reasonable means of assessing and ranking the relative abilities of soil to remove percolating viruses. Once the relative adsorbing capability of the soil was known, the model might also be useful for making some limited extrapolations to field conditions.

While a complete description of the model is beyond the scope of this review, it is important to discuss some aspects of its application. Figure 3 illustrates the predicted adsorption profile of coliphage ΦX174 when it is passed through a column of Kranzburg soil.[111] The curves were generated, with a series of complex equations, by calculating the fraction of unadsorbed (C/Co), or adsorbed viruses ($q/q_{\rightleftharpoons}$) for a specific time (days) and depth (cm). The calculations incorporated values for all the variables listed in Table 9. The series of curves indicated that for a defined set of conditions, virus removal was proportional to column depth. Analysis of the liquid-phase adsorption profile (Figure 3a) showed that after 1.2 days of constant virus loading at a percolation of 0.01 cm/min, all of the phages were not adsorbed (C/Co = 0) until they had penetrated to a depth of 25 cm. If the application were extended to 60 days at this same rate, the model would predict that phages would start to appear in 100 cm column effluents. Reduction of the infiltration rate to 0.004 cm/min would slow the extent of viral penetration such that viruses would move to the 60 cm depth only after 120 days. These projections appear to confirm the experimental findings of Lance and Gerba[129] and Vaughn et al.,[144] which indicated that reduced infiltration rates promoted better viral adsorption. The adsorbent profile (Figure 3b) predicted that the maximum number of viruses would adsorb within the top 40 cm of soil after 12 days of constant percolation. The 120-day curve predicted that all adsorption sites in the top 20 cm of soil would be

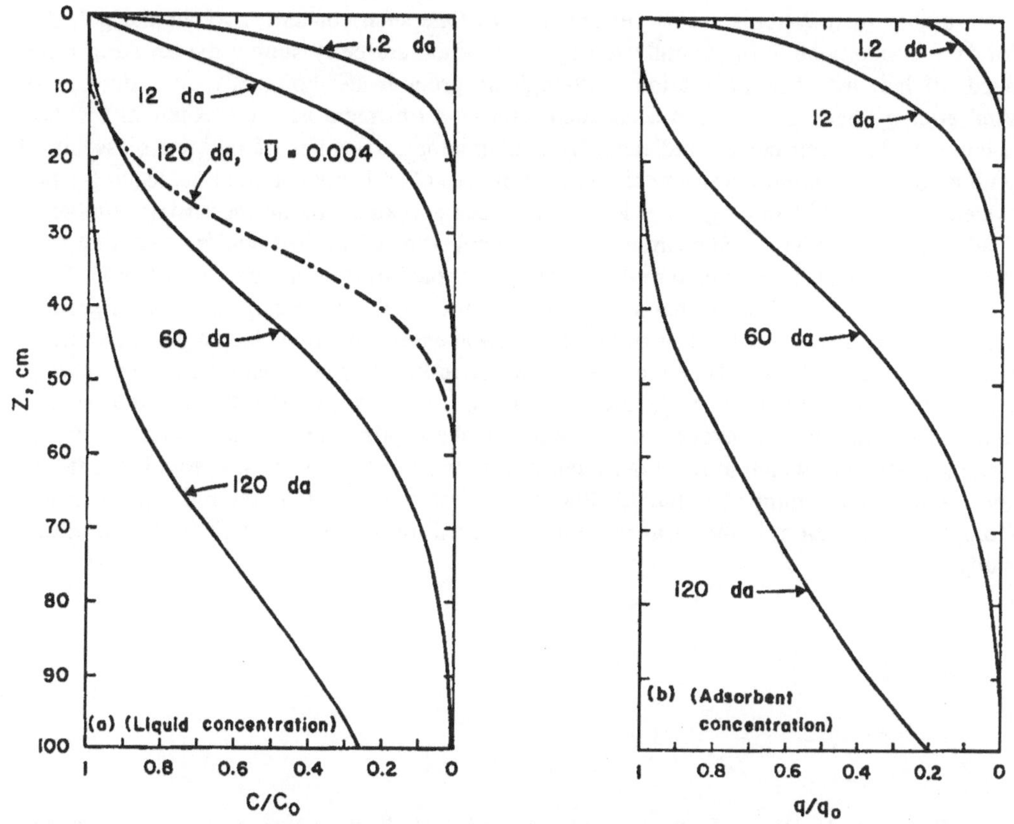

FIGURE 3. Concentration/breakthrough curve for the adsorption of coliphage ϕX174 on Kranzburg soils; (a) liquid-phase virus profile at percolation rate of 0.01 cm/min (———) and 0.004 cm/min (— · — · —), (b) adsorbent-phase virus profile at 0.01 cm/min. C/C_o is the fraction of unadsorbed virus. q/q_o is the fraction of adsorbed virus. Z is the depth of the column penetrated. (From Vilker, V. L. and Burge, W. D., *Water Res.*, 14, 783, 1980. With permission.)

saturated with viruses within this time period, forcing viruses to move to greater soil depths. The curve also predicted that viruses would percolate and appear in the 100 cm column effluent. In addition to predicting virus concentration and breakthrough, the model could be used to rank soil on the basis of its relative ability to adsorb viruses. Vilker[116] illustrated this potential by performing a comparative analysis of the adsorption of coliphage T2 to sandy clay loam soils,[108] coliphage ΦX174 to Kranzburg soil,[115,130] and poliovirus 1 to sandy soil.[132] When these data were standardized and the adsorption profile curve of each calculated, he found that coliphage T2 was retained better on sandy clay loam soil than coliphage ΦX174 was on Kranzburg soil. He also observed that poliovirus 1 suspended in 0.5 *N* NaCl adsorbed poorly to sterile sandy soil. The investigator concluded that while these models would not completely describe the movement of viruses in a field situation, they might play an important role in defining general virus-soil interactions.

VII. SUMMARY AND CONCLUSIONS

Human viruses usually gain access to soil systems through intentional or unintentional discharges of domestic wastewater. Intentional land treatment/disposal systems represent an attractive alternative to surface water discharges, providing both economic and environmental advantages. Major concerns over the possible threat to human health as a result of the large-scale use of such systems are as yet unresolved. One such concern involves the potential for viral contamination of groundwater systems which currently supply the drinking-water needs of half the U.S. population. Although no groundwater-borne disease outbreaks of viral etiology have as yet been associated with land treatment use, the potential for their occurrence has been clearly indicated by epidemiological studies of outbreaks associated with groundwater pollution from unintentional modes of soil application (e.g., leaking septic systems, etc.). Epidemiological evidence has been supported by an increasing number of field studies which have demonstrated viral contamination of shallow aquifers resulting from the use of various land treatment modes (especially rapid infiltration/recharge systems). More recent studies have indicated that soil-associated microbial (i.e., viral) pollution of groundwater may be abated by the use of systems management practices based upon an understanding of the physical and chemical factors which influence viral retention in soil including: temperature; pH; moisture, clay, and ion content; depth to groundwater; infiltration rate; and soil continuity. The proper manipulation of these principles in the operation of land treatment schemes which utilize high-quality wastewater effluents should provide the means for significantly diminishing the likelihood of viral movement to groundwater systems. Complete elimination of the viral pollution potential awaits the result of further scientific study.

REFERENCES

1. **Duboise, S. M., Moore, B. E., Sorber, C. A., and Sagik, B. P.,** Viruses in soil systems, *Crit. Rev. Microbiol.*, 7, 245, 1979.
2. **Sorber, C. A. and Sagik, B. P.,** Health effects of land application of wastewater and sludge: what are the risks?, *Water Sewage Works*, 125, 82, 1978.
3. **Wolman, A.,** Public health aspects of land utilization of wastewater effluents and sludges, *J. Water Pollut. Control Fed.*, 49, 2211, 1977.
4. Environmental Protection Agency, *Environment News*, September 13, 1980.
5. Water and Sewage Works Focus, Land treatment of sewage: EPA's view, *Water Sewage Works*, 125, 64, 1978.
6. **Harris, R. H.,** Water quality management on Long Island, case for recycling municipal wastewater by groundwater recharge, testimony presented at the Tidal Wetlands Hearing, Suffolk County, New York, 1975.
7. **Larkin, E. P.,** Problems associated with the use of sewage wastes on crops, *Assoc. Food Drug Off. U.S., Q. Bull.*, 44, 116, 1980.
8. **Culp, G. L., Culp, R., Hinricks, D., and Williams, R.,** Land treatment and advanced wastewater alternatives at South Tahoe, *J. Water Pollut. Control Fed.*, 51, 2487, 1979.
9. **Reynolds, J. H. and Cissell, J. A.,** Land application of wastewater, *J. Water Pollut. Control Fed.*, 52, 1284, 1980.
10. **Barth, D. S.,** EPA's research and development program on the health effects of land application of municipal wastewater and sludges, in *Proc. Conf. Risk Assess. Health Effects Land Applic. Munic. Wastewater Sludges*, Sagik, B. P. and Sorber, C. A., Eds., Center for Applied Research and Technology, University of Texas, San Antonio, 1978, 1.
11. **Keeley, J. W.,** Meeting the groundwater contamination problem, *Water Sewage Works*, 124, 88, 1977.
12. **Wright, K. R. and Rovey, C. R.,** Land application of waste: state of the art, *Groundwater*, 17, 47, 1979.
13. **Small, M. M.,** Wetlands wastewater treatment systems, in State of Knowledge in Land Treatment of Wastewater, Vol. 2, McKim, H. L., Ed., U.S. Army Cold Regions Research and Engineering Laboratories (CRREL) Hanover, New Hampshire, 1978, 141.
14. **Hunt, P. G.,** Reliability problems with indicator organisms for monitoring overland flow treated wastewater effluent, *J. Environ. Qual.*, 8, 301, 1979.
15. **Vaughn, J. M. and Landry, E. F.,** Data report: the fate of human enteric viruses in a natural sewage recycling system, Rep. 51281, Brookhaven National Laboratory, Upton, N.Y., 1980.

16. **Chase, S. E.,** Nine decades of sanitary engineering, 3. Back to the land, *Water Works Waste Eng.*, 1, 7, 1964.
17. **Bernard, M. A.,** Land disposal and sewage effluent: appraisal of health effects of pathogenic organisms, *J. Am. Water Works Assoc.*, 65, 432, 1973.
18. **Dunlop, S. G.,** Survival of pathogens and related disease hazards, in *Municipal Sewage Effluent for Irrigation*, Wilson, C. W. and Beckett, F. E., Eds., Louisiana Polytechnic Institute Alumni Foundation, Tech Station, Ruston, La., 1968, 107.
19. **Krone, R. B.,** The movement of disease producing organisms through soils, in *Municipal Sewage Effluent for Irrigation*, Wilson, C. W. and Beckett, F. E., Eds., Louisiana Polytechnic Institute Alumni Foundation, Tech Station, Ruston, 1968, 75.
20. Report on United Nation expert group meeting on the achievement of efficiency in the use and reuse of water in cooperation with the Government of Israel, *Water Res.*, 9, 485, 1975.
21. **Uiga, A., Fehrmann, R., and Crites, R. W.,** Relative health factors comparing activated sludge systems to land application systems, in State of Knowledge in Land Treatment of Wastewater, Vol. 1, McKim, H. L., Ed., U.S. Army CRREL Hanover, N.H., 1978, 253.
22. **Wellings, F. M., Lewis, A. L., Mountain, C. W., and Stark, L. M.,** Virus consideration in land disposal of sewage effluents and sludge, *Fla. Sci.*, 38, 194, 1975.
23. **Roy, D.,** Microbiology-detection, occurrence and removal of viruses, *J. Water Pollut. Control Fed.*, 52, 1839, 1980.
24. **Akin, E. W., Jakubowski, W., and Pahren, H. R.,** Health hazards associated with wastewater effluents and sludge: Microbiological considerations, in *Risk Assessment and Health Effects of Land Application of Municipal Wastewater and Sludges*, Sagik, B. P. and Sorber, C. A., Eds., Center for Applied Research and Technology, University of Texas, San Antonio, 1978, 9.
25. **Shuval, H. E.,** Health factors in the re-use of wastewater for agricultural, industrial and municipal purposes, in *Problems in Community Wastes Management*, Publ. Health Papers, No. 38, World Health Organization, Geneva, 1969, 76.
26. **Melnick, J. L.,** Are conventional methods of epidemiology appropriate for risk assessment of virus contamination of water?, in *Risk Assessment and Health Effects of Land Application of Municipal Wastewater and Sludges*, Sagik, B. P. and Sorber, C. A., Eds., Center for Applied Research and Technology, University of Texas, San Antonio, 1978, 61.
27. **Shuval, H. I.,** Land treatment of wastewater in Israel, in State of Knowledge in Land Treatment of Wastewater, Vol. 1, McKim, H. L., Ed., U.S. Army CRREL, Hanover, N.H., 1978, 429.
28. **Schaub, S. A., Kenyon, K. F., Bledsoe, B., and Thomas, R. E.,** Evaluation of the overland runoff mode of land wastewater treatment for virus removal, *Appl. Environ. Microbiol.*, 39, 127, 1980.
29. **Bell, R. G. and Bole, J. B.,** Elimination of fecal coliform bacteria from Reed canary grass irrigated with municipal sewage lagoon effluent, *J. Environ. Qual.*, 5, 417, 1976.
30. **Krishnaswami, S. K.,** Health aspects of land disposal of municipal wastewater effluents, *Can. J. Publ. Health*, 62, 36, 1971.
31. **Moore, B. E., Sagik, B., and Sorber, C.,** Land application of sludges: minimizing the impact of viruses on water resources, in *Risk Assessment and Health Effects of Land Application of Municipal Wastewater and Sludges*, Sagik, B. P. and Sorber, C. A., Eds., Center for Applied Research and Technology, University of Texas, San Antonio, 1978, 154.
32. **Vaughn, J. M. and Landry, E. F.,** Data report: an assessment of the occurrence of human viruses in Long Island aquatic systems, Brookhaven National Laboratory, Rep. 50787, Upton, N.Y., 1977.
33. **Jelnick, C. F.,** Health perspectives, use of municipal sludge on land, in *Proc. 1977 Natl. Conf. Comp. Munic. Residues Sludges*, Information Transfer, Rockville, Md., 1978, 27.
34. **Stokinger, H. E.,** Sanity in research and evaluation of environmental health, *Science*, 174, 662, 1971.
35. **Brown, K. W., Wolf, H. W., Donnelly, K. C., and Slowey, J. F.,** The movement of fecal coliform and coliphages below septic lines, *J. Environ. Qual.*, 8, 121, 1979.
36. **Hurst, C. J. and Gerba, C. P.,** Development of a quantitative method for the detection of enteroviruses in soil, *Appl. Environ. Microbiol.*, 37, 626, 1979.
37. **Landry, E. F., Vaughn, J. M., and Penello, W. F.,** Poliovirus retention in 75 cm soil cores following sewage and rainwater application, *Appl. Environ. Microbiol.*, 40, 1032, 1980.
38. **Kerfoot, W. B., Ketchum, B. H., Kallio, P., Bowker, P., Mann, A., and Scolieri, C.,** Cape Cod wastewater renovation and retrieval system, a study of water treatment and conservation, First year operation, Woods Hole Oceanogr. Rep., 75-32, Woods Hole, Mass., 1975.
39. **Derbyshire, J. B. and Brown, E. G.,** Isolation of animal viruses from farm livestock waste, soil and water, *J. Hyg.*, 81, 295, 1978.
40. **Sagik, B. P., Moore, B. E., and Sorber, C. A.,** Human enteric virus survival in soil following irrigation with sewage plant effluents, Environmental Protection Agency Rep. EPA-600/1-80-004, Cincinnati, Ohio, 1980.

41. **Tierney, J. T., Sullivan, R., and Larkin, E. P.,** Persistence of poliovirus 1 in soil and on vegetables grown in soil previously flooded with inoculated sewage sludge or effluent, *Appl. Environ. Microbiol.*, 33, 109, 1977.
42. **Bagdasar'yan, G. A.,** Sanitary examination of soil and vegetables from irrigation fields for the presence of viruses, *Gig. Sanit.*, 11, 37, 1964.
43. **Artykov, M. S.,** Sanitary-virusological study of the effectiveness of decontamination of wastewater in agricultural irrigation fields, *Gig. Sanit.*, 34, 110, 1973.
44. **Bondarenko, V. I. and Grigor'Eva, L. V.,** The isolation of enteroviruses from the surface of fruits and vegetables, *Microbiol. Zh.*, 37, 244, 1975.
45. **Sadovski, A. Y., Fattal, B., Goldberg, D., Katzenelson, E., and Shuval, H. I.,** High levels of microbial contamination of vegetables irrigated with wastewater by the drip method, *Appl. Environ. Microbiol.*, 36, 824, 1978.
46. **Murphy, W. H. and Syverton, J. T.,** Adsorption and translocation of mammalian viruses by plants. II. Recovery and distribution of viruses in plants, *Virology*, 6, 623, 1958.
47. **Bell, R. G. and Bole, J. B.,** Elimination of fecal coliform bacteria from Reed canarygrass irrigated with municipal sewage lagoon effluent, *J. Environ. Qual.*, 5, 417, 1976.
48. **Gerba, C. P., Wallis, C., and Melnick, J. L.,** Viruses in water: the problem, some solutions, *Environ. Sci. Technol.*, 9, 1122, 1975.
49. **Wellings, F. M., Mountain, C. W., and Lewis, A. L.,** Virus in groundwater, in *Individual Onsite Wastewater Systems*, Proc. 2nd Natl. Conf., McClelland, N. I., Ed., Ann Arbor Science, Ann Arbor, Mich., 1977, 61.
50. **Sobsey, M. D. and Scandura, J.,** Enteric microorganism removal by on-site wastewater disposal systems in coastal plains soils, Interim Project Rep., Sanitary Engineering Section, Division of Health Services, Department of Human Resources, Chapel Hill, North Carolina, 1980.
51. **Vaughn, J. M., Landry, E. F., and Thomas, Mc. H.,** The lateral movement of indigenous enteroviruses in a sandy sole-source aquifer, in Conf. on Microbial Health Considerations of Soil Disposal of Domestic Wastewaters, University of Oklahoma, Norman, 1981.
52. **Kerfoot, W. B. and Ketchum, B. T.,** Cape Cod wastewater renovation and retrieval system, a study of water treatment and conservation, Rep. WHOI-74-13, Woods Hole Oceanographic Institution, Woods Hole, Mass., 1974.
53. **Gilbert, R. G., Rice, R. C., Bouwer, H., Gerba, C. P., Wallis, C., and Melnick, J. L.,** Wastewater renovation and reuse: virus removal by soil filtration, *Science*, 192, 1004, 1976.
54. **Vaughn, J. M., Landry, E. F., Baranosky, J. L., Beckwith, C. A., Dahl, M. C., and Delihas, N. C.,** Survey of human virus occurrence in wastewater-recharged groundwater on Long Island, *Appl. Environ. Microbiol.*, 36, 47, 1978.
55. **Koerner, E. L. and Haws, D. A.,** Long-term effects of land application of domestic wastewater, Environmental Protection Agency Rep. 600/2-79-047, Ada, Okla., 1979.
56. Metropolitan Sanitary District of Greater Chicago and IIT Research Institute, Viral and bacterial levels resulting from the land application of digested sludge, EPA Rep. 600/1-79-015, U.S. Environmental Protection Agency, Cincinnati, Ohio, 1974.
57. **Farrah, S. R. and Bitton, G.,** Fate of viruses following land application of sewage sludge, II. Field experiments, in Int. Symp. Viruses and Wastewater Treatment, University of Surrey, Guildford, U.K., 1980.
58. **Keswick, B. H. and Gerba, C. P.,** State of knowledge and significance of viruses in groundwater, *Environ. Sci. Technol.*, 14, 1290, 1980.
59. **Mack, W. N., Lu, Yue-shoung, and Coohoon, D. B.,** Isolation of poliomyelitis virus from a contaminated well, *Health Serv. Rep.*, 87, 271, 1972.
60. **Wellings, F. M.,** Viral aspects of wetland disposal of effluent, in *Proc. Symp. Freshwater Wetlands and Sewage Effluent Disposal*, Ann Arbor Science, Ann Arbor, Mich., 1976, 297.
61. **Vaughn, J. M. and Landry, E. F.,** The occurrence of human enteroviruses in a Long Island groundwater aquifer recharged with tertiary wastewater effluents, in State of Knowledge in Land Treatment of Wastewater, Vol. 2, McKim, H. L., Ed., U.S. Army CRREL, Hanover, N.H., 1978, 233.
62. **Schaub, S. A. and Sorber, C. A.,** Virus and bacteria removal from wastewater by rapid infiltration through soil, *Appl. Environ. Microbiol.*, 30, 47, 1977.
63. **Marzouk, Y., Goyal, S. M., and Gerba, C. P.,** Prevalence of enteroviruses in groundwater of Israel, *Groundwater*, 17, 487, 1979.
64. **Marzouk, Y., Goyal, S. M., and Gerba, C. P.,** Relationship of viruses and indicator bacteria in water and wastewater of Israel, *Water Res.*, 14, 1585, 1980.
65. **Wellings, F. M., Lewis, A. L., Mountain, C. W., and Pierce, L. V.,** Demonstration of virus in groundwater after effluent discharge onto soil, *Appl. Microbiol.*, 29, 751, 1975.
66. **Van Donsel, D. T.,** Microbiology-waterborne outbreak, *J. Water Pollut. Control Fed.*, 43, 1221, 1971.

67. **McDermott, J. F.,** Virus problems and their relation to water supplies, *J. Am. Water Works Assoc.*, 66, 613, 1974.
68. **Sobsey, M. D.,** Enteric viruses and drinking water supplies, *J. Am. Water Works Assoc.*, 67, 414, 1975.
69. **Taylor, E. B., Eagen, J. H., Smith, H. F. D., and Coene, R. F.,** The case for waterborne infectious hepatitis, *Am. J. Publ. Health.*, 56, 2093, 1966.
70. **Craun, G. F.,** Disease outbreaks caused by drinking water, *J. Water Pollut. Control Fed.*, 52, 833, 1980.
71. **Holmes, I. H.,** Viral gastroenteritis, *Prog. Med. Virol.*, 25, 1, 1979.
72. **Richmond, S. J., Caul, E. O., Dunn, S. M., Ashley, C. R., Clarke, S., and Seymour, N. R.,** An outbreak of gastroenteritis in young children caused by adenovirus, *Lancet*, 1, 1178, 1979.
73. **Craun, G. F.,** Disease outbreaks caused by drinking water, *J. Water Pollut. Control Fed.*, 51, 1751, 1979.
74. **Craun, G. F.,** Disease outbreaks caused by drinking water, *J. Water Pollut. Control Fed.*, 50, 1362, 1978.
75. **Craun, G. F.,** Microbiology-waterborne outbreaks, *J. Water Pollut. Control Fed.*, 45, 1265, 1973.
76. National Academy of Sciences, Drinking water and health, *Waterborne Microorganisms*, 1977, chap. III.
77. **Allen, M. J.,** Microbiology of groundwater, *J. Water Pollut. Control Fed.*, 50, 1342, 1978.
78. **Burge, W. D. and Marsh, P. B.,** Infectious disease hazards of landspreading sewage wastes, *J. Environ. Qual.*, 7, 1, 1978.
79. **Hallgren, R.,** Epidemic hepatitis in the County of Vasterbotten in northern Sweden, an epidemiological and clinical study, *Acta Med. Scand. Suppl.*, 140, 1942.
80. **Eliassen, R. and Cummings, R. H.,** Analysis of waterborne disease outbreaks 1935—45, *J. Am. Water Works Assoc.*, 40, 509, 1949.
81. **Mosley, J. W.,** Waterborne infectious hepatitis, *N. Engl. J. Med.*, 261, 703, 1959.
82. **Garibaldi, R. A.,** Infectious hepatitis outbreak associated with cafe water, *Health Serv. Mental Health Adm. Health Rep.*, 87, 164, 1972.
83. Center for Disease Control, EIS Conf., Atlanta, Ga., 16, 1972.
84. **Vilim, V.,** Viral Hepatitis A — water epidemic in a bungalow community, *Cesk. Epidemiol. Mikrobiol. Immunol.*, 26, 46, 1977.
85. **Baer, G. M., Walker, J. A., and Yager, P. A.,** Studies of an outbreak of acute hepatitis A: 1. Complement level fluctuations, *J. Med. Virol.*, 1, 1, 1977.
86. Center for Disease Control, *Morb. Mort. Weekly Rep.*, 21, 119, 1972.
87. Center for Disease Control, *Morb. Mort. Weekly Rep.*, 27, 183, 1978.
88. Center for Disease Control, *Morb. Mort. Weekly Rep.*, 29, 585, 1980.
89. **Morens, D. M., Zweighaft, R. M., Vernon, T. M., Gary, G. W., Eslien, J. J., Wood, B. T., Holman, R. C., and Dolin, R.,** A waterborne outbreak of gastroenteritis with secondary person-to-person spread, *Lancet*, 964, 1979.
90. Center for Disease Control, *Morb. Mort. Weekly Rep.*, 27, 403, 1978.
91. **Lycke, E.,** Epidemic acute diarrhea in adults associated with infantile gastroenteritis in adults, *Lancet*, 2, 1056, 1978.
92. **Berg, G.,** Removal of viruses from sewage, effluents and waters, *Bull. W.H.O.*, 49, 451, 1973.
93. **Sproul, O. J.,** Removal of viruses by treatment processes, in *Viruses in Water*, Berg, G., Bodily, H. L., Lennette, E. H., Melnick, J. L., and Metcalf, T. G., Eds., American Public Health Association, Washington, D.C., 1976, 167.
94. **Foster, D. H. and Engelbrecht, R. S.,** Microbial hazards in disposing of wastewater on soil, in *Recycling Treated Municipal Wastewater and Sludge through Forest and Cropland*, Sopper, W. E. and Kardos, L. T., Eds., Pennsylvania State University Press, University Park, Penn., 1973, 247.
95. **Sagik, B. P. and Sorber, C. A., Eds.,** *Risk Assessment and Health Effects of Land Application of Municipal Wastewater and Sludges*, University of Texas, San Antonio, 1978.
96. **McKim, H. L., Ed.,** State of Knowledge in Land Treatment of Wastewater, Vol. 1 and 2, U.S. Army CRREL, Hanover, N.H. 1978.
97. **Gerba, C. P., Wallis, C., and Melnick, J. L.,** Fate of wastewater bacteria and viruses in soil, *J. Irrig. Drain. Div., Am. Soc. Civ. Eng.*, 101, 157, 1975.
98. **Bitton, G.,** Adsorption of viruses onto surfaces in soil and water, *Water Res.*, 9, 473, 1975.
99. **Vilker, V. L., Kamdar, R. S., and Frommhagen, L. H.,** Capacity of activated sludge solids for virus adsorption, *Chem. Eng. Commun.*, 4, 569, 1980.
100. **Wellings, F. M., Lewis, A. L., and Mountain, C. W.,** Demonstration of solids-associated virus in wastewater and sludge, *Appl. Environ. Microbiol.*, 31, 354, 1976.
101. **Schaub, S. A. and Sagik, B. P.,** Association of enteroviruses with natural and artificially introduced colloidal solids in water and infectivity of solids-associated virions, *Appl. Microbiol.*, 30, 212, 1975.
102. **Taylor, D. H.,** Interpretation of the adsorption of viruses by clays from their electrokinetic properties, in *Chemistry of Water Reuse*, Cooper, W. J., Ed., Ann Arbor Science, Ann Arbor, Mich., 1981.
103. **Murray, J. P.,** Physical Chemistry of Virus Adsorption and Degradation on Inorganic Surfaces, Environmental Protection Agency Rep. 600/2-80-134, Cincinnati, Ohio, 1980.

104. **Murray, J. P. and Parks, G. A.,** Poliovirus adsorption on oxide surfaces: correspondence with the DLVO-Lifshitz theory of colloidal stability, in *Particulates in Water: Characterization, Fate, Effects and Removal,* Kavanaugh, M. C. and Leckie, J. Q., Eds., American Chemical Society, Adv. Chem. Ser. 189, Washington, D.C., 1980, 97.
105. **Mandel, B.,** Characterization of type 1 poliovirus by electrophoretic analysis, *Virology,* 4, 554, 1971.
106. **Bitton, G., Pancorbo, O., and Gifford, G. E.,** Factors affecting the adsorption of poliovirus to magnetite in water and wastewater, *Water Res.,* 10, 973, 1976.
107. **Taylor, D. H., Bellamy, A. R., and Wilson, A. T.,** Interaction of bacteriophage R17 and reovirus type III with the clay mineral allophane, *Water Res.,* 14, 339, 1980.
108. **Drewry, W. A. and Eliassen, R.,** Virus movement in groundwater, *J. Water Pollut. Control Fed.,* 40, R257, 1968.
109. **Gerba, C. P. and Lance, J. C.,** Poliovirus removal from primary and secondary sewage effluent by soil filtration, *Appl. Environ. Microbiol.,* 36, 247, 1978.
110. **Gerba, C. P., Sobsey, M. D., Wallis, C., and Melnick, J. L.,** Adsorption of poliovirus onto activated carbon in wastewater, *Environ. Sci. Technol.,* 9, 727, 1975.
111. **Vilker, V. L. and Burge, W. D.,** Adsorption mass transfer model for virus transport in soils, *Water Res.,* 14, 783, 1980.
112. **Vilker, V. L.,** Simulating virus movement in soils, in *Modeling Wastewater Renovation Land Treatment,* Iskandar, K. I., Ed., Wiley-Interscience, New York, 1981, chap. 9.
113. **Vilker, V. L.,** An adsorption model for prediction of virus breakthrough from fixed beds, in State of Knowledge in Land Treatment of Wastewater, Vol. 2, McKim, H. L., Ed., U.S. Army CRREL, Hanover, N.H., 1978, 381.
114. **Cookson, J. T. and North, W. J.,** Adsorption of viruses on activated carbon: equilibrium and kinetics of the attachment of *Escherichia coli* bacteriophage T4 on activated carbon, *Environ. Sci. Technol.,* 1, 46, 1967.
115. **Burge, W. D. and Enkiri, N. K.,** Virus adsorption by five soils, *J. Environ. Qual.,* 7, 73, 1978.
116. **Vilker, V. L.,** Virus interactions with soil components: measurements and use for predicting movement during land application of municipal wastewater, in Proc. 3rd Annu. Madison Conf. Appl. Res. Proc. Munic. Ind. Waste, Madison, Wisc., September 1980.
117. **Koya, K. V. A. and Chaudhuri, M.,** Virus retention by soil, *Prog. Water Technol.,* 43, 1977.
118. **Moore, B. E., Sagik, B. P., and Sorber, C. A.,** Viral transport to groundwater at an operational wastewater land application site, presented at Annu. Water Pollut. Control Fed. Symp., Houston, 1979.
119. **Gilbert, R. G., Gerba, C. P., Rice, R. C., Bouwer, H., Wallis, C., and Melnick, J. L.,** Virus and bacteria removal from wastewater by land treatment, *Appl. Environ. Microbiol.,* 32, 333, 1976.
120. **Wellings, F. M., Lewis, A. L., and Mountain, C. W.,** Survival of viruses in soil under natural conditions, in, *Wastewater Renovation and Re-use,* D'Itri, F. M., Ed., Marcel Dekker, New York, 1977, 453.
121. **Bitton, G.,** Fate of viruses in land disposal of wastewater effluents, in *Introduction to Environmental Virology,* Bitton, G., Ed., Wiley-Interscience, New York, 1980, 200.
122. **Chaudhuri, M., Koya, K. V. A., and Sriramulu, N.,** Some notes on virus retention by sand, *J. Gen. Appl. Microbiol.,* 23, 337, 1977.
123. **Gerba, C. P. and Goyal, S. M.,** Adsorption of selected enteroviruses to soils, in State of Knowledge on Land Treatment of Wastewaters, Vol. 2, McKim, H. L., Ed., U.S. Army CRREL, Hanover, N.H., 1978, 225.
124. **Goyal, S. M. and Gerba, C. P.,** Comparative adsorption of human enteroviruses, simian rotavirus and selected bacteriophage to soils, *Appl. Environ. Microbiol.,* 38, 241, 1979.
125. **Landry, E. F., Vaughn, J. M., Thomas, M. Z., and Beckwith, C. A.,** Adsorption of enteroviruses to soil cores and their subsequent elution by artificial rainwater, *Appl. Environ. Microbiol.,* 38, 680, 1979.
126. **Sobsey, M. D., Dean, C. H., Knuckles, M. E., and Wagner, R. A.,** Interactions and survival of enteric viruses in soil materials, *Appl. Environ. Microbiol.,* 40, 92, 1980.
127. **Gerba, C. P., Goyal, S. M., Hurst, C. J., and LaBelle, R. L.,** Type and strain dependence of enterovirus adsorption to activated sludge, soils and estuarine sediments, *Water Res.,* 14, 1197, 1980.
128. **LaBelle, R. L. and Gerba, C. P.,** Influence of pH, salinity and organic matter on the adsorption of enteric viruses to estuarine sediment, *Appl. Environ. Microbiol.,* 38, 93, 1979.
129. **Lance, J. C. and Gerba, C. P.,** Poliovirus movement during high rate land filtration of sewage water, *J. Environ. Qual.,* 9, 31, 1980.
130. **Burge, W. D. and Enkiri, M. D.,** Adsorption kinetics of bacteriophage ΦX174 on soil, *J. Environ. Qual.,* 7, 536, 1978.
131. **Carlson, J. F., Woodard, F. E., Wentworth, D. F., and Sproul, O. J.,** Virus inactivation on clay particles in natural waters, *J. Water Pollut. Control Fed.,* 40, R89, 1968.
132. **Lefler, F. and Kott, Y.,** Virus retention and survival in sand, in *Virus Survival in Water and Wastewater Systems,* Malina, J. F. and Sagik, B. P., Eds., Center for Research in Water Resources, University of Texas, Austin, 1974, 84.

133. **Moore, B. E., Sagik, B. P., and Malina, J. F.,** Viral association with suspended solids, *Water Res.*, 9, 197, 1975.
134. **Stotsky, G., Schiffenbauer, M., Lipson, S. M., and Yu, B. H.,** Surface interactions between viruses and clay minerals and microbes: mechanisms and implications, presented at Int. Symp. on Viruses and Wastewater Treatment, University of Surrey, Guildford, U.K., 1980.
135. **Lance, J. C., Gerba, C. P., and Melnick, J. L.,** Virus movement in soil columns flooded with secondary sewage effluent, *Appl. Environ. Microbiol.*, 32, 520, 1976.
136. **Kessick, M. A. and Wagner, R. A.,** Electrophoretic mobilities of virus adsorbing filter materials, *Water Res.*, 12, 263, 1978.
137. **Bitton, G., Masterson, N., and Gifford, G. E.,** Effect of a secondary treated effluent on the movement of viruses through a cypress dome soil, *J. Environ. Qual.*, 5, 370, 1976.
138. **Scheuerman, P. R., Bitton, G., Overman, A. R., and Gifford, G. E.,** Transport of viruses through organic soils and sediments, *J. Irrig. Drain. Div., Am. Soc. Civ. Eng.*, 105, 629, 1979.
139. **Lo, S. H.,** Inactivation of Virus in Natural Water and Wastewater by Silicate Minerals, M.S. thesis, University of Maine, Orono, 1970.
140. **Bixby, R. L. and O'Brien, D. J.,** Influence of fulvic acid on bacteriophage adsorption and complexation in soil, *Appl. Environ. Microbiol.*, 38, 840, 1979.
141. **Lo, S. H. and Sproul, O. J.,** Polio-virus adsorption from water onto silicate minerals, *Water Res.*, 11, 653, 1977.
142. **Gerba, C. P. and Lance, J. C.,** Poliovirus removal from primary and secondary sewage effluent by soil filtration, *Appl. Environ. Microbiol.*, 36, 247, 1978.
143. **Lance, J. C., Rice, R. C., and Gilbert, R. G.,** Renovation of wastewater by soil columns flooded with primary effluent, *J. Water Pollut. Control Fed.*, 52, 381, 1980.
144. **Vaughn, J. M., Landry, E. F., Beckwith, C. A., and Thomas, M. Z.,** Virus removal during groundwater recharge: effects of infiltration rate on the adsorption of poliovirus to soil, *Appl. Environ. Microbiol.*, 41, 139, 1981.
145. **Duboise, S. M., Moore, B. E., and Sagik, B. P.,** Poliovirus survival and movement in a sandy forest soil, *Appl. Environ. Microbiol.*, 31, 536, 1976.
146. **Green, K. M. and Cliver, D. O.,** Removal of virus from septic tank effluent by sand columns, presented at Natl. Home Sewage Disposal Symp., Chicago, December 1974.
147. **Funderburg, S. W., Moore, B. E., Sagik, B. P., and Sorber, C. A.,** Viral transport through soil columns under conditions of saturated flow, *Water Res.*, 15, 703, 1981.
148. **Bitton, G., Farrah, S. R., Pancorbo, O. C., and Davidson, J. M.,** Fate of viruses following land disposal of sewage sludge, I. Survival and transport patterns in cores: studies under natural conditions, presented at Int. Symp. Viruses and Wastewater Treatment, University of Surrey, Guildford, U.K., 1980.
149. **Damgaard-Larsen, S., Jensen, K. O., Lund, E., and Nissen, B.,** Survival and movement of enterovirus in connection with land disposal of sludges, *Water Res.*, 11, 503, 1977.
150. **Sobsey, M. D., Wallis, C., and Melnick, J. L.,** Studies on the survival and fate of enteroviruses in an experimental model of a municipal solid waste landfill and leachate, *Appl. Microbiol.*, 30, 565, 1975.
151. **Engelbrecht, R. S., Weber, M. J., Amirhor, P., Foster, D. H., and LaRossa, D.,** Biological properties of sanitary landfill leachate, in *Virus Survival in Water and Wastewater Systems*, Malina, J. F. and Sagik, B. P., Eds., Center for Research in Water Resources, University of Texas, Austin, 1974, 201.
152. **Cooper, R. C., Potter, J. L., and Leong, C.,** Virus survival in solid waste treatment systems, in *Virus Survival in Water and Wastewater Systems*, Malina, J. F. and Sagik, B. P., Eds., Center for Research in Water Resources, University of Texas, Austin, 1974, 218.
153. **Cooper, R. C., Potter, J. L., and Leong, C.,** Virus survival in solid waste leachates, *Water Res.*, 9, 733, 1975.
154. **Hori, D. H., Burbank, N. C., Young, R. H., Law, L. S., and Klemmer, H. W.,** Migration of poliovirus type 2 in percolating water through selected Oahu soils, Tech. Rep. 36, Water Resources Research Center, University of Hawaii, Honolulu, 1970.
155. **Berg, G.,** Reassessment of the virus problem in sewage and in surface and renovated waters, *Prog. Water Technol.*, 3, 87, 1973.
156. **Bitton, G., Davidson, J. M., and Farrah, S. R.,** On the value of soil columns for assessing the transport pattern of viruses through soils: a critical outlook, *Water, Air Soil Pollut.*, 12, 449, 1979.
157. **Duboise, S. M., Sagik, B. P., and Malina, J. M.,** Virus migration through soils, in *Virus Survival in Water and Wastewater Systems*, Malina, J. M. and Sagik, B. P., Eds., Center for Research in Water Resources, University of Texas, Austin, 1974, 233.
158. **Wellings, F. M., Lewis, A. L., and Mountain, C. W.,** Virus survival following wastewater spray irrigation of sandy soils, in *Virus Survival in Water and Wastewater Systems*, Malina, J. F. and Sagik, B. P., Eds., Center for Research in Water Resources, University of Texas, Austin, 1974, 253.
159. **Funderburg, S. W., Moore, B. E., Sorber, C. A., and Sagik, B. P.,** Method of soil column preparation for the evaluation of viral transport, *Appl. Environ. Microbiol.*, 38, 102, 1979.

160. **Hurst, C. J., Gerba, C. P., Lance, J. C., and Rice, R. C.,** Survival of enteroviruses in rapid-infiltration basins during the land application of wastewater, *Appl. Environ. Microbiol.*, 40, 192, 1980.
161. **Gerba, C. P., Farrah, S. R., Goyal, S. M., Wallis, C., and Melnick, J. L.,** Concentration of enteroviruses from large volumes of tapwater, treated sewage and seawater, *Appl. Environ. Microbiol.*, 35, 540, 1978.
162. **Jakubowski, W., Hill, W. F., and Clarke, N. A.,** Comparative study of four microporous filters for concentration of viruses from drinking water, *Appl. Microbiol.*, 30, 58, 1975.
163. **Bitton, G., Charles, M. J., and Farrah, S. R.,** Virus detection in soils: a comparison of four recovery methods, *Can. J. Microbiol.*, 25, 874, 1979.
164. **Sobsey, M. D. and Glass, J. S.,** Poliovirus concentration from tapwater with electropositive adsorbent filters, *Appl. Environ. Microbiol.*, 40, 201, 1980.
165. **Noonan, M. J. and McNabb, J. F.,** Contamination of Canterbury groundwater, in *The Quality and Movement of Groundwater in Alluvial Aquifers of New Zealand*, Noonan, M. J., Ed., Tech. Publ. No. 2, Lincoln College, Christ Church, New Zealand, 1979, 195.
166. **Pyle, B. H., Sinton, L. W., Noonan, M. J., and McNabb, J. F.,** The movement of micro-organisms in groundwater, presented at New Zealand Inst. Eng. Tech. Group on Groundwater Symp., Victoria University, Wellington, February 1979.
167. **Larkin, E. P., Tierney, J. T., and Sullivan, R.,** Persistence of poliovirus in soil and on vegetables irrigated with sewage wastes: potential problems, in State of Knowledge in Land Treatment of Wastewaters, Vol. 2, McKim, H. L., Ed., U.S. Army CRREL, Hanover, 1978, 119.
168. **Larkin, E. P., Tierney, J. T., Lovett, J., Van Donsel, D., and Francis, D. W.,** Land application of sewage wastes: potential for contamination of foodstuffs and agricultural soils by viruses and bacterial pathogens in *Risk Assessment and Health Effects of Land Application of Municipal Wastewater and Sludges*, Sagik, B. P. and Sorber, C. A., Eds., University of Texas, San Antonio, 1978, 102.
169. **Bitton, G. B., Fraxedas, R., and Gifford, G. E.,** Effect of solar radiation on poliovirus: preliminary experiments, *Water Res.*, 13, 225, 1979.
170. **Foster, D. M., Emerson, M. A., Buck, C. F., Walsh, D. S., and Sproul, O. J.,** Ozone inactivation of cell- and fecal-associated viruses and bacteria, *J. Water Pollut. Control Fed.*, 52, 2174, 1980.
171. **Hurst, C. J., Gerba, C. P., and Cech, I.,** Effects of environmental variables and soil characteristics on virus survival in soil, *Appl. Environ. Microbiol.*, 40, 1067, 1980.
172. **Liew, P. F. and Gerba, C. P.,** Thermostabilization of enteroviruses by estuarine sediment, *Appl. Environ. Microbiol.*, 40, 305, 1980.
173. **LaBelle, R. L. and Gerba, C. P.,** Influence of estuarine sediment on virus survival under field conditions, *Appl. Environ. Microbiol.*, 39, 749, 1980.
174. **Hurst, C. J., Gerba, C. P., and Lance, J. C.,** Effect of environmental conditions on enterovirus survival in soil, *Abstr. Annu. Meet. Am. Soc. Microbiol.*, Los Angeles, 1979, 227.
175. **Yeager, J. G. and O'Brien, R. T.,** Enterovirus inactivation in soil, *Appl. Environ. Microbiol.*, 38, 694, 1979.
176. **Yeager, J. G. and O'Brien, R. T.,** Structural changes associated with poliovirus inactivation in soil, *Appl. Environ. Microbiol.*, 38, 702, 1979.
177. **Ward, R. L. and Ashley, C. S.,** Inactivation of enteric viruses in wastewater sludge through dewatering by evaporation, *Appl. Environ. Microbiol.*, 34, 564, 1977.
178. **Hurst, C. J., Farrah, S. R., Gerba, C. P., and Melnick, J. L.,** Development of quantitative methods for the detection of enteroviruses in sewage sludges during activation and following land disposal, *Appl. Environ. Microbiol.*, 36, 81, 1978.
179. **Farrah, S. R., Goyal, S. M., Gerba, C. P., Wallis, C., and Melnick, J. L.,** Concentration of enteroviruses from estuarine water, *Appl. Environ. Microbiol.*, 33, 1192, 1977.
180. **Murray, J. P. and Laband, S. J.,** Degradation of poliovirus by adsorption on inorganic surfaces, *Appl. Environ. Microbiol.*, 37, 480, 1979.
181. **Cheo, P. C.,** Antiviral factors in soil, *Soil Sci. Soc. Am. J.*, 44, 62, 1980.
182. **Kott, H. and Fishelson, I.,** Survival of enteroviruses on vegetables irrigated with chlorinated oxidation pond effluents, *Isr. J. Technol.*, 12, 290, 1974.

Chapter 10

INDICATORS OF VIRUSES

Sagar M. Goyal

TABLE OF CONTENTS

I. INTRODUCTION

The potential of polluted water to cause illness has been recognized for a long time, as evidenced by the implementation of water treatment practices 4000 years ago.[1] It was empirically discovered that the disease rate in communities downstream on a river was much higher compared to that in upstream communities. The conclusive scientific proof that human diseases could be transmitted by water was provided by John Snow's classical studies of cholera outbreaks in London.[2] Since then, a number of bacterial, viral, and parasitic diseases have been shown to be transmitted by water. The institution of water and wastewater treatment practices has greatly reduced — and virtually eliminated in many areas — waterborne diseases in the U.S. and other developed countries, but water-related illness, particularly viral, may still be of concern.

Feces of man and other warm-blooded animals are the major sources of pathogens that are carried in water. Human pathogens which can be transmitted by water include *Salmonella, Shigella, Pasteurella, Yersinia, Leptospira, Campylobacter, Mycobacterium, Pseudomonas, Vibrio,* enteropathogenic *Escherichia coli,* hookworm larvae, *Giardia, Entamoeba,* and more than 100 different types of viruses.[3] To test for the presence of all these pathogens in water is neither feasible nor cost-effective. Also, methods are not yet available for detecting the presence of all pathogens with any reasonable degree of accuracy. It is logical, therefore, to seek an indicator organism(s) whose presence may indicate fecal pollution of a given water source. One such group of indicator organisms is the coliform group of bacteria, which have been used to indicate the sanitary quality of water for more than 75 years. The presence of these indicator organisms in water suggests that fecal pollution of that water source has occurred and that pathogens of public health importance may be present.

Since microbiological hazards associated with consumption of water originate from fecal contamination, the search for indicator organisms has logically been associated with organisms in the microbial flora of feces.[4] *Escherichia coli,* the major component of the coliform group, was described by Escherich[5] in 1885; he believed that it was a predominant organism of human feces and that it might be used as an indicator of the sanitary quality of water. The U.S. Public Health Service adapted coliform organisms as indicators of fecal pollution and published bacteriological standards for drinking water in 1914. Since then, these organisms have been used to monitor the bacteriological quality of drinking water.

The presence of coliform organisms, however, does not signal the presence of fecal pollution exclusively, because the coliform group comprises bacteria of both fecal and nonfecal origin. The term "total coliform" is used to indicate the coliform group as a whole, whereas "fecal coliforms" are the bacteria that are definitely fecal in origin. The numerous reports of recovery of coliform organisms from nonfecally contaminated environments, together with reports of multiplication of coliforms in some natural environments, served to stimulate researchers in the separation of fecal from nonfecal coliforms. The test for fecal coliform bacteria is based on the observation of Eijkmann[6] that fecal bacteria are capable of fermenting lactose at an elevated temperature (44.5°C), whereas nonfecal bacteria are not able to do so. Many international and national standards now incorporate both of these indicators and tolerate a higher number of coliforms than previously permitted, provided that the numbers of fecal coliforms are strictly limited.[7,8] The total coliform group is comprised of *E. coli,* other fecal coliforms, nonfecal lactose fermenting bacteria, and certain other gram negative bacilli; the presence of total coliforms is still widely used as a sanitary index for drinking water because it affords a greater margin of safety than fecal coliforms alone.

II. THE COLIFORM CONTROVERSY

Since the turn of the century, coliforms have played a significant role in predicting the presence of enteric bacterial pathogens in water. In fact, the rates of waterborne bacterial disease outbreaks in the U.S. are low today primarily because of faithful implementation of indicator bacteria as an index of the sanitary quality of water. Their use to predict the virological quality of water has been questioned on several occasions, however, because the behavior and fate of animal viruses have been shown to differ significantly from that of indicator organisms. It has been repeatedly established that viruses are more resistant to environmental conditions and sewage treatment processes than the coliform organisms.[9] Thus, little or no removal of enteroviruses was demonstrated during passage of wastewater through a trickling filter plant and a stabilization pond, although coliform populations were reduced to 8% of the original level.[10] The increase in the number of cases of hepatitis A in the U.S. seems to be inversely related to the decrease in the number of cases of typhoid fever.[11] A causative organism is never isolated from more than 50% of waterborne disease outbreaks and it is suspected that most of these cases of undetermined etiology are caused by enteric viruses.[12]

The occurrence of viruses in sewage varies greatly as opposed to that of indicator bacteria, because the latter are regularly excreted in the feces of all warm-blooded animals whereas the excretion rate of viruses depends upon the number of infected individuals in a community. It follows, therefore, that there is no constant ratio of fecal indicator bacteria to viruses in either sewage or in receiving bodies of water and hence the indicators cannot be relied upon to predict the occurrence of viruses in water and wastewater. Thus, Shuval[10] demonstrated a ratio of one virus particle per 10^6 to 10^7 total coliform organisms in raw wastewater in five Israeli cities, whereas in the U.S. Clarke and Kabler[13] calculated a ratio of 1 virus particle per 65,000 total coliforms on the average. Moreover, fecal coliform bacteria appear to survive better in water during winter than in summer.[14] This may be another reason why the virus:coliform ratio is not constant.

Due to the explosive nature of hepatitis A outbreaks and the characteristic symptomatology of the disease, a few waterborne outbreaks of hepatitis A have been documented by several investigators. Waterborne outbreaks of other virus diseases are difficult to recognize because of the inadequacy of epidemiological methods for detecting low-level transmission of these diseases through water. Also, many of these viruses cause inapparent or latent infections. Melnick and Gerba[3] have noted that a person coming in contact with contaminated water (via contact or ingestion) may ingest a virus which may actively multiply in his gastrointestinal and respiratory tracts, and yet that person may not show signs of overt illness because of the low-dose infection. Such a person will, however, serve as a carrier of virus infection and will continue shedding the virus in his secretions and excretions, which may aid in the secondary spread of disease to his contacts. Since the spectrum of diseases caused by such viruses is broad and because such secondary spread results in scattered cases of acute illness, the current epidemiological tools are not sensitive enough to trace the source of infection to contaminated water.[14a] Only when the water is grossly contaminated and a large number of people come into contact with that water at the same time and become ill at the same time can epidemiological proof of waterborne illness be established.

Apart from the question of indicator-pathogen relationships, the reliability of techniques and procedures used to estimate populations of indicator bacteria has been frequently questioned.[15] Highly significant differences were found among various brands of membranes in their ability to recover bacteria from natural waters and sewage.[16] Another problem with estimation of indicator bacterial density is that bacteria which are injured, but not killed, may not resuscitate on selective media[17] and may thus give a false-negative result.

Disinfection of water by chlorine is considered by many to be a panacea for all water pollution problems. Numerous reports, however, indicate that certain viruses are more resistant to chlorination than bacteria.[9,10,18,19] Shuval[20] reported that whereas a chlorine dosage of 2 mg/ℓ killed 99.9% of total coliforms in 60 min, 20 mg/ℓ was required to achieve the

Table 1
RATIO OF INDICATOR BACTERIA TO VIRUSES BEFORE AND AFTER CHLORINATION OF PRIMARY SEWAGE EFFLUENT

Ratio[a] of	Before chlorination	After chlorination
Fecal coliforms:virus	8.9×10^1—7.2×10^4:1	<0.06—3.4:1
Total coliforms:virus	1.5×10^4—>5.7×10^5:1	0.35—16:1
Fecal streptococci:virus	5.2×10^2—8.1×10^3:1	<1—1:1

[a] Ratios were calculated as most probable number (MPN) or CFU (colony-forming units)/100 mℓ for indicator bacteria: plaque-forming units (PFU)/ℓ for virus.

Data from Berg, G., Dahling, D. R., Brown, G. A., and Berman, D., *Appl. Environ. Microbiol.*, 36, 880, 1978.

same kill for polioviruses in sewage effluents. In a recent study, Berg et al.[21] recovered viruses from five of eight chlorinated primary sewage effluents that were negative for fecal coliforms. They also demonstrated that the level of combined chlorine that usually destroyed more than 99.999% of the indigenous total coliforms, fecal coliforms, and fecal streptococci in primary sewage effluents could destroy only 85 to 99% of the indigenous viruses present. They further pointed out that chlorine may destory all vegetative bacteria and viruses, if given sufficient time to act. However, it is a practicality that at a certain point in time, some viruses will survive when all indicator bacteria have been destroyed by chlorine, because of the difference in the kill rate. A summary of their findings (Table 1) clearly indicates that viruses can survive a chlorine dosage that kills virtually all indicator bacteria.

Polioviruses have also been recovered from finished waters containing free chlorine residuals in excess of 1 mg/ℓ and turbidities less than 1 turbidity unit. Bacterial indicator organisms, shown as total plate counts, were absent in each case.[22] Two of the viral isolates were compared with laboratory strains of poliovirus 1 for their susceptibility to chlorine. The surviving fractions of viruses derived from these two natural isolates was orders of magnitude greater than that of standard laboratory strains.[23] Repeated exposures of poliovirus 1 to sublethal doses of chlorine, alternating with virus propagation in cell cultures, lead to a significantly more resistant strain in as few as five cycles, and additional exposures increased resistance of the virus still further.[24] These data suggest that natural viruses may survive modern water treatment practices, including chlorination.

In a recent waterborne outbreak of gastroenteritis and hepatitis in Georgetown, Tex. coxsackieviruses B2 and B3 were isolated from two of the city's wells and from one sample of tapwater.[25] Hepatitis A virus was also detected in the city's sewage and in one sample of wellwater. Some of the virus isolations were made from water that met current bacteriological standards and contained an adequate chlorine level (0.8 mg/ℓ). A preliminary study of two strains of coxsackievirus B3 which were isolated from tapwater showed that those two isolates were more resistant to chlorine than a laboratory strain of coxsackievirus B3.[26] We have also demonstrated the presence of rotaviruses and enteroviruses in potable water supplies in Mexico, some of which met the current bacteriological, free chlorine, and turbidity standards for potable waters.[27]

At least two other outbreaks have been related to potable water which met bacteriological standards. Wellings et al.[28] isolated echovirus 22/23 from a disinfected groundwater supply during an outbreak at a migrant labor camp in Florida. More recently, an iodinated (0.7 to 1.0 ppm) groundwater supply was implicated in an outbreak of viral gastroenteritis at a summer camp in Maryland, even though drinking water samples examined before, during and after the episode were bacteriologically safe.[29]

In a recent study, we examined 155 samples of groundwater, surface water, potable water, and swimming pool water in Israel; of these, 45 (29%) yielded viruses.[30] Enteroviruses were isolated on several occasions when the water in question met current bacteriological standards. In some virus-positive samples, we found no fecal or total coliform bacteria. When the data were analyzed statistically, no significant correlation was found between the occurrence of bacterial indicators and the presence of viruses. The isolation of viruses from 10 of 23 (43.5%) samples of "bacteriologically safe" potable water indicates again that bacteriological indicators are not adequate to monitor the occurrence of viruses in water.

It is still not entirely clear how adequate current water treatment processes are in the control or importance of waterborne viral disease. It has been estimated that the average enteric virus concentration in domestic sewage in the U.S. is about 7000 plaque-forming units (PFU)/ℓ. In South Africa and Israel, concentrations as high as 500,000 PFU/ℓ have been detected, and this number probably only represents the concentration of enteroviruses.[31] Given the high concentration of viruses in the wastewater of these nations, sewage and water treatments must be more effectively applied than in the U.S. to ensure the absence of a viral disease hazard. Enteric viruses are known to be more resistant to commonly used wastewater and water treatment methods (including disinfection) than are enteric bacteria.[21] The inactivation rate in these processes is much greater for bacteria than for viruses. In a country like the U.S., where the virus concentration in sewage is low, the bacterial index of water quality may more closely reflect the occurrence of viruses in potable or other waters. Even if such were not the case, viruses would still be present in relatively low concentrations. However, in other nations a disparity may exist in the application of bacterial standards largely developed by nations with relatively low incidences of viral disease. Indicator bacteria would be expected to be present in domestic wastes at similar concentrations in both populations, while viruses would occur at concentrations 100 to 500 times greater in the waters of developing nations. Thus, reduction of the number of indicator bacteria by treatment processes to levels considered safe or natural die-off during self-purification in natural waters would still leave large numbers of pathogenic enteric viruses. This fact may to some degree explain the relatively large number of times that viruses have been isolated from water meeting acceptable bacterial indicator standards. In developed nations it has often been recognized that the fecal coliform test is not always a suitable indicator for detecting viral contamination; this problem is probably much greater and probably has more significance in less developed nations and in areas with a high incidence of enteric viral disease.

In addition to being inadequate for predicting viral pollution of water, the fecal coliform bacteria, under certain circumstances, may also be inadequate for determining the presence of bacterial pathogens. A classical example of the failure of fecal coliforms as indicators of fecal pollution occurred in Riverside, Calif., in 1965 when a waterborne outbreak of *Salmonella typhimurium* afflicted more than 16,000 individuals.[32] Although *S. typhimurium* was recovered from the incriminated water supply, fecal coliforms were absent. In another study, Seligman and Reitler[33] isolated *Salmonella* and *Arizona* organisms from water in Israel when most probable numbers (MPN) of *E. coli* were low or zero.

In addition to being falsely negative on occasions, the coliforms have been known to give false-positive reactions also. Thus, Presnell and Meiscer[34] found high total coliform levels in an oyster-growing Gulf area where sanitary surveys argued against indictment of human or animal wastes. It was thought that the oyster-growing area was influenced by tides from the Gulf of Mexico and that calcium and magnesium of seawater made the water hard, thus prompting multiplication of the coliform indicators. Dutka[15] cited many studies showing the ability of coliform bacteria to reproduce in enriched waters and thus falsely indicate an elevated health hazard. Dufour[35] also reported elevated fecal coliform counts in the absence

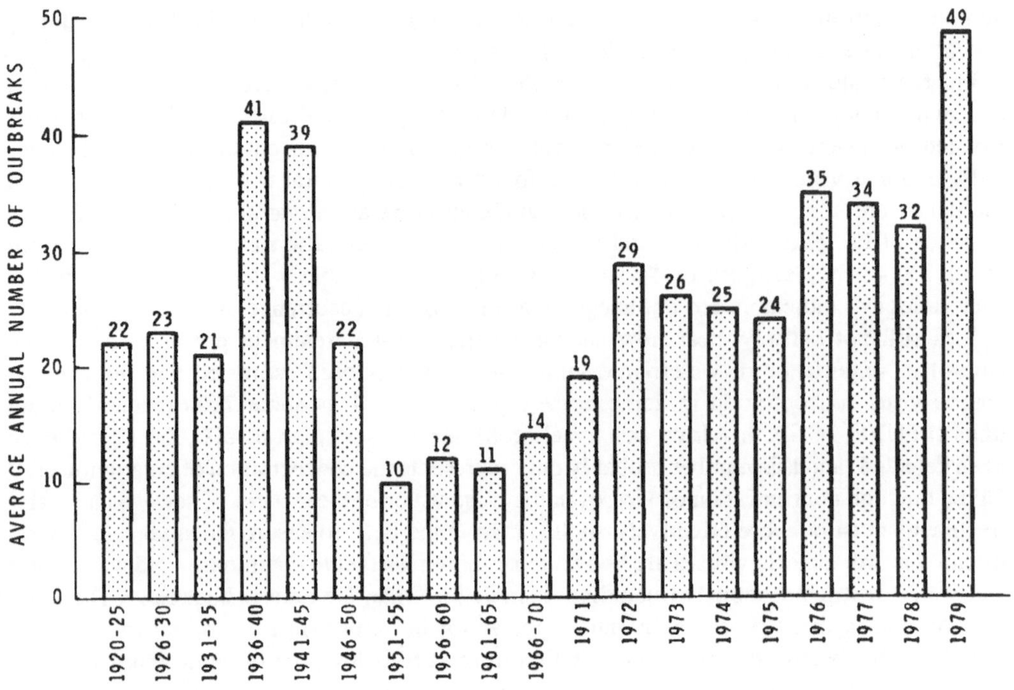

FIGURE 1. Average annual number waterborne disease outbreaks, 1920 to 1979.

of any evidence of fecal pollution from man or animals. The occurrence of this phenomenon may set off a false alarm which might result in closure of clean shellfish beds or impoundment of a good-quality water, resulting in an economic loss to the seafood industry.

The low incidence of waterborne enteric bacterial disease outbreaks is often cited as evidence of the unviolability of coliform indicators. It should be kept in mind that, although the incidence of typhoid fever has been greatly reduced in this country, the number of hepatitis cases has increased over the years. Also, the number of waterborne disease outbreaks has continued to increase since the 1970s (Figure 1). More than 50% of the waterborne outbreaks in the U.S. during 1971 to 1977 were caused by an unknown agent(s).[12] It is suspected that most of these outbreaks of "unknown etiology" were caused by viruses such as rotaviruses and the Norwalk agent.[12] It is also likely that a greater incidence of waterborne illness caused by viruses occurs but passes unrecognized because of inadequacies of epidemiologic surveillance methods.[36] Also, the recent detection of human enteric viruses in treated drinking waters containing free residual chlorine[22,23] raises doubts about the adequacy of current bacterial standards.

III. INDICATORS OF VIRUSES IN THE MARINE ENVIRONMENT

Bivalve mollusks are filter-feeding organisms, i.e., they sieve out suspended food particles from a current of water passing through the shell cavity. In its quest for food, an oyster may filter as much as 1500 ℓ (396 gal) of water per day.[37] If the water in which shellfish feed contains pathogenic bacteria or viruses, they may become entrapped on the mucous membranes and be transferred to the digestive tract. Since the entire shellfish is usually consumed along with the gastrointestinal tract and since the shellfish is usually eaten raw, it may act as a passive carrier of human pathogenic microorganisms. Because shellfish concentrate pathogens from water during feeding, there is a greater potential risk associated with shellfish consumption than with recreational use of the same water. That shellfish can spread enteric bacterial disease was known as early as 1895.[38] Among viral diseases, hepatitis A has been shown unequivocally to be transmitted by the consumption of raw or inadequately cooked shellfish.[39] The occurrence of other waterborne viruses in shellfish has been well-established, but it has been difficult to document their transmission by shellfish ingestion because of the occult nature of these viruses.[39] However, the potential for disease transmission always exists, as demonstrated recently by oyster-associated gastroenteritis in Australia. More than 2000 individuals were affected in that outbreak, which was shown to be caused by Norwalk virus.[40]

It is usually difficult and time-consuming to routinely determine the presence of pathogenic bacteria and viruses in wastewater and shellfish. The fecal coliform index is therefore used as an indicator of the sanitary quality because they are normal inhabitants of the gastrointestinal tract of warm-blooded animals and are excreted in feces in large numbers. The presence of fecal coliforms in a certain sample is then considered as evidence of recent fecal pollution of that sample. Standards exist in the U.S. for shellfish and shellfish-growing waters which state that "Most probable numbers of (total) coliforms in water should not exceed 70 per 100 mℓ and no more than 10% of samples should exceed 230 (total) coliforms per 100 mℓ."[41] Shellfish meat in itself is required to contain no more than 230 fecal coliforms per 100 g.[41] On the basis of these standards, oyster beds have been classified into three categories, i.e., closed, approved, and conditional. Enforcement of these standards has resulted in the apparent absence of shellfish-associated typhoid fever outbreaks in the U.S. since 1959.[42] However, outbreaks of shellfish-associated hepatitis A and nonspecific gastroenteritis continue to occur.[39,40]

Recent studies have indicated that these standards cannot be relied upon because coliform bacteria are more sensitive to treatment processes and natural inactivation factors than some of the more resistant enteroviruses.[9] Morris et al.[43] calculated that enteric viruses may survive in mussel tissue three to six times longer than coliform bacteria.

In several studies, enteroviruses have been isolated from shellfish that met coliform standards. Metcalf and Stiles[44] were among the first investigators who showed that a satisfactory coliform index in seawater did not necessarily mean that the area was free of enteroviruses. In a later study, Metcalf et al.[45] carried out parallel examinations of oysters and overlying seawater in Galveston Bay, Tex. for 3 months for the presence of fecal coliforms and enteroviruses. They isolated viruses from oysters on two occasions when the MPN of fecal coliforms per 100 mℓ of water and oyster meat was 7 and 20, respectively, on one occasion and 9 and 79, respectively, on another.

The inadequacy of bacterial indicators for accurately predicting the virological quality of water or shellfish has also been demonstrated by an epidemic of hepatitis A in the southcentral U.S. that was caused by the consumption of oysters from the Gulf coast.[46,47] The area from which the shellfish were obtained had been closed temporarily to commercial shellfish harvesting because a high total coliform count had been detected. When the number of total coliform bacteria reached a level considered to be safe, the area was reopened. The subsequent outbreak indicated that hepatitis A virus can survive as long as 6 weeks in shellfish and that accepted bacteriological surveillance methods cannot always ensure that a particular water source is free from viral pathogens. Enteric viruses also have been detected in oysters taken from Galveston Bay, which met current bacteriological standards for shellfish harvesting[48,49] and at beaches with no detectable fecal coliforms.[11]

We have been conducting studies along the upper Texas coast to determine the occurrence and fate of human enteric viruses in estuarine water, bottom sediments, and shellfish. Another major objective has been to study the relationship between viruses and bacteriological indicators. In one such study, Goyal et al.[50] monitored the seasonal occurrence and distribution of enteroviruses in several coastal canal communities. Of a total of 78 water samples (400 ℓ size) examined, 44 yielded enteroviruses. Of the virus-positive samples, 26 were positive when the MPN of total coliforms was <70, a figure used to indicate sanitary acceptability of water for harvesting shellfish.[41] Some samples yielded viruses when the MPN of total coliforms was less than 40 (Table 2). Statistically, there was no correlation among environmental factors, bacterial indicators, and the occurrence of viruses, except for a high correlation between viruses in water and total coliforms in sediment. However, isolation of viruses from sediments was not attempted in that study.

In another study, we isolated viruses from water and shellfish samples taken from approved areas[49] (Table 3). Viruses were isolated from water and oyster samples on occasions when the MPN of total and fecal coliforms were as low as 3/100 mℓ of water. Viruses were isolated from 12 water samples and 8 oyster samples when total coliform counts in water samples met the accepted standards. Similarly, 14 samples of water and 7 samples of oysters were positive for viruses when the MPN of fecal coliforms in oysters met the current bacteriological standards for shellfish harvesting.

In yet another study, LaBelle et al.[51] isolated viruses from sediments when the overlying water met recreational water standards (Table 4). Statistical analysis showed that the number of viruses isolated from sediment was associated with only 1 of the 12 factors measured, i.e., fecal coliform numbers in sediment. Viruses were also isolated from seawater when bacterial counts were below the recommended standard. However, the number of viruses in seawater showed no correlation with the number of bacteria in seawater.

Each of these three studies was limited in scope and was conducted in a limited geographical area. Therefore, we combined the data generated from all three studies and analyzed it statistically to determine the relationship of viruses with bacterial indicators and environmental factors.[51a] Only those variables common to all of the studies were considered. Thus, the data analyzed represented a wide range of marine environments (i.e., shallow waterways, open bay waters, etc.) and included areas both heavily polluted and nonpolluted as judged by bacteriological indicators. Study sites included both recreational areas and areas open to shellfish harvesting. The analyses were performed on the largest amount of quantitative virological data that has been reported for marine waters.[51] Although the multivariate regression analysis showed that the number of viruses detected in water correlated significantly with the number of total coliforms in water, the amount of variation in the number of viruses accounted for by this factor (4%) was not large enough to make it a good predictor. Correlations with other bacterial indicators were not statistically significant. Also, viruses were detected 43% of the time in recreational waters considered acceptable as judged by total coliform standards and 44% of the time when judged by fecal coliform standards. Viruses were detected in waters which met acceptable standards for shellfish harvesting 35% of the time. Our failure to correlate the occurrence of viruses in marine waters with indicator bacteria and their occurrence with high frequency in waters which met current bacteriological standards, indicate that these standards do not reflect the occurrence of enteroviruses, and perhaps other human pathogenic viruses, in marine waters.

Without epidemiological data, it is difficult to assess the meaning and importance of the possible failure of indicator bacteria to reflect the presence of low levels of viruses. Epidemiological studies to establish a relationship between viral disease and the presence of viruses in water would be a formidable task and might not yield meaningful results, because clinically observable illness occurs only in a small number of people who become infected and because of the widely varying incubation periods.[36] This consideration, as well as the low infective dose of viruses,[52] has led some to suggest that the presence of enteric viruses in any water is indicative of a potential viral disease hazard.[53]

The need for standards governing the sanitary quality of marine waters used for recreation has been recognized by public health officials for many years. In response to this need, most states have adopted standards based on federal recommendations for the sanitary quality

Table 2
MOST PROBABLE NUMBERS (MPN) OF INDICATOR ORGANISMS ON OCCASIONS WHEN VIRUSES WERE ISOLATED FROM WATER SAMPLES

			MPN of indicator organisms per 100 mℓ			
			Total coliforms		Fecal coliforms	
Site	Month	Number of PFU/ 400 ℓ of water	Water	Sediment	Water	Sediment
H-81	May	2	2,400	—[a]	79	—
	October	31	11,000	35,000	11,000	16,000
	November	120	35,000	540,000	940	17,000
	December	6	790	160,000	490	1,100
	January	48	16,000	—	1,800	28,000
	February	14	540	160,000	110	3,500
	March	17	79	70,000	49	2,400
	April	245	23	110,000	13	120
H-82	May	4	2,400	—	2,400	—
	June	5	24,000	—	16,000	—
	July	5	5,400	—	1,600	—
	September	11	170	—	79	—
	November	108	35,000	540,000	270	1,700
	March	3	540	35,000	130	460
	April	52	350	24,000	17	49
K-27	June	13	24,000	—	1,100	—
	July	2	17,000	—	280	—
	September	22	—	—	540	—
	October	35	540	35,000	70	79
	November	125	35,000	2,400,000	170	2,600
	December	6	1,100	26,000	1,100	7,000
	January	1	24,000	920,000	260	2,800
318	April	2	20	—	20	—
	May	12	49	—	8	—
	June	10	33	—	5	—
	October	2	11	940	2	20
	February	10	33	35,000	2	20
	March	18	23	17,000	8	50
	April	6	79	35,000	4	50
323	May	2	49	—	8	—
	June	4	240	—	11	—
	July	2	8	—	8	—
	September	4	79	—	14	—
	October	23	17	940	7	9
	February	3	23	5,400	8	50
	March	4	23	1,800	2	20
324	April	2	40	—	40	—
	May	12	49	—	5	—
	June	2	540	—	17	—
	July	1	14	—	2	—
	October	16	8	17,000	5	14
	December	2	170	7,000	17	630
	February	30	33	46,000	4	2
	March	8	33	17,000	5	20

[a] Not done.

From Goyal, S. M., Gerba, C. P., and Melnick, J. L., *J. Water Pollut. Control. Fed.*, 50, 2447, 1978. With permission.

Table 3
RECOVERY OF VIRUSES, FECAL COLIFORMS, AND TOTAL COLIFORMS FROM WATER, SEDIMENTS, AND OYSTERS IN APPROVED AND NONAPPROVED SHELLFISH HARVESTING AREAS

	Mean MPN[a] of total coliforms per 100 mℓ or 100 g			Mean MPN of fecal coliforms per 100 mℓ or 100 g			Virus isolations		
Station	Water	Sediment	Oysters	Water	Sediment	Oysters	Water (number positive/number examined)	Oysters (number positive/number examined)	Both positive
Closed to shellfish harvesting									
April Fool's Point	472	33,086	8,170	46	354	1,758	6/9	6/9	4
Mowle's Bait Camp	1,179	91,000	22,930	421	6,445	7,817	3/6	2/6	0
Red Bluff Reef	23	430	93	6	650	480	2/2	0/2	0
Eagle's Point	704	54,600	17,880	285	9,300	8,222	5/6	1/6	1
Tiki Island	219	27,880	5,628	127	3,782	1,133	3/6	2/6	2
Yacht Club	9	11,000	70	0	0	0	0/1	1/1	0
Total							19/30 (63%)	12/30 (40%)	
Open to shellfish harvesting									
North Redfish	23	Not done	210	2	Not done	40	2/3	0/1	0
South Redfish	14	11,000	95	3	30	40	0/4	0/2	0
Reef 59	21	1,387	773	4	47	218	3/4	1/4	1
Reef 63	4	1,500	930	3	90	40	2/3	1/3	1
Total							7/14 (50%)	2/10 (20%)	
All sites	412	36,855	8,729	130	3,308	3,100	26/44 (59%)	14/40 (35%)	9

[a] Most probable number.

From Goyal, S. M., Gerba, C. P., and Melnick, J. L., *Appl. Environ. Microbiol.*, 37, 572, 1979. With permission.

Table 4
RELATIVE RECOVERIES OF TOTAL COLIFORMS, FECAL COLIFORMS, AND VIRUSES FROM WATER AND SEDIMENTS

Site	Date	PFU/20 ℓ of seawater[a]	PFU/20 ℓ of sediment[b]	MPN[c] of indicator organisms per 100 mℓ: Total coliforms, Water	Total coliforms, Sediment	Fecal coliforms, Water	Fecal coliforms, Sediment
H-86	11/15/77	6	200	24,000	46,000	24,000	46,000
	11/22/77	15	0	11,000	240,000	11,000	46,000
	4/12/78	374	0	24,000	240,000	11,000	110,000
	4/12/78	528	140	4,600	210,000	4,600	15,000
	2/27/78	0	350	200	460,000	90	460,000
	2/27/78	0	280	200	460,000	90	460,000
H-86a	11/15/77	0	50	460	11,000	460	2,400
	11/22/77	5	50	24,000	240,000	24,000	21,000
	12/01/77	30	480	24,000	240,000	11,000	240,000
T-50	11/15/77	0	170	23	930	<3	90
H-86b	11/22/77	1	0	24,000	7,500	24,000	7,500
H-81	11/22/77	8	0	24,000	110,000	11,000	110,000
K-27	12/09/77	2	0	24,000	240,000	4,600	46,000
	12/21/77	1	0	4,600	1,100,000	210	2,100
H-86[d]	2/27/78	0	50	50	1,100	6	200
Oyster bed no. 1	3/03/78	7	0	4	400	4	<300

[a] Plaque-forming units of viruses isolated per 20 ℓ of seawater.
[b] Plaque-forming units of viruses isolated per 20 ℓ of sediment.
[c] Most probable number
[d] Sample taken 300 ft downstream.

From LaBelle, R. L., Gerba, C. P., Goyal, S. M., Melnick, J. L., Cech, I., and Bogdan, G. F., *Appl. Environ. Microbiol.*, 39, 588, 1980. With permission.

of waters used for recreational bathing and shellfish harvesting. Total and fecal coliform bacteria are in general use in the U.S. for judging the acceptability of contact recreational waters and shellfish-harvesting waters. Water is generally not considered acceptable for contact recreation when the total and fecal coliform densities of 1000 and 200/100 mℓ, respectively, are exceeded.

These standards were based on epidemiologic studies on Lake Michigan and the Ohio River, where detectable health effects were associated with a total coliform density of 2000/100 mℓ. This was extrapolated to a fecal coliform density of 400/100 mℓ. This value was reduced to 200/100 mℓ on the assumption that the quality of direct-contact recreational waters should be better than that which produced a demonstrable health effect.[42]

Such standards have been applied universally to both marine and fresh water bathing areas. More recent epidemiological studies among swimmers at marine bathing beaches have also indicated a relationship between indicator bacteria and gastrointestinal illness.[54] The Environmental Protection Agency is currently conducting a program to develop such water quality criteria for recreational waters.[55] The result of these findings thus far is that a swimming-associated gastroenteritis, primarily in children, can be quantitatively associated with the quality of the bathing water as measured by *E. coli* or enterococcus densities. The gastroenteritis typically has a short incubation period, an acute onset, a short period of relatively benign symptoms, and no sequelae, although in some individuals the symptoms are disabling enough for them to remain home, remain in bed, or seek medical advice. The association between illness and the presence of as few as 10 *E. coli* per 100 mℓ suggests that the agent(s) responsible for the observed illness is (are) highly infectious, present in sewage in large numbers, and/or survives much longer than *E. coli* in the marine environment. These characteristics, along with the nature of the illness, suggest a viral etiology, probably rotavirus or the Norwalk virus.[56]

Recently, standards were proposed for viral quality of recreational water. Melnick[57] recommended consideration of a limit of 1 infectious unit of virus per 10 gal of recreational water. This standard was not met in about one third of the samples referred to in our studies.[49-51] Because of the lack of epidemiologic data, such a standard is arbitrary and reflects limitations of current detection methodology for enteric viruses in water rather than disease risk. It may be conservative since the efficiency of concentration methodology of enteroviruses is under 50%, and current concentration and detection methods are optimized for enteroviruses and are not capable of recovering rotaviruses, hepatitis A virus, adenoviruses, or the Norwalk virus (which may also be present in wastewater discharges[56]); in fact, current methodology is only optimized for detection of less than 40% of the enteric viruses which may be present in sewage-contaminated waters.

Of the environmental factors examined, the only ones for which a clear relationship with virus isolation could be demonstrated were the occurrence of rainfall within 24 hr of sampling and salinity of water. There was also some indication that a logarithmic correlation existed between the presence of viruses and turbidity, which was probably influenced by the rainfall. The increase in the number of viruses in water after periods of heavy rainfall could result from disturbance of sediments containing viruses, runoff, or flushing of sewage-laden waterways which empty into the bay. Although the statistical model, which estimates the number of viruses detected in water based on three factors — presence or absence of rain, water salinity, and number of total coliforms in water — fits well to the observed data, the amount of variance described by this model is only about 16%. Thus, a large amount of variance remains unexplained and is likely to be due to other factors not considered here. Such variance could be related in part to errors encountered in measurement of the parameters studied, i.e., sampling variations due to differences in the efficiency of virus detection from one sample to another.

The effect of environmental factors controlling enteric viruses in marine water may be greatly influenced by geophysical parameters (i.e., bottom topography, shoreline contours, water depth, inflow changes, etc.), and it may be diffiult to apply findings of this study to other coastal areas. Clearly, more work is needed on factors controlling the occurrence of viruses in marine waters for the effective management of marine water quality.

In a study by Goyal et al.,[49] attempts were made to statistically correlate the occurrence of viruses in shellfish and their overlying waters to the indicator bacteria in water, shellfish, and sediment. A multivariate regression analysis of the data showed that although the number of viruses detected in water correlated significantly with total coliforms in oysters, the amount of variation in the number of viruses explained by this variable (25%) did not permit one to be a good predictor of the other. Even if such a relationship was predictive, such an association would be of little value in monitoring efforts, but again indicates that some relationship exists between the presence of enteroviruses and total coliform bacteria in marine water and oysters.

It would appear that further study of this relationship may be warranted. For example, perhaps increasing the volume of seawater analyzed for coliforms or the number of samples analyzed may increase the usefulness of the total coliform or even fecal coliform index in predicting the occurrence of enteroviruses in marine waters.

IV. ALTERNATE INDICATORS

In addition to total coliform and fecal coliform bacteria, several other organisms and several nonorganismic substances have been suggested as indicators of fecal pollution. These include fecal streptococci, *Clostridium perfringens,* Bifidobacterium, coliphages, chlorine residual, adenosine triphosphate, coprostanol, endotoxin assay by limulus amoebocyte lysate, heterotrophic bacterial counts at 20 and 35°C, and polioviruses. The selection of a particular organism as an indicator must take into consideration the fact that the numbers of normal flora in feces are relatively constant whereas the numbers of pathogens vary. An ideal indicator should fulfill the following criteria:

1. An indicator should always be present when pathogens are present and should be absent when pathogens are absent.
2. The persistence and growth characteristics of both indicator and pathogens should be similar.
3. The pathogens and indicator should occur in a constant ratio so that counts of the indicator give a good estimate of the number of pathogens present.
4. Preferably, the indicator should be present in the source of pollution at levels far in excess of the pathogen concentration.
5. The indicator should be resistant to the environment and disinfectants at the same rate as pathogens.
6. The indicator should be nonpathogenic and easily quantifiable.
7. The test for the indicator organism should be simple, quick, and economical and should be applicable to all types of water.
8. The test should detect only the indicator organism and should not give false-positive reactions.

As of now, no single organism or nonorganismic substance fulfills all these criteria. However, total coliform and fecal coliform bacteria fulfill most of them, have been commendable in the past for indicating fecal pollution of water, and should continue to be effective indicators of bacterial pathogens. For monitoring the virological quality of water, however, coliforms are clearly inadequate. A discussion of suggested alternate indicators follows.

A. Fecal Streptococci

Slanetz and Bartley[58] advocated the use of fecal streptococci as indicators of fecal pollution. Although fecal coliforms occur in human feces in larger numbers than fecal streptococci, the latter occur in animal feces in larger numbers than the fecal coliforms.[59] Therefore, fecal coliforms may not give a true picture of both human and animal feces. Kenner[60] listed the following advantages in using fecal streptococci as indicators:

1. They occur in relatively high numbers in excreta of humans and animals.
2. They are present in wastewaters and known polluted waters.
3. They are absent from pure waters, virgin soils, and environments having no contact with human or animal life.
4. They persist without multiplication outside the animal body, except in very rich wastes such as sugar beet waste effluents.
5. They are present in much greater numbers than pathogens.
6. They are generally more resistant than coliforms to the toxic chemical pollutants in certain waters.
7. They occur in higher numbers than fecal coliforms in feces of warm-blooded animals.

Hanes and Fossa[61] reported a positive correlation between bather load and enterococci and *Pseudomonas aeruginosa* levels. Also, coliform bacteria but not enterococci, multiplied in sewage-water mixtures. They argued, therefore, that the enterococci may be better than coliforms.

Cohen and Shuval[14] found fecal streptococci to be more resistant than coliforms to the natural water environment and to the purification processes. The survival of fecal streptococci reflected more closely the survival of enteric viruses than did the survival of coliforms. Also, fecal streptococci were the only indicator organisms isolated at points distant from the original source of pollution. They concluded that fecal streptococci may, in certain cases, provide a better estimate of the probable virus content in lightly contaminated water than total coliforms or fecal coliforms. More recently, Cabelli et al.[54] found a correlation between the quality of bathing water as measured by *E. coli* and enterococcus densities and the incidence of swimming-associated gastroenteritis, particularly among children.

Some workers, on the other hand, found fecal streptococci of limited value as sole indicators of fecal pollution. Geldreich and Kenner[62] were of the view that fecal streptococci may be useful in conjunction with either total coliforms or fecal coliforms in establishing the source of pollution, on the basis of the fecal coliform/fecal streptococci ratio which is 4 in human feces and 0.7 in animal feces. Burman[63] criticized the use of fecal streptococci as pollution indicators because of (1) their lower initial concentration, (2) the variety of methods proposed for their isolation, and (3) uncertainty as to the selectivity of some of the methods.

B. Coliphages

The ubiquity of coliphages in the feces of man and other warm-blooded animals, in sewage, and in sewage-polluted water has led to the suggestion that coliphages may reliably indicate the presence of enteric virus pollution of water. Considerable interest has been shown in the use of f2 and MS2 coliphages because they are more like enteroviruses (single-stranded RNA, icosahedral, 25 nm diameter) than the T phages.

Guelin[64] was the first to advocate the use of bacteriophages as indicators of enteric virus pollution because coliphages occur in the intestinal tract of humans and animals, as do fecal coliforms. More recently, Kott et al.[65] advocated the use of coliphages as indicators because they found that coliphages MS2 and f2 survived as long as poliovirus 1 in an experimental oxidation pond. Also, these coliphages were found to be as resistant to chlorination as poliovirus 1. These observations, together with the ease, simplicity, convenience, economicity, and rapidity of coliphage assays, led Kott et al.[65] to suggest that coliphages may be better able to predict the virological quality of water than coliforms.

Vaughn and Metcalf[66] examined sewage effluents, shellfish, and shellfish-growing waters for coliphages and enteric viruses over a period of 3 years to determine the practicality of a coliphage indicator system for predicting the virological quality of water. They found that coliphages were widely disseminated throughout an estuary, generally occurring in the absence of detectable enteric virus activity. Conversely, a majority of samples which contained enteroviruses did not yield coliphages. They further observed that oysters accumulated more coliphages than enteric viruses; coliphages multiplied in the estuary during the summer months when the proper host cell was present and two major types of coliphages were found in field samples based on their reactivity with different strains of *E. coli*. They further compared the rate of coliphage isolation with three different *E. coli* host cultures. Although the total number of isolations obtained with *E. coli* 9637 was similar to that obtained with *E. coli* 11303 at the end of 3 years, the results of the first year of study showed that *E. coli* 11303 was three times as sensitive as *E. coli* 9637 and six times as sensitive as *E. coli* UNH. However, during the last 2 years of study, the coliphage isolation rate with *E. coli* 11303 was much lower than that with *E. coli* 9637.

The inability to accurately correlate coliphage and enteric virus occurrence in field samples, along with the potential for the presence of more than one predominant coliphage type, led

Vaughn and Metcalf[66] to suggest that coliphage as an indicator of enteric viruses had serious shortcomings. In addition, they cited the following reasons for nonsuitability of coliphages as indicators:

1. Coliphages were consistently present in raw sewage samples which yielded inconsistent enterovirus isolations.
2. Treated effluents were coliphage-positive but enterovirus-negative.
3. Many (63%) enterovirus isolations occurred without any phage isolation.
4. Phages replicated in estuarine water during the summer months (if proper bacterial hosts were present) making them unsuitable for use as an indicator system.
5. In a controlled experiment on comparative uptake of coliphages and enteroviruses by oysters, uptake of coliphages was 5- and 30-fold higher than that of enteroviruses.[67]

The only advantage of coliphages as cited by these authors was that the survival patterns of coliphages and enteroviruses in estuary waters and in shellfish were similar. However, Joyce and Weiser[68] found that coliphage T2 did not survive as long as enteroviruses in untreated farm pond water at 20 to 25 and 4°C.

The status of coliphages as an indicator needs careful assessment before being totally rejected. Since methods are now available to concentrate and detect coliphages from large volumes of water and wastewater,[69] it will be interesting to see if there is any correlation between the occurrence of coliphages and human enteric viruses.

C. Enteroviruses

Coin et al.[70] suggested polioviruses as indicators of enteric virus pollution because of the frequent isolation of these viruses from sewage-polluted waters. Also, in communities where live attenuated poliovirus vaccine is used, polioviruses may be adequate indicator. Payment et al.[71] monitored raw sewage over a 13-month period in Laval, Canada and found viruses in 47 of 53 samples tested and 39 of these samples yielded polioviruses. Katzenelson and Kedmi,[72] however, were unable to detect polioviruses in as many as 50% of the sewage samples of Israel, where live attenuated poliovirus vaccine is routinely administered to all infants three times during the first year of life. This phenomenon has also been observed in our laboratory when viruses other than polioviruses were isolated from several samples of ground and surface waters.

D. Coprostanol

Coprostanol is a fecal sterol (5β-cholestan-3β-01) and occurs only in the feces of man and other higher animals. It is stable, nonpathogenic, unaffected by chlorine, and can be detected in the presence of other lipid-like components in water.[15] The cumbersome method for the detection of coprostanol and nonexistence of a consistent relationship between fecal sterols and traditional indicator bacteria,[73] however, precludes the use of coprostanol as a routine fecal indicator.

E. *Clostridium perfringens*

European workers advocate *C. perfringens* for the examination of chlorinated waters and the detection of remote pollution. It has been suggested also that in terms of survival, distribution of *C. perfringens* spores in the aquatic environment may be more analogous to that of enteroviruses.[74] Several investigators have demonstrated the widespread presence of this organism in coastal waters and their underlying sediments, and it has been suggested that, because of its long survival, *C. perfringens* may be a very good indicator of past pollution.[75-78] We have demonstrated long survival of enteroviruses in sediments.[79] Since *C. perfringens* spores are expected to settle to the bottom, survive, and accumulate,[74] we examined a number of samples of estuarine sediment. However, we did not find any correlation between the presence of enteroviruses and *C. perfringens*.[51]

F. Bifidobacter

Bifidobacter is a major component of human and animal feces and has been advocated as an indicator of fecal pollution.[80,81] The advantages of this organism include its association with feces, its inability to multiply in nature, and the similarity of its survival rate to that of pathogens and other fecal indicators. Enough work has not been done to advocate or discourage the use of this organism as an indicator.

V. SUMMARY AND CONCLUSIONS

For almost a century, water pollution control has been based on (1) secondary treatment of sewage and (2) infinite dilution of wastes in receiving waters. Since the turn of the century, the use of coliform bacteria to indicate fecal pollution has served as an added protection. The rates of enteric bacterial disease in the U.S. are low today because the spread of enteric pathogens has been controlled by means of sanitary engineering and the practice of good sanitation and personal hygiene. The continuing occurrence of outbreaks of hepatitis A and viral gastroenteritis, however, indicates that potential dangers of water pollution still exist, especially with viruses.

High standards of sanitation and the consequent low fecal carriage rates of infectious agents in the U.S. has made the public increasingly susceptible to enteric diseases where contact with reservoirs exists. This is best illustrated by the high susceptibility of the U.S. public to traveller's diarrhea.

Often, the lack of epidemiologic evidence of waterborne viral disease outbreaks has been cited as evidence that the threat to human health is minimal and that the indicator bacteria are doing their job. Current epidemiologic methods, however, are not sensitive enough to detect virus disease transmission through water, because clinically observable illness occurs only in a small number of people who become infected and because of widely varying incubation periods. This fact, and the low infective dose of viruses,[52] has led some to suggest that the presence of enteric viruses in any water is indicative of a potential viral disease hazard.[82]

It has not been the intent of this chapter to discourage the use of bacterial indicators for judging the sanitary quality of water. An attempt has been made, however, to elaborate on the potential pitfalls associated with their use. The isolation of viruses from water and shellfish which met current bacteriological standards points toward the need for the search for a better indicator for viruses. Recently, standards have been proposed for viral quality of recreation water. Melnick[2] recommended consideration of a limit of 1 infectious unit of virus per 10 gal of recreational water, and Shuval[20] proposed a standard of no viruses in 10-gal samples. Recently, the state of Arizona promulgated standards which require that irrigation water and water for full-body recreation contain a mean of less than 1 PFU of enteric viruses per 40-ℓ sample. For partial-body recreation, the mean allowable limit is 125 PFU/40 ℓ in a minimum of 5 samples.[83]

Because of the lack of quantitative epidemiologic data, all such standards are arbitrary and reflect limitations of current detection methodology of enteric viruses in water rather than disease risk. Still, they may be conservative in view of such factors as: (1) the efficiency of our concentration method is less than 50%, (2) very low concentrations of enteroviruses may cause infection in susceptible hosts,[84] and (3) current concentration and detection methods are optimized for enteroviruses and are not capable of recovering reoviruses, adenoviruses, hepatitis A virus, rotaviruses, and other viruses which may be present in wastewater discharges.[85-87]

The foregoing discussion supports the recommendations made by the World Health Organization[88] and by the American Water Works Association[89] which include a standard for viruses in water and possible monitoring of viral contamination in certain situations.

REFERENCES

1. **Baker, M. N.,** *The Quest for Pure Water,* American Water Works Association, New York, 1949.
2. **Melnick, J. L.,** Detection of virus spread by the water route, in *Virus and Water Quality: Occurrence and Control,* Snoeyink, V. and Griffin, V., Eds., University of Illinois Press, Urbana-Champaign, 1971, 114.
3. **Melnick, J. L. and Gerba, C. P.,** The ecology of enteroviruses in natural waters, *Crit. Rev. Environ. Control,* 10, 65, 1980.
4. **Geldreich, E. E.,** Bacterial populations and indicator concepts in feces, sewage, stormwater and solid wastes, in *Indicators of Viruses in Water and Food,* Berg, G., Ed., Ann Arbor Press, Ann Arbor, Mich., 1978, 51.
5. **Escherich, T.,** The intestinal bacteria of infants, *Fortschr. Med.,* 3, 515, 1885.
6. **Eijkman, C.,** The fermentation test at 46°C as a means of testing potable water, *Zentralbl. Bakteriol. Parasitenkd., Infektionskr. Hyg., Abt. I, Orig.,* 37, 742, 1904.
7. Worth Health Organization, *International Drinking Water Standard,* World Health Organization, Geneva, 1963.
8. National Technical Advisory Committee, Water Quality Criteria, Federal Water Pollution Control Administration, Washington, D.C., 1968.
9. **Shuval, H. I., Thompson, A., Fattal, B., Cymbalista, S., and Weiner, Y.,** Natural virus inactivation processes in seawater, *J. Sanit. Eng. Div., Am. Soc. Civ. Eng.,* SA5, 587, 1971.
10. **Shuval, H. I.,** Detection and control of enteroviruses in the water environment, in *Developments in Water Quality Research,* Shuval, H. I., Ed., Ann Arbor-Humphrey, Ann Arbor, Mich., 1970, 47.
11. **Craun, G. F. and McCabe, L. J.,** Review of the causes of waterborne-disease outbreaks, *J. Am. Water Works Assoc.,* 65, 74, 1973.
12. **Craun, G. F.,** Waterborne disease — a status report emphasizing outbreaks in groundwater systems, *Ground Water,* 17, 183, 1979.
13. **Clarke, N. A. and Kabler, P. W.,** Human enteric viruses in sewage, *Health Lab. Sci.,* 1, 44, 1964.
14. **Cohen, J. and Shuval, H. I.,** Coliforms, fecal coliforms, and fecal streptococci as indicators of water pollution, *Water, Air, Soil Pollut.,* 2, 85, 1973.

14a. **Melnick, J. L.,** Are conventional methods of epidemiology appropriate for risk assessment of virus contamination of water?, in *Risk Assessment and Health Effects of Land Application of Municipal Wastewater and Sludges,* Sagik, B. P. and Sorber, C. A., Eds., Center for Applied Research and Education, University of Texas, San Antonio, 1978, 61.

15. **Dutka, B. J.,** Microbiological indicators, problems and potential of new microbial indicators of water quality, in *Biological Indicators of Water Quality,* James, A. and Evison, L., Eds., John Wiley & Sons, New York, 1979, 18.
16. **Tobin, R. S., Lomax, P., and Kushner, D. J.,** Comparison of nine brands of membrane filters and the most-probable-number method for total coliform enumeration in sewage-contaminated drinking water, *Appl. Environ. Microbiol.,* 40, 186, 1980.
17. **Braswell, J. R. and Hoadley, A. W.,** Recovery of *Escherichia coli* from chlorinated secondary sewage, *Appl. Environ. Microbiol.,* 28, 328, 1974.
18. **Weidenkopf, S. J.,** Inactivation of type I poliomyelitis virus with chlorine, *Virology,* 5, 56, 1958.
19. **Clarke, N. A. and Kabler, P. W.,** The inactivation of purified coxsackie virus in water by chlorine, *Am. J. Hyg.,* 59, 119, 1954.
20. **Shuval, H. I.,** Water needs and usage, in *Viruses in Water,* Berg, G., Bodily, H. L., Lennette, E. H., Melnick, J. L., and Metcalf, T. G., Eds., American Public Health Association, Washington, D.C., 1976, 12.
21. **Berg, G., Dahling, D. R., Brown, G. A., and Berman, D.,** Validity of fecal coliforms, total coliforms, and fecal streptococci as indicators of viruses in chlorinated primary sewage effluents, *Appl. Environ. Microbiol.,* 36, 880, 1978.
22. **Hoehn, R. C., Randall, C. W., Bell, S. A., and Shaffer, P. T. B.,** Tri-halomethanes and viruses in a water supply, *J. Environ. Eng. Div., Am. Soc. Civ. Eng.,* 103, 803, 1977.

23. **Shaffer, P. T. B., Metcalf, T. G., and Sproul, O. J.,** Chlorine resistance of poliovirus isolates recovered from drinking water, *Appl. Environ. Microbiol.*, 40, 1115, 1980.
24. **Bates, R. C., Shaffer, P. T. B., and Sutherland, S. M.,** Development of resistant poliovirus by repetitive sublethal exposure to chlorine, in *Water Chlorination — Environmental Impact and Health Effect,* Vol. 2, Jolley, R. L., Ed., Ann Arbor Press, Ann Arbor, Mich., 1978, 471.
25. **Hejkal, T. W., Keswick, B., LaBelle, R. L., Gerba, C. P., Sanchez, Y., Dreesman, G. R., Hafkin, B. A., and Melnick, J. L.,** Viruses in a community water supply associated with an outbreak of gastroenteritis and infectious hepatitis, *J. Am. Water Works Assoc.*, 74, 318, 1982.
26. **Goyal, S. M.,** unpublished data.
27. **Deetz, T. R., Smith, E. M., Goyal, S. M., Gerba, C. P., Vollet, J. J., Tsai, L., and DuPont, H. L.,** Rota- and Enteroviruses in Potable and Environmental Water in Mexico, in preparation.
28. **Wellings, F. M., Lewis, A. L., Mountain, C. W., and Pierce, L. V.,** Demonstration of virus in groundwater after effluent discharge onto soil, *Appl. Microbiol.*, 29, 751, 1975.
29. **Woodward, W. E.,** personal communication.
30. **Marzouk, Y., Goyal, S. M., and Gerba, C. P.,** Relationship of viruses and indicator bacteria in water and wastewater of Israel, *Water Res.*, 14, 1585, 1980.
31. **Buras, N.,** Concentration of enteric viruses in wastewater and effluent: a two year study, *Water Res.*, 10, 295, 1976.
32. **Ross, E. C., Campbell, K. W., and Ongerth, H. J.,** *Salmonella typhimurium* contamination of Riverside, California, supply, *J. Am. Water Works Assoc.*, 58, 165, 1966.
33. **Seligmann, R. and Reitler, R.,** Enteropathogens in water with low *E. coli* titer, *J. Am. Water Works Assoc.*, 57, 1572, 1965.
34. **Presnell, M. W. and Meiscer, J. J.,** Coliforms and fecal coliforms in an oyster-growing area, *J. Water Pollut. Control Fed.*, 43, 407, 1971.
35. **Dufour, A. P.,** *Escherichia coli:* the fecal coliform, in *Bacterial Indicators/Health Hazards Associated with Waters,* ASTM Special Tech. Publ. No. 635, Hoadley, A. W. and Dutka, B. J., Eds., American Society for Testing and Materials, Philadelphia, 1977, 48.
36. **Goldfield, M.,** Epidemiological indicators for transmission of viruses by water, in *Viruses in Water,* Berg, G., Bodily, H. L., Lennette, E. H., Melnick, J. L., and Metcalf, T. G., Eds., American Public Health Association, Washington, D.C., 1976, 70.
37. **Anon.,** Oysters on the move, in *Aquanotes,* Office of Sea Grant Development, Louisiana State University, Baton Rouge, 1976.
38. **Metcalf, T. G.,** Evaluation of shellfish sanitary quality by indicators of sewage pollution, in *Discharge of Sewage from Sea Outfalls,* Gameson, A. L. H., Ed., Pergamon Press, Elmsford, N.Y., 1975, 75.
39. **Gerba, C. P. and Goyal, S. M.,** Detection and occurrence of enteric viruses in shellfish: a review, *J. Food Protec.*, 41, 743, 1978.
40. **Linco, S. J. and Grohmann, G. S.,** The Darwin outbreak of oyster-associated viral gastroenteritis, *Med. J. Aust.*, 1, 211, 1980.
41. **Wood, P. C.,** *Guide to Shellfish Hygiene,* World Health Organization, Geneva, 1976, 68.
42. **Cabelli, V.,** New standards for enteric bacteria, in *Water Pollution Microbiology,* Mitchell, R., Ed., Wiley-Interscience, New York, 1978, 233.
43. **Morris, R. L., Mearus, A. J., and Kim, J.,** Viruses and bacteria in coastal waters and shellfish, Annual Report of the Southern California Coastal Water Research Project, El Segundo, Calif., 1976.
44. **Metcalf, T. G. and Stiles, W. C.,** Viral pollution of shellfish in estuary waters, *J. Sanit. Eng. Div., Am. Soc. Civ. Eng.*, 94, 595, 1968.
45. **Metcalf, T. G., Wallis, C., and Melnick, J. L.,** Virus enumeration and public health assessments in polluted surface water contributing to transmission of virus in nature, in *Virus Survival in Water and Wastewater Systems,* Malina, J. F. and Sagik, B. P., Eds., Center for Research in Water Resources, University of Texas, Austin, 1974, 57.
46. **Portnoy, B. L., Mackowiak, P. A., Caraway, C. T., Walker, J. A., McKinley, T. W., and Klein, C. A., Jr.,** Failure of shellfish certification programs to prevent outbreaks, *JAMA*, 233, 1065, 1975.
47. **Mackowiak, P. A., Caraway, C. T., and Portnoy, B. L.,** Oyster-associated hepatitis: lessons from the Louisiana experience, *Am. J. Epidemiol.*, 103, 181, 1976.
48. **Fugate, K. J., Cliver, D. O., and Hatch, M. T.,** Enteroviruses and potential bacterial indicators in Gulf coast oysters, *J. Milk Food Technol.*, 38, 100, 1975.
49. **Goyal, S. M., Gerba, C. P., and Melnick, J. L.,** Human enteroviruses in oysters and their overlying waters, *Appl. Environ. Microbiol.*, 37, 572, 1979.
50. **Goyal, S. M., Gerba, C. P., and Melnick, J. L.,** Prevalence of human enteric viruses in coastal canal communities, *J. Water Pollut. Control Fed.*, 50, 2247, 1978.
51. **LaBelle, R. L., Gerba, C. P., Goyal, S. M., Melnick, J. L., Cech, I., and Bogdan, G. F.,** Relationships between environmental factors, bacterial indicators, and the occurrence of enteric viruses in estuarine sediments, *Appl. Environ. Microbiol.*, 39, 588, 1980.

51a. **Gerba, C. P., Goyal, S. M., LaBelle, R. L., Cech, I., and Bogdan, G. F.,** Failure of indicator bacteria to reflect the occurrence of enteroviruses in marine waters, *Am. J. Publ. Health,* 69, 116, 1979.
52. **Westwood, J. C. N. and Sattar, S. A.,** The minimal infective dose, in *Viruses in Water,* Berg, G., Bodily, H. L., Lennette, E. H., Melnick, J. L., and Metcalf, T. G., Eds., American Public Health Association, Washington, D.C., 1976, 61.
53. **Berg, G.,** Integrated approach to the problem of viruses in water, *J. Sanit. Eng. Div., Am. Soc. Civ. Eng.,* 97, 867, 1971.
54. **Cabelli, V. J., Dufour, A. P., Levin, M. A., McCabe, L. J., and Haberman, P. W.,** Relationship of microbial indicators to health effects at marine bathing beaches, *Am. J. Publ. Health,* 69, 690, 1979.
55. **Cabelli, V.,** What do indicators indicate?, in *Aquatic Microbial Ecology,* Colwell, R. R., Ed., University of Maryland Sea Grant Program, College Park, 1980, 47.
56. **Melnick, J. L., Gerba, C. P., and Wallis, C.,** Viruses in water, *Bull. W. H. O.,* 56, 499, 1978.
57. **Melnick, J. L.,** Viruses in water, in *Viruses in Water,* Berg, G., Bodily, H. L., Lennette, E. H., Melnick, J. L., and Metcalf, T. G., Eds., American Public Health Association, Washington, D.C., 1976, 3.
58. **Slanetz, L. W. and Bartley, C. H.,** Detection and sanitary significance of fecal streptococci in water, *Am. J. Publ. Health,* 54, 609, 1964.
59. **Kenner, B. A., Clark, H. F., and Kabler, P. W.,** Fecal streptococci. II. Quantification of streptococci in feces, *Am. J. Publ. Health,* 50, 1553, 1960.
60. **Kenner, B. A.,** Fecal streptococcal indicators, in *Indicators of Viruses in Water and Food,* Berg, G., Ed., Ann Arbor Press, Ann Arbor, Mich., 1978, 247.
61. **Hanes, N. B. and Fossa, A. J.,** A quantitative analysis of the effects of bathers in recreational water quality, in *Proc. 5th Int. Water Pollut. Res. Conf.,* Pergamon Press, Elmsford, N.Y., 1971, HA9-1.
62. **Geldreich, E. E. and Kenner, B. A.,** Concepts of fecal streptococci in stream pollution, *J. Water Pollut. Control Fed.,* 41, R336, 1969.
63. **Burman, N. P.,** Indicators, in *Progress in Microbiological Techniques,* Collins, C. H., Ed., Butterworth, London, 1967, 185.
64. **Guelin, A.,** Etude quantitative de bacteriophage de la mer, *Ann. Inst. Pasteur, Paris,* 74, 104, 1948.
65. **Kott, Y., Roze, N., Sperber, S., and Betzer, N.,** Bacteriophages as viral pollution indicators, *Water Res.,* 8, 165, 1974.
66. **Vaughn, J. M. and Metcalf, T. G.,** Coliphages as indicators of enteric viruses in shellfish and shellfish raising estuarine waters, *Water Res.,* 9, 613, 1975.
67. **Metcalf, T. G., Vaughn, J. M., and Stiles, W. C.,** The occurrence of human viruses and coliphage in marine waters and shellfish, in *FAO Tech. Conf. Mar. Pollut. Its Effects on Living Resources and Fishing,* Food and Agriculture Organization, Rome, Italy, December 1970, FIR:MP/70/E-24.
68. **Joyce, G. and Weiser, H. H.,** Survival of enteroviruses and bacteriophage in farm pond waters, *J. Am. Water Works Assoc.,* 59, 491, 1967.
69. **Goyal, S. M., Zerda, K. S., and Gerba, C. P.,** Concentration of coliphages from large volumes of water and wastewater, *Appl. Environ. Microbiol.,* 39, 85, 1980.
70. **Coin, L., Menetrier, M. L., Loboude, J., and Hannoun, M. C.,** Modern microbiological and virological aspects of water pollution, in *Advances in Water Pollution Research,* Vol. 2, Jaag, O., Ed., Pergamon Press, London, 1965, 10.
71. **Payment, P., Larose, Y., and Trudel, M.,** Polioviruses as indicators of virological quality of water, *Can. J. Microbiol.,* 25, 1212, 1979.
72. **Katzenelson, E. and Kedmi, S.,** Unsuitability of polioviruses as indicators of virological quality of water, *Appl. Environ. Microbiol.,* 37, 343, 1979.
73. **Dutka, B. J. and El-Shaarawi, A.,** Relationship between various bacterial populations and coprostanol and cholesterol, *Can. J. Microbiol.,* 9, 1386, 1975.
74. **Cabelli, V. J.,** *Clostridium perfringens* as a water quality indicator, in *Bacterial Indicators/Health Hazards Associated with Water,* ASTM Special Tech. Publ. No. 635, Hoadley, A. W. and Dutka, B. J., Eds., American Society for Testing and Materials, Philadelphia, 1977, 65.
75. **Bonde, G. J.,** Studies on the dispersion and disappearance phenomena of enteric bacteria in the marine environment, *Rev. Int. Oceanogr. Med.,* 9, 17, 1968.
76. **Davies, J. A.,** Isolation and identification of Clostridia from North Sea sediments, *J. Appl. Bacteriol.,* 32, 164, 1969.
77. **Smith, L. D.,** The clostridial flora of marine sediments from a productive and a nonproductive area, *Can. J. Microbiol.,* 14, 1301, 1968.
78. **Matches, J. R. and Liston, J.,** Mesophilic clostridia in Puget Sound, *Can. J. Microbiol.,* 20, 1, 1974.
79. **Smith, E. M., Gerba, C. P., and Melnick, J. L.,** Role of sediment in the persistence of enteroviruses in the estuarine environment, *Appl. Environ. Microbiol.,* 35, 685, 1978.
80. **Evison, L. M. and James, A.,** Bifidobacterium as an indicator of fecal pollution in water, *Prog. Water Technol.,* 7, 57, 1975.

81. **Levin, M. A.,** Bifidobacteria as water quality indicators, in *Bacterial Indicators/Health Hazards Associated with Waters,* ASTM Special Tech. Publ. No. 635, Hoadley, A. W. and Dutka, B. J., Eds., American Society for Testing and Materials, Philadelphia, 1977, 131.
82. **Sobsey, M. D., Carrick, R. J., and Jensen, H. R.,** Improved methods for detecting enteric viruses in oysters, *Appl. Environ. Microbiol.*, 36, 121, 1978.
83. **Anon.,** *Water Quality Standards for Surface Waters,* Arizona Water Quality Control Council, Phoenix, 1979.
84. **Katz, M. and Plotkin, S. A.,** Minimal infective dose of attenuated poliovirus for man, *Am. J. Publ. Health,* 57, 1837, 1967.
85. **Lund, E., Hedstrom, C. E., and Stannegard, O.,** A comparison between virus isolations from sewage and from fecal specimens from patients, *Am. J. Epidemiol.*, 84, 282, 1966.
86. **Sattar, S. A. and Westwood, J. C. N.,** Isolation of apparently wild strains of poliovirus type 1 from sewage in the Ottawa area, *Can. Med. Assoc. J.*, 116, 25, 1977.
87. **Malherbe, H. H. and Strickland-Cholmley, M.,** Quantitative studies on virus survival in sewage purification processes, in *Transmission of Viruses by the Water Route,* Berg, G., Ed., Wiley-Interscience, New York, 1967, 379.
88. World Health Organization, Human viruses in water, wastewater and soil, W. H. O. Tech. Rep. Series No. 639, 1979.
89. American Water Works Association T&P Committee, Viruses in drinking water, committee report, *J. Am. Water Works Assoc.*, 71, 441, 1979.

Index

INDEX

D

E

F

G

H

I

K

L

M

N

O

P

Q

R

S

T

U

V

W

Y

Z